Advances in Mobile Computing and Communications

Perspectives and Emerging Trends in 5G Networks

OTHER COMMUNICATIONS BOOKS FROM CRC PRESS

Advances in Mobile Computing and Communications

Perspectives and Emerging Trends in 5G Networks

M. Bala Krishna • Jaime Lloret Mauri

CRC Press
Taylor & Francis Group
Boca Raton   London   New York

CRC Press is an imprint of the
Taylor & Francis Group, an **informa** business

CRC Press
Taylor & Francis Group
6000 Broken Sound Parkway NW, Suite 300
Boca Raton, FL 33487-2742

© 2016 by Taylor & Francis Group, LLC
CRC Press is an imprint of Taylor & Francis Group, an Informa business

No claim to original U.S. Government works

Printed on acid-free paper
Version Date: 20160225

International Standard Book Number-13: 978-1-4987-0113-6 (Hardback)

Library of Congress Cataloging-in-Publication Data

Names: Krishna, M. Bala, editor. | Lloret Mauri, Jaime, editor.
Title: Advances in mobile computing and communications : perspectives and emerging trends in 5G networks / edited by M. Bala Krishna, Jaime Lloret Mauri.
Description: Boca Raton : Taylor & Francis, 2016. | Includes bibliographical references and index.
Identifiers: LCCN 2015044794 | ISBN 9781498701136
Subjects: LCSH: Mobile computing. | Mobile communication systems.
Classification: LCC QA76.59 .A377 2016 | DDC 004--dc23
LC record available at http://lccn.loc.gov/2015044794

**Visit the Taylor & Francis Web site at
http://www.taylorandfrancis.com**

**and the CRC Press Web site at
http://www.crcpress.com**

Printed and bound in the United States of America by Publishers Graphics,
LLC on sustainably sourced paper.

Contents

Preface

Long-term evolution advanced (LTE-A) and heterogeneous networks (HetNets) are high-speed wireless communication networks compared to traditional, flat IP-based networks. Fourth-generation (4G) heterogeneous networks improve the spectral efficiency with an optimum size of customers experiencing data rates in Gbps. HetNet combines a broad range of technologies such as device-to-device communication, machine-to-machine communication, Cloud RAT/RAN, multiple radio access networks, advanced coding and multiplexing, and multiple access techniques such as OFDM, OFDMA, and WCDMA to provide better coverage for indoor and outdoor applications. HetNet overcomes the theoretical limits of existing radio linking technologies and enhances wide spectrum utilities for high-speed communication in wireless networks. Deployment of macro, micro, pico, and femto base stations and remote relay head nodes exemplifies the wireless network by providing broadband coverage to thousands of users in the network. Enhancements in existing free space propagation models and diffraction and scattering models improve the radio link aspects of 4G and beyond networks. Managing and controlling spectrum usage and networking resources to achieve minimum intercellular interference is a significant factor in multicellular 5G HetNets. Massive MIMO, 3D beamforming and millimeter wave (mmWave) technologies improve the services of WLAN and WPAN networks. Channel aspects such as indoor impulse response, outdoor impulse response, and small-scale characterizations yield better radio propagation in multicellular HetNets.

Physical layer aspects such as channel state information, feedback, and transmission adaptations of single and multiple users improve the performance of communication links. Multicellular networks are affected by intercellular or intracellular interference and hence require efficient power control and subcarrier allocation techniques to improve the fairness of node communication in the network. A game theory approach in physical resource block allocation increases the energy efficiency of radio resource management in multicellular HetNets. mmWave regulation and standardization techniques achieve multigigabit transmission in dense networks. Macrocell antennas mounted on the top of buildings provide coverage for outdoor users, and picocells/femtocells antennas mounted on low height buildings and inside buildings provide coverage for indoor users. Microcell base stations are deployed in dense areas to reduce the effect of cell outage. Advance radio technologies such as software-defined radio and cognitive radio networks address the scarcity of spectral resources and efficiently manage spectral allocation in multicellular networks. High-speed audio and video streaming are the promising aspects of 5G networks that aim at using advanced video coding techniques such as the H.264 standard, optimized resource allocation, layered video coding, and error resilient and concealment procedures. Multicellular HetNets aim at flexible network designs to achieve better performance in Wi-Fi hotspot zones.

This book provides state-of-the-art, novel approaches for advances in mobile computing and communication, covering current and emerging topics and is an excellent reference book for developers, researchers, academicians, and graduate students. The book comprises contributions from prominent researchers and academicians working in the area of mobile communications and wireless networks. It aims to enlighten the readers in developing areas of 5G networks.

Chapter 1 explains the emerging trends in 5G networks based on M-RAT, OFDM, multiple MIMO, SDN, C-RAN, D2D, and M2M systems. This chapter describes the features of LTE/LTE-A (release 8 to release 12), MIMO enhancements based on 3D beamforming, and mmWave technology.

Chapter 2 explains the radio propagation and channel model features for 4G and beyond networks. Channel models based on deterministic, statistical, shadowing, and fading techniques are given. Outdoor channel modeling, indoor channel modeling, and modern broadband communications for 4G and beyond networks are elucidated in this chapter.

Chapter 3 explains channel state information (CSI) for single-user MIMO and multiuser MIMO. CSI feedback for 5G networks based on Grassmannian quantization and Stiefel manifold quantization are elucidated in this chapter.

Chapter 4 explains the game theory approach for power control and subcarrier allocation in OFDMA cellular networks. Basic power control algorithms and capacity maximization techniques are highlighted. Game theory for dynamic resource allocation based on transmit power minimization and fairness are elucidated in this chapter.

Chapter 5 explains mmWave technology (regulation and standardization), single carrier mmWave, multicarrier mmWave, and mmWave antennas for multigigabit transmission. mmWave technology for 5G cellular networks based on channel characterization and massive MIMO are elucidated in this chapter.

Chapter 6 explains the architecture of multicellular HetNets based on macrocells, microcells, picocells, femtocells, and remote relay head notes. Multitier architecture of Cloud RAN and mobile cloud computing in multicellular HetNets are elucidated in this chapter.

Chapter 7 explains energy-efficient operations for 4G and beyond networks using HetNets. Radio resource management based on SMDP model and JCAC optimal policy model are highlighted. Energy efficiency methods based on cognitive radio Hetnets and dynamic coverage management are elucidated in this chapter.

Chapter 8 explains opportunistic multiconnect with P2P Wi-Fi and cellular providers. Multiconnect as network virtualization is highlighted. Group connect methods for mobility traces and opportunistic methods are also elucidated in this chapter.

Chapter 9 explains video streaming over wireless channels for 4G and beyond networks. Advanced video streaming and wireless video streaming techniques are highlighted in this chapter. Streaming video and mobile multimedia communications for 5G networks, MIMO, and home networks are explained. Wireless technology in massive crowds based on femtocell approach and group connect are elucidated.

We take this opportunity to express our sincere gratitude to all the authors and the publishing team. First, we are deeply indebted to our active authors and researchers, who share our vision of this book, *Advances in Mobile Computing and Communications: Perspectives and Emerging Trends in 5G Networks*, and contributed to high-quality chapters. We sincerely thank the authors for their tireless effort in bringing this book to a noble and presentable form. Second, we thank various technical societies for their kind support. Third, we thank the CRC Press editorial team, especially Richard O'Hanley, Karen Schober, Ashley Weinstein, Theresa Gutierrez, and Richard Tressider for their patience, guidance, support, and encouragement during the initial phase, progress phase, and final phase of this book. Finally, we thank our families for their warm support and encouragement.

M. Bala Krishna
Corresponding Editor
University School of Information and Communication Technology
Guru Gobind Singh Indraprastha University
New Delhi, India

Jaime Lloret Mauri
Department of Communications
Polytechnic University of Valencia
Valencia, Spain

Editors

M. Bala Krishna received his bachelor of engineering (BE) in computer engineering from Delhi Institute of Technology (presently Netaji Subhas Institute of Technology), University of Delhi, Delhi, India, and master of technology (MTech) in information technology from University School of Information Technology (presently University School of Information and Communication Technology), Guru Gobind Singh Indraprastha University, Delhi, India. He also holds a PhD in computer engineering from JMI Central University, New Delhi, India. He worked as a senior research associate and project associate in the Indian Institute Technology, Delhi, India, in the areas of digital systems and embedded systems. He worked as a faculty member and handled projects related to networking and communication. He is presently working as an assistant professor in the University School of Information and Communication Technology. His areas of interest include computer networks, wireless networking and communications, mobile and ubiquitous computing, and embedded system design. He has published in international journals and conferences and book chapters. His teaching focuses on wireless networks, mobile computing, data and computer communications, embedded systems, programming languages, etc. His current research work includes wireless ad hoc and sensor networks, green networking and communications, advances in mobile computing and communications, cognitive networks, and software-defined networks. He is a member of IEEE and ACM technical societies. He is a technical program committee member for IEEE and ACM international conferences.

Jaime Lloret Mauri received his MSc in physics in 1997, MSc in electronic engineering in 2003, and PhD in telecommunication engineering (Dr. Ing.) in 2006. He is a Cisco Certified Network Professional Instructor. He worked as a network designer and administrator in several enterprises. He is currently associate professor at the Polytechnic University of Valencia. He is the head of the Communications and Networks Research Group of the Integrated Management Coastal Research Institute and the head of the Innovative Team on Active and Collaborative Techniques and Use of Technological Resources in Education (EITACURTE). He is the director of the Computer Networks and Communications diploma program and of the Digital Post Production master's program, both taught at the Polytechnic University of Valencia. He was Internet Technical Committee chair (a joint committee of the IEEE Communications Society and the Internet Society) for the 2014–2015 term. He has authored 22 book chapters and has more than 350 research papers published in national and international conferences and international journals (more than 130 with ISI Thomson JCR). He has been the coeditor of 38 conference proceedings and guest editor of several international books and journals. He is editor-in-chief of *Ad Hoc and Sensor Wireless Networks* (with ISI Thomson Impact Factor), the international journal *Network Protocols and Algorithms*, and the *International Journal of Multimedia Communications*, is IARIA Journals board chair (8 journals), has been associate editor of 46 international journals (16 of them with JCR). He has been involved in more than 320 program committees of international conferences and more than 130 organization and steering committees. He leads many national and international projects. He is currently the chair of the Working Group of the Standard IEEE 1907.1. He has been general chair (or co-chair) of 26 international workshops and conferences (chairman of SENSORCOMM 2007, UBICOMM 2008, ICNS 2009, ICWMC 2010, eKNOW 2012, SERVICE COMPUTATION 2013, COGNITIVE 2013, and ADAPTIVE 2013, and co-chairman of ICAS 2009, INTERNET

2010, MARSS 2011, IEEE MASS 2011, SCPA 2011, ICDS 2012, 2nd IEEE SCPA 2012, GreeNets 2012, 3rd IEEE SCPA 2013, SSPA 2013, AdHocNow 2014, MARSS 2014, SSPA 2014 IEEE CCAN 2015, 4th IEEE SCPA 2015, IEEE SCAN 2015, ICACCI 2015 and SDRANCAN 2015 and local chair of MIC-WCMC 2013 and IEEE Sensors 2014). He is currently chair of the 12th AICT 2016, 11th ICIMP 2016, and 3rd GREENETS 2016. He is IEEE Senior and IARIA Fellow.

His publications, h-index and quality of research can be found at

http://scholar.google.es/citations?user=ZJYUEGEAAAAJ

https://www.researchgate.net/profile/Jaime_Lloret2/publications/

http://www.informatik.uni-trier.de/~ley/pers/hd/m/Mauri:Jaime_Lloret

http://www.scopus.com/authid/detail.url?authorId=23389476400

http://orcid.org/0000-0002-0862-0533

http://www.researcherid.com/rid/H-3994-2013

Contributors

Alagan Anpalagan received his BASc (H), MASc, and PhD in electrical engineering from the University of Toronto, Toronto, Ontario, Canada. He joined the Department of Electrical and Computer Engineering at Ryerson University, Toronto, Ontario, Canada, in 2001 and was promoted to full professor in 2010. He served the department as the director for the Graduate Program (2004–2009) and the Interim Electrical Engineering Program (2009–2010). During his sabbatical (2010–2011), he was a visiting professor at the Asian Institute of Technology and visiting researcher at Kyoto University. Dr. Anpalagan's industrial experience includes working at Bell Mobility on 1xRTT system deployment studies (2001), Nortel Networks on SECORE R&D projects (1997), and IBM Canada as an IIP intern (1994). Dr. Anpalagan directs a research group working on radio resource management and radio access and networking areas within the WINCORE Lab. His current research interests include cognitive radio resource allocation and management, wireless cross-layer design and optimization, collaborative communication, green communications technologies, machine-to-machine communication, and small cell networks. He is coauthor/editor of the *Handbook of Green Information and Communication Systems* (Wiley, 2012) and *Routing in Opportunistic Networks* (Springer-Verlag, 2013). Dr. Anpalagan serves as associate editor for the *IEEE Communications*

Surveys & Tutorials (2012–), *IEEE Communications Letters* (2010–), and *Springer Wireless Personal Communications* (2009–). He is past editor for EURASIP *Journal of Wireless Communications and Networking* (2004–2009). He also served as guest editor for special issues in ACM/Springer MONET *Green Cognitive and Cooperative Communication and Networking* (2012), EURASIP *Radio Resource Management in 3G+ Systems* (2006), and EURASIP *Fairness in Radio Resource Management for Wireless Networks* (2008). Dr. Anpalagan served as the IEEE Toronto Section chair (2006–2007), ComSoc Toronto Chapter chair (2004–2005), and the IEEE Canada Professional Activities Committee chair (2009–2011). He is the recipient of the Dean's Teaching Award (2011), Faculty Scholastic, Research and Creativity Award (2010), and Faculty Service Award (2010) all from Ryerson University.

Glaucio H.S. Carvalho received his BS, MS, and PhD, all in electrical engineering, from the Federal University of Para (UFPA), Brazil, in 1999, 2001, and 2005, respectively. From 2005 to 2012, he worked as an associate professor at UFPA in the faculties of information systems, statistics, and computing. From January 2013 to June 2014, he was a visiting postdoctoral fellow at the DABNEL lab, Department of Computer Science at Ryerson University. His research interests include the application of operational research methods for modeling and performance optimization of computer and communication networks.

Sanjay Kumar Dhurandher received his MTech and PhD in computer science from the Jawaharlal Nehru University, New Delhi, India. He is an associate professor and head of the Advanced Centre CAITFS, Division of Information Technology, Netaji Subhas Institute of Technology (NSIT), University of Delhi, New Delhi, India. From 1995 to 2000, he worked as a scientist/engineer at the Institute for Plasma Research, Gujarat, India, which is under the Department of Atomic Energy, India. His current research interests include wireless ad hoc networks, sensor networks, computer networks, network security, and underwater sensor networks. He serves as an associate editor for the *International Journal of Communication Systems*, Wiley.

Emmanuel Duarte-Reynoso is pursuing his PhD in electrical engineering at the Center for Research and Advanced Studies of National Polytechnic Institute (IPN), México City, México. He received his BSc in telematics engineering from the Advanced Technologies and Interdisciplinary Engineering Professional Unit of IPN in 2012. His research interests include resource allocation in mobile cellular communications and applications of game theory in wireless networks.

Kazi R. Islam received his MSc in computer science from Ryerson University, Toronto, Canada. He also received an ME in electrical and computer engineering from Ryerson University, Toronto, Canada. He received his BE in computer science and engineering from Shahjalal University of Science and Technology (SUST) in Sylhet, Bangladesh. He is currently pursuing a PhD in computer science under the supervision of Dr. Faisal Qureshi at the University of Ontario Institute of Technology (UOIT), Oshawa, Canada. His current research interests include sensor networks and computer vision.

Domingo Lara-Rodríguez received his BSc in electronics and communications engineering and MSc and PhD from the National Polytechnic Institute (IPN), México City, México. He is currently with the Communication Section, Department of Electrical Engineering, Center for Research and Advanced Studies of IPN (CINVESTAV-IPN), México City. His research interests include radio resource management, performance evaluation, and architectural design in mobile cellular systems.

Narges Noori received her BSc, MSc, and PhD (with honors) from the Iran University of Science and Technology (IUST), Tehran, Iran, all in electrical engineering, in 1998, 2000, and 2006, respectively. From June 2004 to April 2005, she was with the RF/Microwave and Photonics Group, University of Waterloo, Waterloo, Ontario, Canada, as a visiting scholar. In May 2005, she joined the Research Institute for Information and Communications Technology, previously known as Iran Telecommunication Research Center (ITRC), Tehran, Iran, where she is now working as a research assistant professor. In 2008, she received the best researcher award from the Iran Ministry of ICT. She has participated in and managed several research projects of the radio communication group of ITRC. She has been a senior member of the IEEE since June 2013. Her research interests include beamforming and power control in cognitive radio networks, radio propagation modeling, ultra-wide band communications, and numerical methods in electromagnetics.

Md Mizanur Rahman received his MSc in computer science in 2013 from the Department of Computer Science, Ryerson University, Toronto, Ontario, Canada. He is a PhD candidate in the same department. His research interests include radio resource management in heterogeneous wireless networks and cloud computing.

Fernando Ramírez-Mireles received his PhD in electrical engineering from the University of Southern California in Los Angeles. He has more than five years of industrial experience in companies in the Silicon Valley and more than ten years of experience in academia. He has coauthored three IEC comprehensive technology reports and more than 30 technical articles in areas including UWB, DSL, and speech recognition. His work on UWB is frequently cited, and he holds four patents on ADSL and VDSL. He is a full professor at the Instituto Tecnológico Autónomo de México in México City.

Stefan Schwarz received his BSc in electrical engineering and his Dipl-Ing (MSc equivalent) in telecommunications engineering with highest distinctions in 2007 and 2009, respectively, both at the Vienna University of Technology, Wien, Austria. He also received his Drtechn (PhD equivalent) in telecommunications engineering with highest distinctions in 2013 at the Vienna University of Technology. In 2010, he received the honorary prize from the Austrian Federal Ministry of Science and Research for excellent graduates of scientific and artistic universities, and in 2014, he received the INiTS award in the Information and Communication Technologies category for his dissertation, "Limited Feedback Transceiver Design for Downlink MIMO OFDM Cellular Networks." Since 2008, he has been working as a project assistant at the Mobile Communications group of the Institute of Telecommunications, Vienna University of Technology. His research interests lie in the broad fields of wireless communications and signal processing. In his dissertation, he focused on limited feedback single- and multiuser MIMO-OFDM communication systems and multiuser scheduling algorithms. He actively serves as a reviewer for the *IEEE Transactions on Communications, Signal Processing, Wireless Communications,* and *Vehicular Technology* and for EURASIP *Journal on Signal Processing* and *Journal on Wireless Communications and Networking.*

Isaac Woungang received his MSc and PhD in mathematics from the Université de la Méditerranée–Aix Marseille II, Luminy, France, and Université du Sud, Toulon-Var, France, in 1990 and 1994, respectively. In 1999, he received an MASc in telecommunications from the INRS-Energy, Materials and Telecommunications, University of Quebec in Montreal, Canada. From 1999 to 2002, he worked as a software engineer at Nortel Networks. Since 2002, he has been with the Department of Computer Science at Ryerson University, where he is now a full professor. In 2004, he founded the Distributed Applications and Broadband NEtworks Laboratory (DABNEL) R&D group. His research interests include computer communication networks, mobile communication systems, and network security. Dr. Woungang serves as coeditor-in-chief of the *International Journal of Communication Networks and Distributed Systems* (IJCNDS), Inderscience, UK; associate editor of the *International Journal of Communication Systems* (IJCS), Wiley; and associate editor of the *Computers and Electrical Engineering* (C&EE) journal, Elsevier. Dr. Woungang has edited several books in the areas of networks and pervasive computing, published by Springer, Elsevier, and Wiley. Since January 2012, he has been serving as acting chair of the Computer Chapter of the IEEE Toronto Section.

Marat Zhanikeev received his MS and PhD in global information and telecommunications studies from Waseda University in Tokyo, Japan, in 2003 and 2007, respectively. His research interests include network measurement, network monitoring, and network management, but also extend to the practical applications in these topics. End-to-end network performance in clouds is his specific, immediate target at the time of this writing. He is an associate professor at the Kyushu Institute of Technology (Kyutech) and is a regular member of the IEICE and the IPSJ.

Abbreviations

2G	Second-generation networks
3D	Three-dimensional
3G	Third-generation networks
3GPP	Third-Generation Partnership Project
4G	Fourth-generation networks
5G	Fifth-generation networks
A/V	Audio/video
AAS	Active antenna system
ABS	Almost blank subframe
ADC	Analog-to-digital converter
ADI	Active digital identity
AGC	Automatic gain controller
AMC	Adaptive modulation and coding
AOA	Angle of arrival
AOD	Angles of departure
AP	Access point
API	Application program interfaces
ASK	Amplitude shift keying
AVC	Advanced video coding
AWGN	Additive white Gaussian noise
BAN	Body area network
BBU	Base band unit

BD	Block diagonalization
BER	Bit error ratio
BICM	Bit interleaved coded modulation
BLER	Block error ratio
BPSK	Binary phase shift keying
BS	Base station
CAC	Call admission control
CBR	Constant bitrate
CC	Control channels
CCE	Control channel element
CDD	Cyclic delay diversity
CDMA	Code division multiple access
CDN	Content delivery network
CDN2U	CDN-to-users
CEOFDM	Constant envelope OFDM
CIR	Channel impulse response
CLSM	Closed-loop spatial multiplexing
CMOS	Complementary metal-oxide-semiconductor
CMS	Common mode signaling
CoMP	Coordinated multipoint
CP	Cyclic prefix (based on Chapter 5)
CP	Content provider (based on Chapter 9)
CPMA	Code-modulated path-sharing multiantenna
CP-OFDM	Cyclic-prefix-based OFDM
CQI	Channel quality indicator
CR	Cognitive radio
C-RAN	Cloud-radio access network
CRRM	Common radio resource management
CRS	Cell-specific reference signals
CSI	Channel state information
CSIT	Channel state information at the transmitter
D2D	Device-to-device
DAC	Digital-to-analog converter
DAS	Distributed antenna system
DCI	Downlink control information
DCM	Dynamic coverage management
DMCs	Dense multipath components
DOA	Direction-of-arrival

DoF	Degrees of freedom
DoS	Denial of service
DSA	Dynamic spectrum access
E2E	End-to-End
ECMA	European Computer Manufacturers Association
eICIC	Enhanced intercell interference coordination
eMBMS	Evolved Multimedia Broadcast/Multicast Service
eNBs	Evolved NodeBs
ESM	Effective SINR mapping
FBMC	Filter-bank-based multicarrier
FCC	Federal Communication Commission
FDD	Frequency division duplex
FDMA	Frequency division multiple access
FD-MIMO	Full dimension MIMO
FDTD	Finite-difference time-domain
feICIC	Further enhanced ICIC
FFT	Fast Fourier transform
FSK	Frequency shift keying
GNSS	Global navigation satellite system
GOP	Group of pictures
GPS	Global positioning system
GRAN	GSM EDGE radio access network
GSCM	Geometry-based stochastic channel model
GSM	Global system for mobile communications
HARQ	Hybrid automatic repeat request
HA-WLAN	High-speed access WLAN
HD	High definition
HDMI	High-definition multimedia interface
HeNB	Home eNodeB
HetNet	Heterogeneous network
HSI	High-speed interface
HSPA	High-speed packet access
IA	Interference alignment
ICI	Intercell interference
ICIC	Intercell interference coordination
IF	Intermediate frequency
IFFT	Inverse fast Fourier transform
iid	Independent and identically distributed

IMT	International mobile telecommunication
IoT	Internet of Things
IPTV	Internet Protocol television
ISP	Internet service provider
IT	Information theory
ITU	International Telecommunication Union
ITU-R	International Telecommunications Union Radiocommunication Sector
JCAC	Joint call admission control
JRRM	Joint radio resource management
KKT	Karush-Kuhn-Tucker
LNA	Low-noise amplifier
LOS	Line of sight
LTE	Long-term evolution
LTE-A	LTE advanced
LVC	Layered video coding
M2M	Machine-to-machine
MAC	Medium access control
MANET	Mobile ad hoc network
MBSFN	Multimedia broadcast single frequency network
MCS	Modulation and coding scheme
MET	Maximum eigenmode transmission
MIESM	Mutual information ESM
MIMO	Multiple-input, multiple-output
MMSE	Minimum mean squared error
mmW	Millimeter wave
MNO	Mobile network operators
MPC	Multipath component
MPTCP	Multi-Path Transmission Control Protocol
MRC	Maximum ratio combining
MRT	Maximum ratio transmission
MSE	Mean squared error
MVN	Mobile virtual networks
NAICS	Network-assisted interference cancellation and suppression
NBS	Nash bargaining solution
NCP-SC	Null cyclic prefix single carrier
NFV	Network function virtualization

NLOS	Non–line of sight
NP2AN	Network provider to access network
OFDM	Orthogonal frequency division multiplexing
OFDMA	Orthogonal frequency division multiple access
OLSM	Open-loop spatial multiplexing
OPEX	Operating expenses
oQoE	Objective quality of experience
OSA	Opportunistic spectrum access
P2P	Peer-to-peer
P4P	A version of P2P where cache is maintained by service providers
PA	Power amplifier
PAPR	Peak-to-average power ratio
PDCCH	Physical downlink control channel
PDP	Power delay profile
PHY	Physical layer
PM	Physical machine
PMI	Precoding matrix indicator
PRBs	Physical resource blocks
PSNR	Peak signal-to-noise ratio
PU2RC	Per user unitary rate control
QAM	Quadrature amplitude modulation
QBC	Quantization-based combining
QoE	Quality of experience
QoS	Quality of service
QPSK	Quadrature phase shift keying
RAPs	Random access points
RAT	Radio access technology
RB	Resource block
RBD	Regularized block diagonalization
RE	Resource element
RF	Radio frequency
RI	Rank indicator
RRHNs	Remote radio head nodes
RRM	Radio resource management
RS	Relay station
RVQ	Random vector quantization
SC	Single carrier

SC-FDMA	Single carrier frequency division multiplexing
SDMN	Software-defined mobile networks
SDN	Software-defined network
SDR	Software-defined radio
SiGe	Silicon-germanium
SIMO	Single-input, multiple-output
SINR	Signal-to-interference-plus-noise ratio
SISO	Single-input, single-output
SLA	Service level agreement
SLNR	Signal-to-leakage-and-noise ratio
SMDP	Semi-Markov decision process
SNR	Signal-to-noise ratio
SOI	Silicon-on-insulator
SON	Self-organizing network
SP	Service provider
SQBC	Subspace QBC
SVC	Scalable video coding
SVD	Singular value decomposition
TDD	Time division duplex
TDM	Time division multiplex
TDMA	Time division multiple access
TE	Transverse electric
TM	Transverse magnetic
TOA	Time of arrival
TTI	Transmission time interval
TxD	Transmit diversity
UE	User equipment
UEP	Unequal error protection
UL	Uplink
UMTS	Universal Mobile Telecommunications System
UTD	Uniform theory of diffraction
UTRAN	Universal terrestrial radio access network
UWB	Ultra-wide band
V2V	Vehicle-to-vehicle
VBR	Variable bit rate
VCO	Voltage controlled oscillator
VM	Virtual machine
VR	Visibility region

VWU	Virtual wireless user
WCDMA	Wideband Code Division Multiple Access
Wi-Fi	Wireless fidelity
WiMAX	Worldwide interoperability for microwave access
WLAN	Wireless local area network
WN	Wireless network
WPAN	Wireless personal area network
WSP	Wireless service provider
WU	Wireless user
ZF	Zero-forcing

1

ADVANCES IN 4G COMMUNICATION NETWORKS

A 5G Perspective

M. BALA KRISHNA AND STEFAN SCHWARZ

Contents

1.1 Introduction

Long-term evolution advanced (LTE-A) and fourth-generation (4G) communication networks implement a multicellular architecture over traditional IP networks and offload the network traffic from a macro base station to multiple small-cell base stations. This approach improves network performance. Advanced coding, multiplexing (orthogonal frequency division multiplexing [OFDM]), multiple access (orthogonal frequency division multiple access [OFDMA]), intercell interference (ICI) mitigation (enhanced intercell interference coordination [eICIC]), and antenna synchronization (multiple-input-multiple-output [MIMO]) techniques are used in 4G and beyond networks to improve spectrum efficiency and achieve high through-put rates (100 Mbps to 10 Gbps). Even though 4G networks accommodate heterogeneity in network interfaces and device connectivity, challenging aspects such as exploding data, dynamic network traffic technology, efficient spectrum allocation, resource sharing, and energy management are yet to be resolved. Software-defined radio (SDR) and cognitive radio (CR) networking features are yet to be explored for managing spectral and network resources.

The limitations of 4G networks are due to (1) the increasing demands of ever-growing devices and users; (2) requirement of robust, reliable, and high-quality audio and video streaming services; (3) coordinating macro- and small-cell networks; (4) computational complexities in dense deployment of small cells; (5) sustaining high mobility and addressing handover issues; (6) dynamic network topology for varying spectrum allocation; (7) minimizing end-to-end latencies; and (8) data transmission in high-speed vehicular networks.

1.2 Evolution toward 5G Networks

Fifth-generation (5G) networks aim to achieve high-speed data transfer rates (in tens of Gbps) to meet the exploding demands of millions of users per square kilometer. High user density escalates the network traffic to

Table 1.1   Requirements, Technology, and Addressing Issues in 5G Networks

REQUIREMENTS	INTENDED TECHNOLOGY	ADDRESSING ISSUES
Spectrum management and energy efficiency	Cognitive radios (CRs), software-defined radio, and software-defined networks (SDR-SDN) technology [2]	Efficient usage of spectrum and networking resources over the licensed and unlicensed spectrum
Improvement of area spectral efficiency	Massive multiple-input multiple-output (MIMO) [5], spatial MIMO	Intercell interference (ICI), reduction in noise and fading
Dense deployment of small cells	Cloud-RAN, multi-RAT systems	Synchronizing distributed antennas, operational and maintenance cost
Peak data rates reaching 10 Gbps and cost-effective services	Millimeter wave technology at backhaul networks	Standardizing the services of short-range and broadband applications
Reduction in delivery latency for reliable and robust services	Device-to-device and end-to-end systems, machine-type and human-type systems	Providing data, voice, and video services to end users.
High mobility	Mobility models and vehicular systems	Mobility levels reaching up to 500 km/h

Sources: Mumtaz, S. et al., *IEEE Wireless Commun. Mag.*, 21(5), 14, October 2014; Condoluci, M. et al., *IEEE Commun. Mag. Extrem. Dense Wireless Networks*, 53, 134, January 2015; Lai, C.-F. et al., *IEEE Network Mag.*, 29(1), 49, January–February 2015; Wang, R. et al., *IEEE Access*, 2, 1187, October 1, 2014; Wang, C.-X. et al., *IEEE Commun. Mag.*, 52(2), 122, February 2014.

TB/km and user mobility to 400–500 km/h. Latencies are minimized by considering device-to-device (D2D) [1] and machine-to-machine (M2M) supporting systems in 5G networks. Table 1.1 enumerates the requirements, technology, and addressing issues in 5G networks.

1.3 Challenges in 5G Networks

The challenges in 5G networks are as follows:

- Increase in spectrum bandwidth, transmission rate, and traffic density
- Design of small-cell infrastructure [1] (based on femtocells and microcells with eNBs) to increase the spectrum efficiency and decrease the load of base station
- Ultradense networks, multiple radio access technologies (M-RATs), and mobile crowd sensing
- Exponential increase in data traffic based on multimedia live streaming, multivideo conferencing, Internet Protocol television (IP-TV), and so on

- D2D and M2M [2] connectivity
- Mobility management for high-speed vehicular network
- Energy management in cellular communication
- Network virtualization and software-defined network (SDN)
- Mobile vehicular cloud management

1.4 Emerging Trends in 5G Networks

5G communication systems aim to extend the services of wireless technology with respect to cloud services, heterogeneous device connectivity, vehicular mobility, and distinct connectivity for indoor and outdoor users using multiple MIMO antennas. Indoor wireless communication for short-range distance uses Wi-Fi, femtocells, ultra-wideband (UWB), and millimeter wave (mmWave) technology. Outdoor wireless communication for long-range distance uses microcells, macrocells, large antenna arrays, and intermediate relay head nodes to communicate with indoor systems and gateway nodes in the network.

1.4.1 Multiple Radio Access Technology (M-RAT)

M-RAT supports a broad range of radio access technologies [6] such as 2G, 3G, Third Generation Partnership Project (3GPP), and 4G networks under a coexistent networking system [4] known as multiple radio access networks. Traditional high-speed access wireless local area network (WLAN) operates with a number of access points to support a high density of user and devices in the network. The M-RAT system offloads the network traffic from high-density areas to adjacent microcells or picocells to facilitate seamless connectivity and handover management in the cellular network. M-RAT systems use unlicensed and licensed spectrum bands that allow cooperative reuse of Global System for Mobile Communications (GSM) spectrum in LTE networks to achieve a high throughput rate.

1.4.2 OFDM and Multiple MIMO Systems

OFDM and multiple MIMO technologies [1] improve the pairing and spectrum reuse capabilities that allocate an optimum number of physical resource blocks in the communication channel. D2D communication systems upgrade the number of devices through radio link control and packet data control techniques [5]. Communication system is enhanced by using diverse multiple accessing schemes for uplink (single-carrier FDMA) and downlink (OFDMA) signals. Control mechanisms are applied in physical and medium access control (MAC) layers of LTE-A to regulate the network traffic.

1.4.3 Device-to-Device Communication Systems

Device-to-Device (D2D) communication systems are short-range systems that utilize licensed and unlicensed spectrum bands to share the radio access and minimize the load in the base station. D2D communication systems [1] are classified as follows:

1. Inband D2D systems support device communication using the licensed spectrum from adjacent base stations and minimize the intercellular interference. Small-cell base stations use underlay (same resources for all) and overlay (dedicated resources for specified users) techniques for inband D2D communication.

2. Outband D2D systems support device communication using the unlicensed spectrum that is supported by technologies such as Wi-Fi, ZigBee, and Bluetooth. Outband D2D systems considerably reduce the adjacent interference from dedicated inband devices. Outband services are implemented in two modes: (a) In the controlled mode, the central base station single-handedly controls the secondary network interfaces; and (b) in the autonomous mode, the communicating devices choose the secondary interface networks as per their requirements. For outband D2D operations, the user devices must be compatible for switching between LTE and Wi-Fi systems.

1.4.4 Software-Defined Networks

Future international mobile telecommunication aims to achieve high spectrum efficiency, increase throughput rate, high mobility, heterogeneous device connectivity (D2D, M2M, etc.), and network virtualization. The features of SDNs are network configuration [3], network management, network virtualization, access points, servers, routers, and switches in the network. SDNs scan licensed and unlicensed spectrum bands and efficiently utilize available frequencies for devices working on multiple interfaces.

SDNs use scale-up and scale-down modes to enhance network services. Software-defined mobile networks support adaptable and dynamic configuration of network resources to achieve optimal connectivity between the mobile nodes in a network. SDR concentrates on resource and spectrum allocation, and SDN concentrates on network connectivity and path management. The cross-layer approach [7] integrates the services of SDR and SDN in the design of 4G and beyond networks.

1.4.5 Cloud Technologies for 5G RANs

Cloud-RAN, or centralized-RAN (C-RAN), is a scalable and flexible architecture comprising (1) remote radio head nodes (RRHNs), (2) dense deployment of low-power base stations [8] supported by network function virtualization (NFV), and (3) SDN. C-RAN consists of baseband units with centralized processors and RRHNs with remote antennas connected by a front-haul network to achieve high data rates and minimum latency in the network. C-RAN [4] with multiple-RATs enhances radio resource management, load balancing, latency reduction, mobility management, traffic steering, and service innovation schemes in the network. With emerging trends in cloud computing [9], radio access network as a service (RANaaS) (partial centralization of radio access networks) [10] enables the features of NFV and SDN that are used to configure and manage user devices connected to cellular and mobile networks. Random access points (RAPs) are deployed at optimum distances from macrocell and small-cell base stations to offload the network traffic, reduce packet latency, and increase data transfer rate. Deployment of RAPs near picocells and femtocells enables the macro user nodes with uninterrupted services. RANaaS coordinates with the macro base station and local RAPs of the network. The combination of RANaaS with RAPs and SDN-based transport node forms the virtual eNB in 5G cloud networks, where the transport nodes are managed by network controllers and gateway nodes. 3GPP-RAN aims at designing energy-efficient cellular networks supported by eNB and self-organized network technology.

1.4.6 Machine-Type and Human-Type Communications

With enhancements in LTE-A systems, machine-type (MT) [2] and human-type (HT) communication systems are considered in 5G networks. MT communication systems use remote servers such as actuators and smart meters that operate without the intervention of networking operators. MT communication systems enable the macrocells to coordinate with small cells such as femto and pico and micro base stations to allocate resources in the network. HT communication systems coordinate with application servers and networking devices, and the network traffic is rerouted based on macrocell resources. LTE-A systems support MT/HT communication systems to reduce installation and maintenance costs in 5G networks. With LTE-A systems as backbone networks, the MT systems control and coordinate with application server nodes and service capability server nodes. Application servers coordinate with network operators and service providers to monitor the active connections in control plane and the number of closed and open access groups associated with the base station. Application servers allocate resources based on network traffic conditions.

1.4.7 New Carrier-Type Cell Systems

Traditional 3GPP and LTE standard systems use cell-specific reference signals that enable the transmitters to be vigilant even at low-traffic conditions, leading to the deprival of energy resources. Cloud-RAN-based 5G networking systems implement two-plane communication systems, also known as the new carrier-type cell systems [6]. New carrier-type cell systems define different tasks to the control and user planes. The functioning of control and user planes is discriminated as (1) control plane, which addresses the issues related to the connection and mobility of user devices, and (2) user plane, which addresses the issues of data transmission. Simultaneous macro- and small-cell connectivity is provided to the user device. The user equipment (UE) is allowed to connect to a femto-/picocell (for user plane operations) while maintaining connectivity with the macrocell (for control plane operations). The macrocells are allowed to use low-frequency spectral bands for control signals and high-frequency spectral bands for data transmission. This technique considerably reduces interference and improves the throughput levels of 5G networks.

1.5 LTE/LTE-A 4G and Beyond Technology

Cellular networks are currently experiencing a tremendous growth of data traffic. A 10- to 11-fold increase of global mobile data traffic between 2013 and 2018 is predicted by several studies [11,12]. A few notable statements of Cisco's Visual Networking Index [11], illustrating these trends, are as follows:

- Mobile data traffic in 2013 was nearly 18 times the size of the entire Internet in 2000.
- Average smartphone usage grew 50% in 2013.
- Smartphones represented only 27% of total global handsets in use in 2013, but represented 95% of total global handset traffic.

The predicted trends in global mobile traffic, according to Ericsson [12], are shown in Figure 1.1a, illustrating the expected exponential growth in traffic due to data services (video, web browsing, e-mail, etc.). The driving forces behind the expected traffic explosion are new applications enabled by modern devices such as smartphones and tablet computers, while the classical role of the mobile as a phone becomes less and less significant.

To cope with such ever-increasing demands, 3GPP is constantly enhancing its Universal Mobile Telecommunications System (UMTS) cellular network standard. The latest evolution of 3GPP radio access technology is denoted as UMTS LTE. The global acceptance of 3GPP's cellular network standards is illustrated in Figure 1.1b. Currently (2014), GSM is still dominating global mobile subscriptions; however, with increasing application of smartphones, LTE subscriptions are predicted to outnumber other technologies in the mid-2020s.

1.5.1 Overview of LTE Release 8/9 Features

3GPP initiated standardization of LTE in 2004 with the goal to enhance UMTS and to provide a long-term replacement of the High Speed Packet Access radio access technology. The first release of UMTS LTE (Release 8) was finalized in 2008. LTE employs OFDMA as modulation format and multiple access scheme. It utilizes adaptive modulation and coding to adapt the transmission rate to the current channel conditions, exploiting channel state information (CSI)

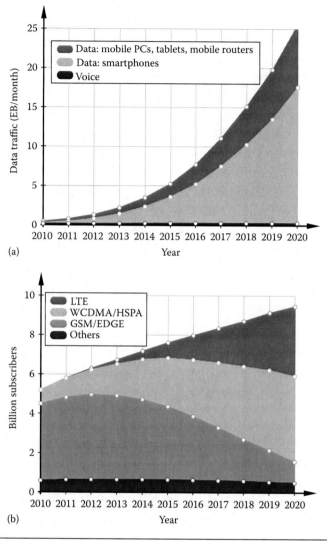

(a)

(b)

Figure 1.1 Estimated growth of (a) global mobile data and voice traffic and (b) global mobile subscriptions per technology. *Note*: 1 EB = 10^{18} bytes. (From Ericsson, Traffic exploration tool, Online, December 10, 2014, http://www.ericsson.com/TET.)

feedback from UE. LTE implements a bit interleaved coded modulation architecture, combining 4/16/64 quadrature amplitude modulation with a powerful turbo code for forward error correction. It applies fast physical layer (PHY) retransmissions in case of erroneous packet reception, employing hybrid automatic repeat request. LTE supports MIMO technology, enabling beamforming, diversity, and spatial multiplexing gains through standard-defined precoding schemes;

transmissions from up to four antenna ports are supported. For more details on LTE PHY, the interested reader is referred to [14,15].

In Release 9 of UMTS, finalized in 2009, several additional features have been added to further enhance the LTE standard. Evolved multimedia broadcast and multicast service (eMBMS) has been standardized to enable broadcast/multicast transmissions over cellular networks, targeting applications such as digital audio and video broadcasting. eMBMS supports multimedia broadcast single-frequency network to substantially improve cell edge performance in broadcast/multicast scenarios. Positioning methods have been introduced to pinpoint the location of mobile devices, facilitating not only commercial location-based services but also safety-relevant positioning. Also, the Home eNodeB or LTE femtocell has been added in Release 9, which operates as residential, enterprise, or outdoor access point to locally provide better coverage and higher capacity. Finally, self-organizing network capabilities have been extended to further reduce OPEX and automatically improve network quality. A detailed discussion of Release 9 features is available in [16].

1.5.2 LTE Release 10 (LTE-A) Enhancements

Legacy LTE was not able to meet the official 4G requirements as specified by the International Telecommunication Union [17], mostly due to missing MIMO support for uplink transmissions. Hence, LTE was enhanced by the 3GPP in 2011 to not only meet but even surpass these 4G requirements in Release 10, also known as LTE-A. This was achieved through several enhancements, such as higher-order MIMO (8-layer downlink and 4-layer uplink), carrier aggregation, support of relay nodes, and heterogeneous networks (HetNets), as well as eICIC; see [18].

Carrier aggregation in Release 10 enables bandwidth expansions of up to 100 MHz by serving users over multiple carriers as shown in Figure 1.2a. This figure illustrates the three different aggregation concepts that have been standardized, that is, *contiguous aggregation* of neighboring carriers; *noncontiguous aggregation* of carriers in the same band that are, however, separated in frequency; and *interband aggregation*, stretching over multiple frequency bands. Potential application scenarios for carrier aggregation are illustrated in Figure 1.2b. The upper-left eNodeB employs two frequency bands $F1 < F2$ to

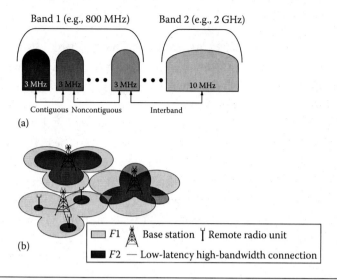

Figure 1.2 Illustration of (a) carrier aggregation concepts and (b) potential carrier aggregation scenarios.

serve the same area; the lower-left eNodeB utilizes frequency band $F2$ to provide local hot-spot capacity, and the right eNodeB applies the two carrier frequencies to cover each other's cell edge.

Relaying techniques are employed to enhance the coverage and capacity of cellular networks; they are mostly applied to improve coverage in indoor locations, rural areas, or otherwise dead zones. Relays in LTE are distinguished by the functionality they implement:

- Layer 1 or *amplify and forward* relays simply retransmit signals in the radio-frequency (RF) domain.
- Layer 2 or *decode and forward* relays decode and reencode data before retransmission.
- Layer 3 relays appear to UE as normal eNodeBs and, conversely, to eNodeBs as UE.

HetNets enable the deployment of low-power base stations (small cells) and other radio equipment, such as RRHs and relays, to primarily improve network capacity in hot-spot locations, such as city centers, shopping malls, and airports. Small cells mostly distinguish from macro base stations due to their much reduced transmit power and much smaller equipment size, providing the advantage of easier and cheaper site acquisition. The LTE femtocell, introduced in Release 9, is one example of a small cell; in Release 10, micro- and picocells have been added. Micros, picos, and femtos are distinguished by their corresponding cell sizes (in decreasing order). Such a mix of different cell sizes complicates network planning and often requires coordination of multiple access points to mitigate interference.

ICIC has already been included in Release 8 of LTE as simple means to reduce intercell interference specifically at the cell edge. This is achieved by coordination of traffic channel resource allocations in the time–frequency domain via the X2 interface. eICIC extends such coordination principles to deal with interference issues in HetNet scenarios. Specifically, the almost blank subframe (ABS) has been introduced to mute macrocells in specific subframes, such as to avoid excessive interference to small cells in the same area; in ABSs, only control information is radiated from macrocells.

1.5.3 Further Enhancements of Releases 11, 12, and Beyond

The main feature introduced in Release 11 of LTE in 2013 is coordinated multipoint transmission (CoMP), which enables dynamic coordination of transmission and reception over multiple base stations, to mitigate or even exploit interfering signals. CoMP schemes are classified into three categories, as illustrated in Figure 1.3a:

1. *Coordinated scheduling* schemes coordinate the assignment of time–frequency domain resources among the coordination set (set of coordinated base stations); ICIC methods belong to this category.

2. *Coordinated beamforming* schemes exploit multiple antennas at the base stations to form favorable transmission beams. A performance criterion that is frequently optimized is the signal to leakage and noise ratio [19], that is, the ratio of signal power arriving at the intended user and interference leakage caused to other users.

3. *Joint transmission* techniques utilize antennas at multiple spatially distributed transmission points (e.g., base stations) to form a single virtual MIMO antenna array. Such methods promise the largest gains in terms of performance; however, they also impose the strictest requirements for backhaul connections and available CSI at the transmitter.

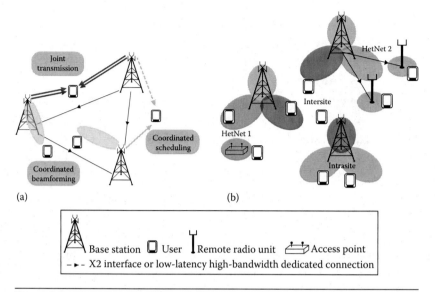

Figure 1.3   Illustration of (a) different CoMP concepts and (b) distinction of CoMP scenarios.

Different CoMP scenarios are distinguished depending on the network equipment involved, as shown in Figure 1.3b. With intrasite CoMP, cells belonging to the same eNodeB are coordinated, requiring no extra backhaul connection. Intersite CoMP involves several spatially distributed macro base stations that are coordinated, for example, over the X2 interface. HetNet CoMP specifies coordination in heterogeneous networks, that is, coordination of small cells, relay nodes, as well as RRHs.

Another feature of Release 11 is further enhanced ICIC (feICIC), dealing with interference control for cell-specific reference signals (CRS); such reference signals cannot be muted within the ABS of Release 10 to ensure backward compatibility with Release 8/9. In Release 11, transmit- and receive-side methods are proposed to either avoid or remove interference due to CRS. Transmit-side feICIC relies on muting data channels on resource elements that experience strong interference due to CRS. Receive-side feICIC, on the other hand, applies interference cancellation at the UE to eliminate the dominant CRS interference.

With Release 12, dual connectivity has been included into the LTE standard in 2014, which enables simultaneous connection of UEs with two different network access points, that is, a master eNodeB and a secondary eNodeB. This is especially advantageous for improved mobility support in HetNet scenarios. Here, the macro base station can act as master eNodeB and provide continuous control information, while data are provisioned by small cells without requiring frequent handovers in between small cells.

Finally, network-assisted interference cancellation and suppression has been studied in Release 12 to evaluate the performance of advanced interference cancellation and suppression receivers with and without network assistance. It has been concluded that such advanced interference-aware receivers can substantially outperform simple interference-rejection combining, as investigated in Release 11. Furthermore, it has been shown that receiver complexity can be reduced if some of the interference parameters, such as applied modulation and coding schemes, are signaled to the UE, thus avoiding blind detection of all interference parameters. Further studies are, however, required before specifying the necessary network assistance signaling.

Several other study items are currently being discussed within 3GPP for potential future inclusion in the LTE standard, such as elevation beamforming and full-dimension MIMO (FD-MIMO), enhancements for MT communications, and operation of LTE in unlicensed spectrum; some of these topics are discussed further later. A more detailed discussion of the features of different LTE-A releases is provided in [20].

1.6 MIMO Enhancements: 3D Beamforming, Full-Dimension MIMO, and Massive MIMO

1.6.1 3D Beamforming and Full-Dimension MIMO

Macro base stations in current cellular networks mostly apply MIMO antenna arrays that enable controlling the beam pattern radiation in the horizontal plane only. Although most such antenna arrays consist of several antenna elements that are stacked vertically to obtain a narrow vertical beam width (e.g., 10°–20°), these vertically stacked antenna elements are all driven with the same signal, hence not allowing for controllable beam patterns in the vertical plane. The elevation domain is exploited only by (mechanically or electrically) tilting the antenna with respect to the vertical axis as illustrated in Figure 1.4a, with the aim to improve the coverage of the served area and to reduce

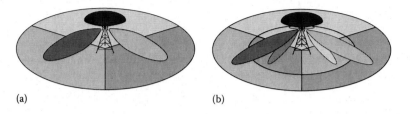

(a)   (b)

Figure 1.4   Comparison of (a) conventional horizontal sectorization and (b) mixed horizontal–vertical sectorization.

interference to other cells. If multiple such antenna arrays are stacked horizontally, the azimuthal radiation pattern can be changed by feeding the individual arrays with appropriate signals, forming one or several beams that point into different angles in the horizontal plane. The idea of 3D beamforming, also known as elevation beamforming, is to extend such 2D beamforming approaches to the elevation domain, employing vertically stacked antennas. This can increase signal strength at the receiver by pointing the vertical main lobe toward the user, and it can reduce intercell interference, especially when serving users that are close to the center of the cell.

A simple static method to exploit the elevation domain is to apply vertical sectorization in addition to horizontal sectorization as proposed, for example, in [21,22] and illustrated in Figure 1.4b. With optimized beam pattern designs and sectorization into three horizontal times three vertical sectors, gains in the order of 200% in average cell throughput compared to traditional horizontal-only sectorization have been observed in [21]. By employing state-of-the-art antenna patterns, improvements in spectral efficiency and cell edge user throughput are also reported in [22]; however, it is not that significant.

Static elevation beamforming can be further extended to dynamic elevation beamforming if the electrical down-tilt of the antenna arrays can be controlled dynamically, enabling to point the vertical main lobe directly toward the intended receiver. The electrical down-tilt is in many cases controlled in the RF domain; hence, such dynamic elevation beamforming does not necessarily require an individual RF chain for each vertical antenna element but can also be achieved with analogue beamforming. As shown in [22], dynamic elevation beamforming outperforms the static approach; however, the minimum down-tilt must be limited to avoid excessive intercell interference otherwise caused by transmissions to users that are located close to the cell edge.

If each antenna element in the horizontal and the vertical plane is active, that is, is equipped with its own RF chain, the 3D beam pattern can be adjusted dynamically by employing adaptive baseband precoding across all antenna elements. Such FD-MIMO active antenna systems [23] enable the application of sophisticated multiuser MIMO and spatial coordination schemes over 2D antenna arrays, as illustrated in Figure 1.5. As shown in [22], the elevation domain is especially advantageous in urban scenarios, when serving users in different floors of high-rise buildings from a common base station outside the buildings. In comparison to 1D antenna arrays, the application of 2D arrays has the practical advantage of reducing the form factor of the antenna; at 2.6 GHz, an 8 × 8 planar array with 0.5λ spacing has a size of approximately 50 cm × 50 cm, whereas a linear array of 64 antennas is around 4 m wide. However, when serving multiple users in parallel, with the users being approximately located on the same level, the performance of linear arrays is better because much sharper azimuthal beams can be formed, thus reducing the multiuser interference [23].

Due to the gains achieved by 3D beamforming, 3GPP has opened study items on this topic in Release 12. As a first step to obtain realistic performance investigations, a 3D channel model has been specified in [24]; further specifications building on this channel model are expected in Releases 13 and beyond.

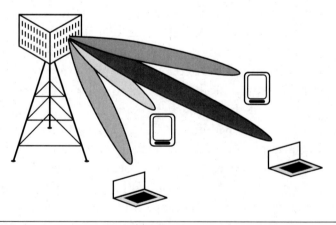

Figure 1.5   Illustration of full-dimension MIMO using a 2D array of active antennas.

1.6.2 Massive MIMO

Massive MIMO systems employ very large antenna arrays at the base station, consisting of several hundreds to even thousands of (collocated or distributed) active antenna elements, to serve a much smaller number of several tens of users equipped with few antennas by means of spatial multiplexing [25]. This operating condition brings along several advantages as detailed in [26,27]:

- The network capacity and energy efficiency can be drastically improved. Capacity gains are obtained by simultaneously serving a large number of users. Energy efficiency is increased because the transmit signal can be focused very precisely on the intended receiver, avoiding waste of signal energy.

- In the limit of infinite number of antennas at the base station, but with only a single antenna per user, maximum ratio transmission (MRT) in the downlink, respectively maximum ratio combining (MRC) in the uplink, becomes optimal. These strategies can be implemented very efficiently, simply by multiplying the received signal on each antenna by the conjugate of the channel response. This effect is known as *favorable channel conditions* in which the channels of different users become asymptotically orthogonal. Even with a finite number of antennas, say, 100, MRT/MRC performs very well, provided the number of served users is around one magnitude below the number of base station antennas.

- Expensive high-power amplifiers and RF components can be replaced with corresponding inexpensive low-power equipment. This is because with increasing number of transmit antennas, the transmit power per antenna element shrinks proportionally. Such low-power equipment is much cheaper and energy efficient than high-power components.

There are, however, several issues that need to be addressed and resolved to make massive MIMO practically feasible:

- Spatial multiplexing of users requires accurate CSI at the base station. While this can be accomplished easily in the uplink, by inserting uplink pilot signals, downlink CSI is harder to obtain. Most investigations in literature assume time division duplex (TDD) transmission with a small enough duty cycle, such that uplink CSI can be reused for the downlink transmission, exploiting channel reciprocity. Reciprocity, however, in general, only holds for the transmission medium and not for the transceiver chains; hence, careful transceiver calibration is required to guarantee reciprocity. TDD is not the preferred operation mode of most cellular network deployments nowadays as it requires discontinuous transmission, degrading the performance of RF power amplifiers. Furthermore, base stations need to be accurately synchronized with respect to uplink/downlink transmissions to avoid uplink–downlink intercell interference.

• Uplink channel estimation relies on orthogonal uplink pilot sequences to identify the channels of different users. The number of such sequences is fundamentally limited by the coherence time–bandwidth product of the channel [25]. In many cases, the number of users in the network will be substantially larger than the number of available orthogonal pilot sequences, thus requiring either to reuse the same sequences in different cells or to employ not perfectly orthogonal sequences. Both solutions lead to pilot contamination, causing interference at those terminals that apply mutually nonorthogonal sequences. Solutions to this problem rely either on clever assignment of pilot sequences among users exploiting, for example, second-order statistics of the channels [28], or on sophisticated blind channel estimation algorithms [29].

1.7 mmWave Communication Technology

1.7.1 mmWave for 5G Cellular

Today's mobile communication networks operate mostly in frequency bands below 3 GHz, typically around 800 MHz and 2 GHz. This frequency operating regime has not changed much since first-generation cellular systems as it is well suited to achieve good coverage with relatively few macro base stations. The ever-increasing demand for larger network capacity, however, requires aggressive densification of cellular networks to achieve a spatial reuse gain of the scarce resource bandwidth. In ultradense networks, with small-cell access points on every "lamp post," only a small area needs to be covered by each small cell, which can be achieved with much higher operating frequencies. Hence, research interest has recently shifted to mobile communication systems operating at centimeter wavelengths (3–30 GHz) and even at millimeter wavelengths (30–300 GHz) [30]. The major advantage of using frequencies above 10 GHz is the availability of large amounts of continuous spectrum chunks in the order of GHz [31], enabling transmission rates of multiple Gbps and user. A further advantage is the reduction of antenna dimensions proportionally to the wavelength, enabling the placement of large numbers of antennas even in mobile devices and making mmWave a natural candidate for massive MIMO systems [32].

The mmWave regime, however, has several drawbacks that have to be addressed to enable practical mmWave cellular systems. According to the Friis transmission equation, keeping the gains of transmit and receive antennas constant, the path loss increases with decreasing wavelength [33]. This effect, however, does not mean that the electromagnetic wave itself is stronger attenuated at higher frequencies. It just appears like this in the Friis equation due to shrinking antenna sizes with decreasing wavelength because the antenna gains are assumed as fixed. On the other hand, keeping the antenna size constant, the received signal strength can even be improved at mmWave under line-of-sight conditions because of increasing transmit and receive antenna gains.

Yet achieving these gains requires well-aligned transmit and receive beams and thus necessitates accurate CSI at the transmitter and the receiver. More problematic than line-of-sight propagation, though, is non-line-of-sight propagation, which is a typical situation in mobile communications. Due to reduced diffraction, worse reflection, and larger material attenuation, shadowing effects are much more severe in the mmWave-band than below 3 GHz [34]. Still, non-line-of-sight propagation does occur and can be exploited in mobile communications to improve coverage. Nevertheless, outages are more likely in mmWave propagation and have to be counteracted by means of macroscopic diversity techniques.

Due to these restrictions, mmWave systems have to be considered as an add-on to existing macro cellular networks, locally providing

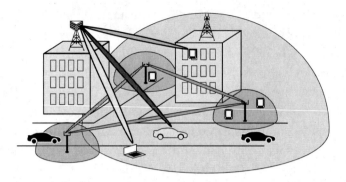

Figure 1.6   Illustration of a heterogeneous cellular network employing several 5G technologies.

large bandwidth and capacity in hot-spot locations. The future cellular network will most likely be a heterogeneous mixture of many different types of network access nodes as illustrated in Figure 1.6 [35,36]. Global coverage and mobility management will be provided and maintained by macro base stations, while small cells and massive MIMO systems supply network capacity and bandwidth. mmWave will be applied mostly on small cells as well as to establish wireless backhaul connections, avoiding the need for expensive wired backhaul.

1.7.2 60 GHz WLAN and WPAN

WLAN and wireless personal area network (WPAN) communities have also recognized the advantages of mmWave communication and published the IEEE 802.11ad amendment of the Wi-Fi standard, also known as WiGig, in 2012 [37]. WiGig operates in the 60 GHz band on multiple 2 GHz bandwidth data channels that each supports 7 Gbps. 802.11ad is intended for short-range (<20 m) indoor wireless communication, with applications such as uncompressed high-definition (HD) video streaming for HD television, instant wireless file synchronization between computers and handheld devices, wireless Gigabit Ethernet, and wireless gaming, ensuring very low latency. The 60 GHz band offers a total bandwidth in the order of 4–7 GHz depending on national regulations. In comparison to mmWave cellular systems, application of mmWave in WLAN/WPAN has the advantage that only very small areas (typically single rooms) have to be covered; thus, path loss and shadowing effects are not significant issues. The increased path loss can even be an advantage as it facilitates dense deployments of WiGig devices with low interference levels [38].

Other standardization groups have taken up the mmWave topic as well. The IEEE 802.15.3c amendment of the UWB WPAN standard provides an mmWave-based PHY alternative to existing 802.15.3 specifications, which allows for data rates over 1 Gbps [39]. Also, the European Computer Manufacturers Association standardized PHY and MAC for 60 GHz wireless networks [40] and the WirelessHD Consortium developed 60 GHz technology for wireless HD video links, supporting data rates of up to 28 Gbps [41]. A comprehensive overview of 60 GHz technology for WLAN/WPAN is available in [42].

1.8 Conclusions

Multicellular architecture over traditional IP networks offloads the network traffic from a macro base station to multiple small-cell base stations in LTE-A and emerging 4G networks. Carrier aggregation enables bandwidth expansions of 100 MHz by serving users over multiple carriers. HetNets enable the deployment of low-power base stations (small cells) and other radio equipment, such as RRHs, to primarily improve network capacity in hot-spot locations, such as city centers, shopping malls, and airports. CoMP mitigates interfering signals based on coordinated scheduling and beamforming, and massive MIMO systems increase the coverage area of 4G networks. Master eNodeB and secondary eNodeB improve mobility support in HetNets. Fifth generation (5G) networks aim to achieve high-speed data transfer rates (Gbps) to meet the exploding demands of millions of users per square kilometer. The main challenges in 5G networks are ultradense networks, multiple radio access technologies (M-RATs), and mobile crowd sensing.

1.9 Future Research Directions

Future research includes the study of spectral bandwidth and usage of white spaces through cognitive radio technology in HetNets and designing of flexible networking topologies with public–private partnerships. Improving the deployment strategies of small cells such as pico- and femtocells for indoor applications and addressing the issue of intercellular interference and power losses need to be considered. Mobility management in ultradense networks and traffic engineering in HetNets need to be developed. Massive MIMO and millimeter wave technologies with adaptive technologies need to be considered for multicellular HetNets. Optimal allocation of small-cell base station to conserve energy and minimize signal interference needs to be considered in 5G networks. Cell range expansion based on resource sharing and traffic conditions and cross-tier channel allocation methods need to be designed for HetNets.

List of Abbreviations

2G	Second-Generation Networks
3D	Three-Dimensional
3G	Third-Generation Networks
3GPP	Third-Generation Partnership Project
4G	Fourth-Generation Networks
5G	Fifth-Generation Networks
A/V	Audio/video
AAS	Active Antenna System
ABS	Almost Blank Subframe
AP	Access Point
API	Application Program Interfaces
CoMP	Coordinated Multi Point
CR	Cognitive Radio
C-RAN	Cloud-Radio Access Network
CRS	Cell-specific Reference Signals
CSI	Channel State Information
D2D	Device-to-Device
E2E	End-to-End
ECMA	European Computer Manufacturers Association
eICIC	Enhanced Intercell Interference Coordination
eMBMS	Evolved Multimedia Broadcast and Multicast Service
eNBs	Evolved NodeBs
FDD	Frequency Division Duplex

FDMA	Frequency Division Multiple Access
FD-MIMO	Full-Dimension MIMO
feICIC	Further enhanced ICIC
GSM	Global System for Mobile Communications
HA-WLAN	High-Speed Access WLAN
HARQ	Hybrid Automatic Repeat Request
HD	High Definition
HDMI	High-Definition Multimedia Interface
HeNB	Home eNodeB
HetNet	Heterogeneous Network
HSPA	High Speed Packet Access
ICI	Intercell Interference
ICIC	Intercell Interference Coordination
ITU	International Telecommunication Union
LTE	Long-Term Evolution
LTE-A	LTE Advanced
M2M	Machine-to-Machine
MAC	Medium Access Control
MANET	Mobile ad hoc network
MIMO	Multiple-Input Multiple-Output
mmW	millimeter Wave
MRC	Maximum Ratio Combining
MRT	Maximum Ratio Transmission
NAICS	Network-Assisted Interference Cancellation And Suppression
NFV	Network Function Virtualization
OFDM	Orthogonal Frequency Division Multiplexing
OFDMA	Orthogonal Frequency Division Multiple Access
PHY	Physical layer
QAM	Quadrature Amplitude Modulation
RAPs	Random Access Points
RAT	Radio Access Technology
RF	Radio Frequency
RRHNs	Remote Radio Head nodes
SDMN	Software-Defined Mobile Networks
SDN	Software-Defined Network
SDR	Software-Defined Radio
SLNR	Signal to Leakage and Noise Ratio

SON   Self-Organizing Network
TDD   Time Division Duplex
TDM   Time Division Multiplex
TDMA   Time Division Multiple Access
UE   User Equipment
UMTS   Universal Mobile Telecommunications System
UWB   Ultra-wideband
Wi-Fi   Wireless Fidelity
WiMAX   Worldwide Interoperability for Microwave Access
WLAN   Wireless Local Area Network
WN   Wireless Network
WPAN   Wireless Personal Area Network

References

1. Mumtaz, S., Huq, K. M. S., and Rodriguez, J., Direct mobile-to-mobile communication: Paradigm for 5G, *IEEE Wireless Communications Magazine*, October 2014, 21(5), 14–23.
2. Condoluci, M., Dohler, M., Araniti, G., Molinaro, A., and Zheng, K., Toward 5G DenseNets: Architectural advances for effective machine-type communications over femtocells, *IEEE Communications Magazine (Topic: Extremely Dense Wireless Networks)*, January 2015, 53(1), 134–141.
3. Lai, C.-F., Hwang, R.-H., Chao, H.-C., Hassan, M. M., and Alamri, A., A buffer-aware HTTP live streaming approach for SDN-enabled 5G wireless networks, *IEEE Network Magazine*, January–February 2015, 29(1), 49–55.
4. Wang, R., Hu, H., and Yang, X., Potentials and challenges of C-RAN supporting multi-RATs toward 5G mobile networks, *IEEE Access*, October 1, 2014, 2, 1187–1195.
5. Wang, C.-X., Haider, F., Gao, X., You, X.-H. et al., Cellular architecture and key technologies for 5G wireless communication networks, *IEEE Communications Magazine (Topic: 5G Wireless Communication Systems: Prospects and Challenges)*, February 2014, 52(2), 122–130.
6. Chin, W. H., Fan, Z., and Haines, R., Emerging technologies and research challenges for 5G wireless networks, *IEEE Wireless Communications Magazine*, April 2014, 21(2), 106–112.
7. Cho, H.-H., Lai, C.-F., Shih, T. K., and Chao, H.-C., Integration of SDR and SDN for 5G, *IEEE Access*, September 11, 2014, 2, 1196–1204.
8. Hossain, E., Rasti, M., Tabassum, H., and Abdelnasser, A., Evolution toward 5G multi-tier cellular wireless networks: An interference management perspective, *IEEE Wireless Communications Magazine*, June 2014, 21(3), 118–127.

9. Rost, P., Bernardos, C. J., Domenico, A. D., Girolamo, M. D., Lalam, M., Maeder, A., Sabella, D., and Wübben, D., Cloud technologies for flexible 5G radio access networks, *IEEE Communications Magazine* (*Topic: 5G Wireless Communication Systems: Prospects and Challenges*), May 2014, 52(5), 68–76.

10. Sabella, D., Domenico, A. D., Katranaras, E., Imran, M. A., Girolamo, M. D., Salim, U., Lalam, M., Samdanis, K., and Maeder, A., Energy efficiency benefits of RAN-as-a-Service concept for a cloud-based 5G mobile network infrastructure, *IEEE Access*, December 18, 2014, 2, 1586–1597.

11. Cisco Systems Inc., Global mobile data traffic forecast update 2014–2019. http://www.cisco.com/c/en/us/solutions/collateral/service-provider/visual-networking-index-vni/white_paper_c11-520862.html (accessed January, 2016).

12. Ericsson, Ericsson mobility report, White Paper, June 2014. http://www.ericsson.com/mobility-report (accessed January, 2016).

13. Ericsson, Traffic exploration tool, Online, December 10, 2014, http://www.ericsson.com/TET.

14. Dahlman, E., Parkvall, S., and Skold, J., *4G LTE/LTE-Advanced for Mobile Broadband*, Elsevier Academic Press, Oxford, U.K., 2011.

15. Caban, S., Mehlführer, C., Rupp, M., and Wrulich, M., *Evaluation of HSDPA and LTE: From Testbed Measurements to System Level Performance*, John Wiley & Sons Ltd., Chichester, U.K., 2012.

16. Rohde and Schwarz, 1MA191: LTE Release 9 technology introduction, White Paper, December 8, 2011, https://www.rohde-schwarz.com/en/applications/lte-release-9-technology-introduction-white-paper-application-note_56280-15541.html (accessed December, 2014).

17. International Telecommunication Union (ITU), Requirements related to technical performance for IMT-advanced radio interface(s), Online: ITU-RM.2134, 2008, http://www.itu.int/pub/R-REP-M.2134-2008/ (accessed December, 2014).

18. Agilent Technologies, Introducing LTE-advanced: Application note. Printed in USA, March 8, 2011. http://cp.literature.agilent.com/litweb/pdf/5990-6706EN.pdf (accessed January, 2016).

19. Sadek, K., Tarighat, A., and Sayed, A. H., A leakage-based precoding scheme for downlink multi-user MIMO channels, *IEEE Transactions on Wireless Communications*, May 2007, 6(5), 1711–1721.

20. Keysight Technologies, LTE-Advanced: Technology and Test Challenges, Application Notes. Published in USA, July 31, 2014. http://literature.cdn.keysight.com/litweb/pdf/5990-6706EN.pdf (accessed January, 2016).

21. Lee, C.-S., Lee, M.-C., Huang, C.-J., and Lee, T.-S., Sectorization with beam pattern design using 3D beamforming techniques, *IEEE Asia-Pacific Signal and Information Processing Association Annual Summit and Conference* (*APSIPA*), Kaohsiung, Taiwan, October 29–November 1, 2013, pp. 1–5.

22. Halbauer, H., Saur, S., Koppenborg, J., and Hoek, C., 3D beamforming: Performance improvement for cellular networks, *Bell Labs Technical Journal*, September 2013, 18(2), 37–56.

23. Kim, Y., Ji, H., Lee, J., Nam, Y.-H., Ng, B. L., Tzanidis, I., Li, Y., and Zhang, J., Full dimension MIMO (FD-MIMO): The next evolution of MIMO in LTE systems, *IEEE Wireless Communications*, June 2014, 21(3), 92–100.

24. 3GPP: The Mobile Broadband Standard, TSG RAN study on 3D channel model for LTE (release 12), September 2014, http://www.3gpp.org/DynaReport/36873.htm (accessed December, 2014).

25. Marzetta, T. L., Noncooperative cellular wireless with unlimited numbers of base station antennas, *IEEE Transactions on Wireless Communications*, November 2010, 9(11), 3590–3600.

26. Rusek, F., Persson, D., Lau, B. K., Larsson, E. G., Marzetta, T. L., Edfors, O., and Tufvesson, F., Scaling up MIMO: Opportunities and challenges with very large arrays, *IEEE Signal Processing Magazine*, January 2013, 30(1), 40–60.

27. Larsson, E., Edfors, O., Tufvesson, F., and Marzetta, T., Massive MIMO for next generation wireless systems, *IEEE Communications Magazine*, February 2014, 52(2), 186–195.

28. Yin, H., Gesbert, D., Filippou, M., and Liu, Y., A coordinated approach to channel estimation in large-scale multiple-antenna systems, *IEEE Journal on Selected Areas in Communications*, February 2013, 31(2), 264–273.

29. Muller, R. R., Cottatellucci, L., and Vehkapera, M., Blind pilot decontamination, *IEEE Journal of Selected Topics in Signal Processing*, October 2014, 8(5), 773–786.

30. Rappaport, T. S., Sun, S., Mayzus, R., Zhao, H., Azar, Y., Wang, K., Wong, G. N., Schulz, J. K., Samimi, M., and Gutierrez, F., Millimeter wave mobile communications for 5G cellular: It will work!, *IEEE Access*, 2013, 1, 335–349.

31. Pi, Z. and Khan, F., An introduction to millimeter-wave mobile broadband systems, *IEEE Communications Magazine*, June 2011, 49(6), 101–107.

32. Swindlehurst, A. L., Ayanoglu, E., Heydari, P., and Capolino, F., Millimeter-wave massive MIMO: The next wireless revolution? *IEEE Communications Magazine*, September 2014, 52(9), 56–62.

33. Friis, H. T., A note on a simple transmission formula, *Proceedings of the IRE*, May 1946, 34(5), 254–256.

34. Rangan, S., Rappaport, T. S., and Erkip, E., Millimeter-wave cellular wireless networks: Potentials and challenges, *Proceedings of the IEEE*, March 2014, 102(3), 366–385.

35. Hur, S., Kim, T., Love, D. J., Krogmeier, J. V., Thomas, T. A., and Ghosh, A., Millimeter wave beamforming for wireless backhaul and access in small cell networks, *IEEE Transactions on Communications*, October 2013, 61(10), 4391–4403.

36. Dehos, C., González, J. L., De Domenico, A., Kténas, D., and Dussopt, L., Millimeter-wave access and backhauling: The solution to the exponential data traffic increase in 5G mobile communications systems? *IEEE Communications Magazine*, September 2014, 52(9), 88–95.
37. IEEE, IEEE standard for information technology—Telecommunications and information exchange between systems—Local and metropolitan area networks—Specific requirements—Part 11: Wireless LAN medium access control (MAC) and physical layer (PHY) specifications amendment 3: Enhancements for very high throughput in the 60 GHz band, IEEE Std 802.11ad-2012 (Amendment to IEEE Std 802.11-2012, as amended by IEEE Std 802.11ae-2012 and IEEE Std 802.11aa-2012), December 2012, pp. 1–628.
38. Nitsche, T., Cordeiro, C., Flores, A. B., Knightly, E. W., Perahia, E., and Widmer, J. C., IEEE 802.11ad: Directional 60 GHz communication for multi-Gigabit-per-second Wi-Fi, *IEEE Communications Magazine*, December 2014, 52(12), 132–141.
39. Baykas, T., Sum, C.-S., Lan, Z., Wang, J., Rahman, M. A., Harada, H., and Kato, S., IEEE 802.15.3c: The first IEEE wireless standard for data rates over 1 Gb/s, *IEEE Communications Magazine*, July 2011, 49(7), 114–121.
40. ECMA International, ECMA standard 387, High rate 60 GHz PHY, MAC and HDMI PAL, December 2008, http://www.ecma-international.org (accessed December, 2014).
41. WirelessHD, WirelessHD specification version 1.1 overview, May 2010, http://www.wirelesshd.org/about/specification-summary.
42. Yong, S.-K., Xia, P., and Valdes-Garcia, A., *60 GHz Technology for Gbps WLAN and WPAN: From Theory to Practice*, John Wiley & Sons Ltd., Chichester, U.K., 2011.

2

RADIO PROPAGATION AND CHANNEL MODELING ASPECTS OF 4G AND BEYOND NETWORKS

NARGES NOORI

Contents

2.1 Introduction

The growing request for mobile broadband has motivated research activities for the next generation of cellular networks to increase data rates and achieve a higher quality of service. The long term evolution advanced (LTE-Advanced) networks fully meet 4G requirements and were commercially launched in 2012 to satisfy user demand. However, the rapid growth of data traffic as a result of revolution at the user devices and increased connection between vehicles, sensors, and health devices are driving the advent of the new generation wireless mobile 5G networks. Research activities on 5G communication technologies have been recently started in both the academic and industrial communities [1,2].

The LTE-Advanced (4G) and beyond cellular communication networks offer new techniques to help achieve very high data rates even in mobility conditions. However, the performance of such networks depends intensively on the propagation channel. The ability to predict the minimum required transmit power for an acceptable coverage over the network service area is necessary at the network planning stage.

It is helpful for estimating the effect of such transmissions on existing services and is essential for the improvement of frequency reuse and interference management. That is why propagation and channel characterization are vital for designing modern broadband mobile communication networks [3].

In this chapter, after exploring the general principles of wave propagation and channel modeling, the special aspects of 4G and beyond channels are considered. Then, the channel modeling aspects of very large multiple-input, multiple-output (MIMO) systems are investigated. Path loss prediction models and large- and small-scale fading are considered for different scenarios of indoor, outdoor, and outdoor-to-indoor wave propagation in 4G and beyond networks. The characterization of the propagation environment as an important issue of upcoming millimeter wave mobile broadband networks is also considered. Device-to-device (D2D) radio channel characterization, especially for vehicle-to-vehicle scenarios, is discussed as well. Moreover, body area network channel modeling for modern health monitoring systems is studied. Finally, initial METIS channel models for 5G networks are briefly described.

2.2 Principles and Background of Propagation and Channel Modeling

The wireless channel is referred to as a wireless propagation environment where an electromagnetic wave travels from a transmitter antenna to a receiver point. If detailed characteristics of the wireless propagation channel are available, it is possible to successfully design wireless systems and achieve good system performance. Therefore, a better knowledge of the propagation mechanisms is crucial to deploy a reliable wireless system. According to Figure 2.1, there are many complex interactions between electromagnetic waves and the wireless channel. All of these interactions must be taken into account to analyze the performance of deployed communication systems. In a wireless channel, the received signal may be a summation of direct line of sight (LOS) and numerous multipath components (MPCs) arriving from different paths with different amplitudes and delays. A multipath phenomenon is the result of the main propagation mechanisms, known as reflection, diffraction, and scattering. These mechanisms cause signal fades and additional signal propagation losses. However,

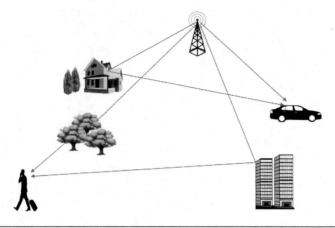

Figure 2.1   Different mechanisms of radio propagation in wireless channels.

many different types of simplifications are required to develop a simple and accurate model of the channel.

2.2.1 Free Space Propagation Model

Loss of transmitted power is mainly caused by the wavefront as the wave propagates away from the transmitter. This means that the total electromagnetic power within any unit area on the wavefront will decay with distance. In free space, the received power can be expressed by the well-known Friis transmission formula as follows:

$$P_r = \frac{P_t G_t G_r \lambda^2}{\left(4\pi d\right)^2},$$ (2.1)

where

P_t is the transmitted power

G_t and G_r are antenna gain of the transmitter and the receiver, respectively

λ is the wavelength

d is the distance between the transmitter and the receiver

2.2.2 Reflection, Refraction, and Transmission

When an electromagnetic wave is incident onto the boundary between two different electromagnetic media, some energy is reflected back and some energy is transmitted. Accordingly, two new waves are generated. The first wave, which returns to the originating medium, is called the reflected wave. The second wave passes through the boundary and travels into the other medium and is called the transmitted or refracted wave, which results from the mechanism of refraction. The relation between the reflection angle, θ_r, and the incidence angle, θ_i, is determined by Snell's law of refraction:

$$\theta_r = \theta_i. \tag{2.2}$$

Thus Snell's law of refraction, which relates the angle of the transmitted wave, θ_t, to the incidence angle, is given as

$$\frac{\sin\theta_i}{\sin\theta_t} = \frac{n_2}{n_1}, \tag{2.3}$$

where n_1 and n_2 are the refractive indexes of the media in which the incident and transmitted waves travel, respectively.

The amplitudes of both reflected and transmitted waves are related to the amplitude of the incident wave through the Fresnel reflection and transmission coefficients. These coefficients are different for transverse electric (TE) and transverse magnetic (TM) waves. The reflection and transmission coefficients are represented by Γ and τ, respectively

$$\Gamma_{TE} = \frac{\eta_2 \cos\theta_i - \eta_1 \cos\theta_t}{\eta_2 \cos\theta_i + \eta_1 \cos\theta_t} \quad \text{and} \quad \tau_{TE} = \frac{2\eta_2 \cos\theta_i}{\eta_2 \cos\theta_i + \eta_1 \cos\theta_t}, \tag{2.4}$$

$$\Gamma_{TM} = \frac{\eta_1 \cos\theta_i - \eta_2 \cos\theta_t}{\eta_1 \cos\theta_i + \eta_2 \cos\theta_t} \quad \text{and} \quad \tau_{TM} = \frac{2\eta_2 \cos\theta_i}{\eta_1 \cos\theta_i + \eta_2 \cos\theta_t}, \tag{2.5}$$

where η_1 and η_2 are the wave impedances of the media in which the incident and transmitted waves travel, respectively.

2.2.3 Diffraction

Propagating waves tend to spread out or bend around corners when they are incident at the edges of obstacles. This phenomenon, known as diffraction, can be expressed by using Huygen's principle. This principle states that all points located on a wavefront can be considered a point source for spherical wavelets, which form the secondary wavefront. Accordingly, when a wave diffracts over an edge, the resulted curved wavefront helps to fill the geometrical shadow region with diffracted rays. This is highly advantageous in mobile communication, since even when a user moves into deep shadow, they can receive the signal because of the diffraction effect.

2.2.4 Scattering

The reflection process discussed previously is only applicable to smooth surfaces. Scattering, often called diffuse reflection, occurs when the incident wave encounters a surface that is not perfectly smooth. In this case, the wave gets scattered from many points on the surface. The diffuse reflected terms are usually weaker than those returning from specular reflection. The degree of scattering is related to the incidence angle and the surface roughness. Surface roughness is determined by a critical height defined by a Rayleigh criterion as

$$h_c = \frac{\lambda}{8 \cos \theta_i}, \qquad (2.6)$$

where λ is the wavelength. A surface is considered smooth if the maximum height difference between various points on the surface, Δh, is smaller than h_c and rough if Δh is more than h_c. For rough surfaces, the Fresnel reflection coefficient is multiplied by the following scattering loss factor ρ_S:

$$\rho_S = \exp\left(-8\left(\frac{\pi \sigma_h \sin \theta_i}{\lambda}\right)^2\right), \qquad (2.7)$$

where σ_h is the standard deviation corresponding to the Gaussian height distribution.

2.3 Classification and Principles of the Channel Models

Generally, propagation channel models are categorized as deterministic (or site-specific) and statistical (or empirical) [4]. Deterministic channel modeling is more accurate and needs a huge amount of data related to the geometry of the propagation environment and requires more computations. Statistical channel models are developed based on a statistical description of the propagation environment. Their implementation is easier as they require less computational resources and are less sensitive to the channel's geometry.

2.3.1 Deterministic Models

Deterministic models are developed based on the concept of electromagnetic wave propagation. The propagation of electromagnetic waves can be studied by solving Maxwell's equations. In these models, a large amount of complex mathematical operations are executed by considering a detailed geometrical database of the physical environment. The electromagnetic characteristics of the materials must be considered in the database. A key advantage of deterministic models is their high accuracy.

The ray tracing technique is a deterministic method based on the geometrical optics to model wave propagation in different environments. According to geometrical optics, energy is radiated through infinitesimally small tubes known as rays. The rays propagate along straight lines in the direction of wave propagation. This technique assumes that the scatterers act as simple reflectors or refractors [5]. Since signal propagation is modeled via ray propagation, simple geometric equations can be employed as an alternative to the complicated Maxwell's equations to compute the received signal by means of ray superposition. The ray tracing method along with the uniform theory of diffraction (UTD) can be used to model diffracted rays.

2.3.2 Statistical Models

Statistical channel models are presented based on the statistical characterization of a received signal. These models are more computationally efficient than deterministic ones with less complexity and aim at providing accurate channel predictions. Statistical models can be modified for different propagation environments by tuning the channel parameters. A statistical distribution of these parameters can be extracted from real-time measurement data.

2.3.3 Path Loss, Shadowing, and Multipath Fading

Path loss denotes attenuation suffered by the signal as it travels from the transmitter to the receiver. It can be given by the difference of the transmitted, P_t, and received, P_r, power in dB as

$$PL(\text{dB}) = 10 \log_{10} P_t / P_r. \tag{2.8}$$

The free space loss model mentioned earlier describes signal attenuation if the influences of all absorption, diffraction, obstruction, refraction, reflection, and scattering effects are neglected on propagation. Extensive efforts have been made to develop mathematical path loss models for various rural, urban, suburban, and indoor environments. Shadowing or large-scale fading is the result of obstructions that a signal encounters due to different objects in the environment. This causes slow variations in the received signal level in comparison to the nominal value given by the path loss model. The mentioned random variations should be modeled statistically. Based on the measurement results reported in the literature, the large-scale fading follows a log-normal distribution. By taking into account shadowing, path loss at a distance d (in m) is given by

$$PL(\text{dB}) = PL(d_0) + 10n \log_{10} \left(\frac{d}{d_0} \right) + X, \tag{2.9}$$

where
   $PL(d_0)$ is the path loss in dB at a reference distance of $d_0 = 1$ m
   n is the path loss exponent
   X is the log-normal large-scale fading

Besides path loss and shadowing, some significant variations are observed in the received signal as a result of multipath phenomenon. The multipath components arrive at the receiver with various delays and intensities. The total received power may increase or decrease significantly along very small distances depending on the phase of each individual component. This phenomenon is called small-scale or multipath fading, and the rapid signal fluctuation can be predicted by statistical distributions. The amplitude fading statistics of mobile channels in non-line of sight (NLOS) scenarios are described by the Rayleigh distribution. Under the LOS condition, the dominant signal component is much stronger than the others. In this situation, the Rice distribution gives a better description of amplitude fading.

2.3.4 Small-Scale Channel Characterization

Measurement results show that MPC rays arrive at the receiver in clusters. Each cluster consists of a collection of MPCs with similar time of arrival (TOA), angles of arrival (AOAs) and angles of departure (AODs. The channel impulse response (CIR) is expressed by the sum of the contributions from the MPC clusters:

$$h(t) = \sum_{n=0}^{N} \sum_{m=0}^{M_n} a_{m,n} \delta(t - T_n - \tau_{m,n}), \tag{2.10}$$

where
  $\delta(\cdot)$ is the Dirac delta function
  N is the number of clusters
  M_n is the number of MPCs within the nth cluster
  $a_{m,n}$ is the gain of the mth MPC in the nth cluster
  T_n is the TOA of the first MPC arriving within the nth cluster
  $\tau_{m,n}$ is the TOA of the mth MPC relative to the TOA of the nth cluster

The parameters T_l and $\tau_{k,l}$ are defined by two independent exponential distributions as follows:

$$p(T_n | T_{n-1}) = \Lambda \exp[-\Lambda(T_n - T_{n-1})], \quad n > 0 \tag{2.11}$$

$$p(\tau_{m,n} | \tau_{(m-1),n}) = \lambda \exp[-\lambda(\tau_{m,n} - \tau_{(m-1),n})], \quad m > 0 \tag{2.12}$$

where Λ and λ are the mean cluster and ray arrival rates, respectively. Usually, each cluster contains a large number of rays where $\lambda \gg \Lambda$.

The average power of the clusters and their corresponding rays is considered to decay exponentially, such that

$$\overline{a_{m,n}^2} = \overline{a_{0,0}^2} \, e^{-T_m/\Gamma} e^{-\tau_{m,n}/\gamma} \tag{2.13}$$

where

$\overline{a_{0,0}^2}$ is the average gain of the first ray within the first cluster

Γ and γ are the clusters' and the rays' power delay time constants, respectively

2.4 4G and Beyond Channel Modeling Challenges

Technology requirements to offer end users desired services along with environment scenarios are the main factors determining the necessities of propagation modeling [6]. From a technology viewpoint, the propagation challenges are predominantly due to higher frequencies, broader bandwidths, and arrays with considerably greater physical sizes and more number of elements. Millimeter wave frequencies can provide a considerable amount of spectrum for high speed data transmission. The fundamental aspects of millimeter wave propagation must be considered to develop a channel model. Furthermore, very large or massive MIMO approach to improve radio link efficiency in 5G mobile communications was recently proposed [7,8]. Massive MIMO involves base stations equipped with a large number (a few hundreds) of antennas extending over many wavelengths. In massive MIMO systems, correlation properties are more crucial than conventional MIMO because of the possibility to achieve very high spatial diversity gains [9]. On the other hand, channels of massive MIMO systems are different from the usual MIMO channels because they experience large-scale fading over the antenna array. The proposed channel models also require improvement in angular resolution and spherical wave modeling.

From an environment viewpoint, the model shall cover all relevant propagation scenarios, such as urban macro- and microcells, suburban, rural, indoor, and outdoor-to-indoor [10]. The model parameters and their statistical distributions for different scenarios must be extracted from the measurements.

2.5 MIMO Channel Modeling for 4G and Beyond

MIMO technology can support the deployment of faster and more reliable transmissions. In MIMO systems, multiple antenna arrays are deployed in both transmitter and receiver, allowing the system to exploit the spatial dimension of the propagation environment. This technology is a crucial component in 4G and beyond networks.

MIMO channel modeling has been carried out in European research projects and actions like COST 259 [11], COST 273 [12], and IST-WINNER [13]. The COST 259 model was the first geometry-based stochastic channel model (GSCM) for multi-antenna base stations. Full MIMO systems were later considered in the COST 273 model. The COST 2100 model was then built upon the framework of the COST 259 and 273 models for MIMO channels [14].

The WINNER channel, closely related to the COST 2100 model, is also a GSCM and allows separation of antennas and propagation parameters. This model considers a wide range of mobile communication scenarios. In WINNER I work package 5 (WP5), a wideband MIMO channel model was developed at 2 and 5 GHz. It permits creation of an arbitrary geometry-based antenna independent channel model. WINNER II [15] extended the model features of the WINNER I project at a frequency range of 2–6 GHz.

2.5.1 COST 2100 MIMO Channel Model

In the COST 2100 channels, the MPCs are mapped to their related scatterers and are categorized according to their delay, azimuth and elevation of departure, and azimuth and elevation of arrival [14]. Clusters are molded by a combination of scatterers that generate MPCs with analogous delays and similar azimuth and elevation directions. The scattering phenomena from the base station to the mobile user may involve one interaction with an object (single-bounce), or a few successive interactions with various objects (multiple-bounce). Local clusters are located around the base station or mobile user and are characterized by single-bounce scatterers, while twin clusters enable multibounce modeling. A circular visibility region (VR) is defined for each cluster. If the mobile user is within the VR of a cluster, the signal will propagate through that cluster. When the mobile user is in an area with several VR overlaps, the corresponding clusters are observable instantaneously. The COST 2100 model supports spherical wave modeling considering the polarization performance of the channel. The generic multilink advanced version of the COST 2100 channel model is applied to investigate transmissions between multiple base stations and mobile users [16].

Several research groups have made considerable effort to extract the parameters of the COST 2100 model for various scenarios. The time-varying CIR of the COST 2100 model along with a whole list of model parameters and their recommended distributions is presented in Verdone and Zanella [16] for different propagation environments.

2.5.2 WINNER II MIMO Channel Model

The WINNER II MIMO channel model considers N paths with specific delay, AOD, AOA, gain, and cross-polarization power ratio. There are three types of paths: direct or LOS path, single-bounce path, and multibounce path. Each NLOS path consists of numerous subpaths called rays. The time-varying impulse response of the channel is represented by the following matrix:

$$\mathbf{H}(t;\tau) = \sum_{n=1}^{N} \mathbf{H}_n(t;\tau), \tag{2.14}$$

where n is the path index. This matrix is composed of the transmitter and the receiver antenna array response matrices $\mathbf{F}_{tx}$ and $\mathbf{F}_{rx}$, respectively, as well as the channel response matrix $\mathbf{h}_n$ related to cluster n as follows:

$$\mathbf{H}_n(t;\tau) = \iint \mathbf{F}_{rx}\mathbf{h}_n(t;\tau,\phi,\varphi)\mathbf{F}_{tx}^T(\phi)\,d\phi d\varphi. \tag{2.15}$$

In the case of a dynamic radio channel, all the parameters of this channel response will vary with time [17].

2.5.3 Characterization of Dense Multipath Components

Various measurements have shown that in the channel impulse response there are some strong paths corresponding to specular MPCs and many weak paths corresponding to dense or diffuse MPCs [3,18], as follows:

$$H = H_{SC} + H_{DMC}, \qquad (2.16)$$

where H_{SC} and H_{DMC} correspond to the specular and diffuse MPCs, respectively. However, the previously mentioned GSCMs are only based on a superposition of specular MPCs.

Dense multipath components (DMCs), originating from, for example, scattering due to different objects that are small in size compared to the wavelength or comprise rough surfaces cannot be characterized through superposition of individual plane waves. In Poutanen [18] and Quitin et al. [19], the characteristics of DMCs along with their angular and delay domain distributions have been studied based on an extensive set of channel sounding and measurements.

2.6 Outdoor Channel Modeling for 4G and Beyond

2.6.1 Outdoor Channel Impulse Response and Small-Scale Characterizations

The WINNER II MIMO channel model is applicable for all urban, suburban, and rural environments [20]. It has been accepted by ITU-R M.2135 for 4G networks because of its high accuracy in modeling real propagation channels. The time-varying random behavior of the channel parameters makes the channel response a random complex process. Consequently, statistical representations should be employed to describe outdoor radio channel responses. A complete set of model parameters used in system simulations for 4G and beyond are presented in Kyösti et al. [15]. These parameters are entirely calculated using statistical methods and are estimated based on the data obtained from a huge amount of field measurements from different outdoor scenarios [20]. An interested reader can refer to De La Roche et al. [3] for a detailed statistical characterization of the outdoor channels.

2.6.2 Outdoor Path Loss Models and Shadowing

The COST 231 Hata model can be applied to different macrocell environments in the planning of 4G and beyond networks [21]. This model can be expressed by the following formula [22]:

$$L = \left(46.3 - 13.82 \log_{10} h_t + 33.9 \log_{10} f_c\right)$$
$$+ \left(44.9 - 6.55 \log_{10} h_t\right) \log_{10}\left(d\right) - K + K_M \quad \text{(dB)}, \quad (2.17)$$

where

f is the working frequency of the system in MHz

h is the base station antenna height in m

d is the distance between the mobile user and the base station in km

K_M is 0 dB for suburban regions and medium-sized cities, and 3 dB for metropolitan cities

K is the correction factor for mobile antenna height, which is calculated according to Cost 231 [22] for different propagation environments—from open areas to large cities

The COST 231 Walfisch–Ikegami model distinguishes between LOS and NLOS situations [22]. This model is applicable for micro-cellular environments. In LOS and NLOS conditions, the path loss is given by the following formulas, respectively

$$L_{\text{LOS}} = 42.64 + 26 \log_{10}\left(d\right) + 20 \log_{10}\left(f\right) \quad \text{(dB)}, \quad (2.18)$$

$$L_{\text{NLOS}} = L_0 + L_{rts} + L_{msd} \quad \text{(dB)}, \quad (2.19)$$

where

d is in km

f is in MHz L_0 is the attenuation in free-space

L_{rts} is diffraction from rooftop to street

L_{msd} represents diffraction loss due to multiple obstacles [22]

Different path loss models have been proposed in WINNER II for various propagation scenarios. The general form of WINNER II path loss model is typically expressed by Kyösti et al. [15]:

$$PL = K_1 \log_{10} d + K_2 + K_3 \log_{10} \left(\frac{f_c}{5.0} \right) + X, \qquad (2.20)$$

where

    d is the transmitter–receiver distance in m
    f_c is the frequency in GHz

The parameter K_1 contains the path loss exponent, K_2 is the intercept, K_3 defines frequency dependency of the path loss and X is an optional parameter related to the environment. The models can be used from 2 to 6 GHz and for various antenna heights. The shadow fading follows a log-normal distribution. The outdoor propagation scenarios of the WINNER II model and its corresponding path loss formulas and channel parameters are described in full detail in Kyösti et al. [15].

2.7 Indoor Channel Modeling for 4G and Beyond

Since the demand for mobile communications services in indoor offices and hotspot environments, such as shopping malls, factories, train stations, and airports, is high, indoor propagation prediction for broadband mobile communication is extremely important [3,23–26]. Some radio propagation models for 4G and beyond cellular networks that describe the behavior of the signal in different outdoor environments have been the subject of various research works [3,6,21,26,27].

2.7.1 Indoor Channel Impulse Response and Small-Scale Characterizations

Indoor environments can also be described by the WINNER II MIMO channel model [20]. The channel impulse response is expressed by Equation 2.16 where its statistical parameters should be extracted from the measurements executed in indoor channels. The model parameters have been extracted in ITU-R M.2135 [20] for indoor hotspots. A detailed statistical characterization of the indoor channels is presented in Verdone and Zanella [16].

2.7.2 Indoor Path Loss Models and Shadowing

The COST 231 multiwall indoor propagation model [22] incorporates a linear component of loss related to the number of penetrated walls and a complex term corresponding to the number of penetrated floors as

$$L_T = L_F + L_C + \sum_{i=1}^{W} L_{wi} n_{wi} + L_f n_f^{\left(\left(n_f+2\right)/\left(n_f+1\right)-b\right)}, \qquad (2.21)$$

where

L_F is the free space loss along the direct transmitter–receiver path
W is the number of wall types
n_{wi} is the number of walls penetrated by the direct path of type i
L_{wi} is the penetration loss corresponding to a wall of type i
n_f is the number of floors penetrated by the path
b and L_C are two empirical constants
L_f is the loss per floor

Specific suggested values are L_w = 1.9 dB at 900 MHz and 3.4 dB at 1800 MHz in the case of light walls; 6.9 dB at 1800 MHz in the case of heavy walls; L_f = 14.8 dB at 900 MHz and 18.3 dB at 1800 MHz; and b = 0.46 [28].

The WINNER II indoor path loss models are also expressed by Equation 2.33, the corresponding parameters of which are presented in Kyösti et al. [15]. The shadow fading also has a log-normal distribution.

2.7.3 Ray Tracing Models for Indoor Environments

The ray tracing modeling technique may be applied for 4G and beyond network planning in indoor environments if sufficient data and computing time are available [29,30]. In the ray launching method the rays are sent at numerous distinct angles from the transmitter. These rays interact with different objects as they travel through the propagation environment. The propagation of each ray is stopped when its power drops below an assumed threshold. The ray tracing model takes into account all reflection and transmission effects, as well as diffraction mechanisms.

The electric field related to the ith received ray path is given by Saunders and Aragón-Zavala [28]

$$E_i = E_0 f_{ti} f_{ri} L_F(r) \left[\prod_j R_j \prod_m T_m \prod_l D_l A_l(S_l, S_l') \right] e^{-jkr}, \quad (2.22)$$

where

E_0 stands for the reference field

f_{ti} and f_{ri} represent radiation patterns of the transmitting and receiving antennas, respectively

L_F is the free-space loss

R_j is the reflection coefficient corresponding to the jth reflection

T_m is the transmission coefficient corresponding to the mth transmission

D_l and A_l are the diffraction coefficient and the spreading attenuation corresponding to the lth diffraction, respectively

e^{-jkr} is the propagation phase factor (r represents the unfolded ray path length)

k is the wave number

S_l and S_l' are the lth diffraction point distances from the source and field points of the lth diffracted ray, respectively

This model can be used with complete three-dimensional (3D) data for AOA, AOD, power delay profile, and coverage predictions.

2.8 Outdoor to Indoor Channels

One of the main objectives of 4G and beyond technologies is to extend their wireless broadband indoor coverage. However, the poor radio coverage inside buildings caused by wall penetration losses makes it difficult for indoor cellular users to receive high-speed wireless broadband services. A comprehensive description of the outdoor to indoor radio channels and the related path loss is needed to inquire about the network coverage. Furthermore, the outdoor to indoor channel modeling is also important for managing intercell interference between the outdoor macro/microcells and the indoor pico/femtocells [3,16,26,31,32]. Actually, the dominant propagation mechanisms are different at various frequencies because of the frequency dependency of the electrical properties of the building materials. At lower frequencies, radio waves experience deeper penetration into buildings than at higher frequencies.

If an LOS path occurs between one building face and an external antenna, the COST 231 model expresses the total outdoor to indoor path loss as [22]

$$L_T = L_F + L_e + L_g\left(1-\cos\theta\right)^2 + \max\left(L_1, L_2\right), \qquad (2.23)$$

where

r_e and r_i are the external and the internal straight path distances between the transmitter and receiver

$\theta = \cos^{-1}(r_p/r_e)$ is the angle of incidence

L_F is the free space loss for the whole path length $(r_i + r_e)$

L_e is the path loss corresponding to the external wall at $\theta = 0°$

L_g is the additional external wall loss experienced at $\theta = 90°$

$$L_1 = n_i L_i, \quad L_2 = \alpha\left(r_i - 2\right)\left(1-\cos\theta\right)^2, \qquad (2.24)$$

where

n_i stands for the number of walls along the internal path

L_i is the loss corresponding to the internal wall

α is the specific attenuation (dBm^{-1}) of unobstructed internal paths

All these distances are in meters. The model is applicable at distances up to 500 m.

In the NLOS condition, the COST 231 model presents an expression for the loss in a building from an outside transmitter to the outside reference loss, L_{out}, related to a point on the nearest side of the wall of interest at a height of 2 m above the ground. The total loss is given by

$$L_T = L_{out} + L_e + L_{ge} + \max(L_1, L_3) - G_{fh}, \qquad (2.25)$$

where

L_3, L_e, α and L_1 are all defined in the COST 231 LOS model expressed in Equation 2.24

G_{fh}, known as the floor height gain, is represented by

$$G_{fh} = \begin{cases} nG_n \\ hG_h \end{cases}, \qquad (2.26)$$

where

h is the floor height in m
n is the floor number (for the ground floor $n = 0$)

Shadowing is log-normal with a standard deviation of 4–6 dB. Typical model parameters are given in Torres et al. [30] for both LOS and NLOS conditions.

The WINNER II model considers some scenarios for outdoor to indoor channels. The path loss model expressed in Equation 2.20 is also applicable for these scenarios. The corresponding parameters for the path loss models are presented in Kyösti et al. [15]. The shadowing is expressed by the log-normal distribution.

Deterministic methods, for example, ray tracing and finite-difference time-domain (FDTD) and hybrid techniques (combine two or more propagation modeling approaches), can offer more accurate outdoor to indoor propagation modeling [33–36].

2.9 Millimeter Wave Channel Measurement and Characterization

In order to overcome the global bandwidth shortage facing wireless broadband communication, the millimeter wave frequency spectrum is proposed for 5G cellular communications. Understanding the radio channel is an essential requirement to develop millimeter wave mobile networks [37,38]. In Rappaport et al. [37], extensive propagation measurements in indoor, urban, and suburban environments have been conducted to understand the millimeter wave channel.

Furthermore, the penetration and reflection characteristics of the building materials have been measured in Rappaport et al. [37]. Tinted glass and brick pillars, found on the exterior surfaces of buildings, represent high penetration losses such as 28.3 and 40.1 dB, respectively, at 28 GHz. This provides a high isolation between both outdoor and indoor networks. However, indoor building materials such as drywall and clear, nontinted glass have relatively low losses of 3.6 and 6.8 dB, respectively. The penetration measurements have also been made through several obstructions in typical office environments. The results show that the type and number of obstructions affect the penetration loss more than the transmitter–receiver separation distance. Moreover, the outdoor materials represent greater reflection coefficients of 0.815 and 0.896 for concrete and tinted glass, respectively, at an incident angle of 10° as compared to drywall and clear, nontinted glass, which have reflection coefficients of 0.704 and 0.740, respectively. The urban propagation measurements show numerous MPCs with large excess delay for LOS and NLOS conditions. In LOS condition, the average number of MPCs and its standard deviation is 7.2 and 2.2, respectively for a transmitter-receiver distance of less than 200 m. While, in NLOS case, with a separation distance of less than 100 m, the average number of MPCs is 6.8 with the same standard deviation. Furthermore, in an LOS case, with a 52 m separation, an excess delay of 753.5 ns is observed, while, in NLOS conditions and distance of over 423 m, the excess delay is extended to 1388.4 ns. The average path loss exponents that result from all the urban measurements are 2.55 and 5.76 for LOS and NLOS conditions, respectively. The distributions of AOA and AOD have also been presented for the urban measurements. The complete results of 38 GHz suburban channel measurements are also presented in Rappaport et al. [37,38].

The millimeter wave bands are more affected by atmospheric losses due to water vapor and oxygen absorption. The 22 GHz water vapor and 60 GHz oxygen absorptions are serious limiting factors for the coverage range of millimeter wave spectrum. However, millimeter wave links particularly the links work at 22 and 60 GHz face lower interference than the other links. The millimeter wave links at these frequencies are not expected to interfere since their signal will be largely degraded by atmospheric attenuation.

2.10 Channel Modeling in Other Applications of Modern Broadband Communications

2.10.1 Device-to-Device Channel Measurement and Modeling

One of the important goals of 5G mobile communication systems is to deliver the technical requirements for D2D scenarios as any direct connection between two mobile devices, including any human-to-human, machine-to-machine, and vehicle-to-vehicle (V2V) links [39]. D2D propagation models should be compatible with the corresponding cellular channel ones.

D2D propagation scenarios are different from the cellular cases. In D2D communications, both terminals are at a low height (e.g., at the street level) and can be moving. Human interaction may be present at both ends of the communication link and the path loss is higher than in the cellular scenarios. Furthermore, the link is symmetric. This means that both transmitter and receiver are in similar environments [39]. The proposed model should also cover the different cases of outdoor, outdoor-to-indoor, and indoor links. A detailed description of the channel impulse response for D2D propagation channels is presented in Nurmela et al. [39].

In Roivainen et al. [40], the channel measurement results of V2V case are presented at 2.3 and 5.25 GHz frequencies. The measurements are performed for a single-input multiple-output (SIMO) structure. Some channel features for example path loss, delay spread and standard deviation of shadowing are extracted from the measurement data.

2.10.2 Body Area Network Channel Measurement and Modeling

In 4G and beyond networks, wearable and implanted devices play an important role in people's lives, especially for health monitoring [41,42]. A body area network (BAN) consists of a central unit worn on the body, and some devices or sensors are wirelessly connected to this unit. The devices may be implanted in or surface-mounted on the body. The data collected from the devices or sensors is routed to localized base stations via the central unit using 4G and beyond cellular technologies. This may involve three kinds of off-body (from off-body to another on-body device), on-body (within on-body devices and wearable networks). and in-body (from an on-body device to an implant) communication channels [41]. Consequently, it is necessary to explore how the electromagnetic wave that originates from mobile phones interacts with various parts of the human body such as, skin, tissue, and bones. This is the main reason for the development of wireless body area channel models [3,43,44].

Since the dielectric characteristics of body tissue vary significantly with frequency, the following expression can be used for their permittivity [41]:

$$\varepsilon(\omega) = \varepsilon_\infty + \sum_{m=1}^{4} \frac{\Delta\varepsilon_m}{1+\left(j\omega\tau_m\right)^{(1-\alpha_m)}} + \frac{\sigma_j}{j\omega\varepsilon_0}, \qquad (2.27)$$

where
   ε_∞ is the material permittivity at terahertz frequency
   ε_0 is the free-space permittivity
   σ_j is the ionic conductivity
   ε_m, τ_m, and α_m are material parameters for each dispersion region

A complete set of various parameters needed to find $\varepsilon(\omega)$ at any frequency, is given in Hall and Hao [41] for a range of body tissue types. Furthermore, relative permittivity, conductivity, and penetration depth have been reported in Hall and Hao [41] for muscle and fat across a wide frequency spectrum, from 1 MHz to 100 GHz.

The off-body channel studies include investigation of the propagation features of indoor and outdoor environments in addition to the performance of the body-worn antennas for various body proximity and alignment. At microwave frequencies, the human body makes a deep null in the antenna radiation pattern as the electromagnetic wave cannot penetrate through the body [44]. In this case, electrically large ground planes are preferred for the antennas to improve stability and protect the human body from electromagnetic radiation [3].

A measurement campaign was conducted in two laboratory and corridor indoor environments at 5.5 GHz with three different types of antennas (one for transmitters and two for receivers) [44].

The test devices were placed on three subjects in various points on their bodies, including shoulder, chest, belt, and ankle. The path-loss exponent ranged from 1 to 2.3 for the various points of the transmitter and receiver devices. Body shadowing showed a zero mean log-normal distribution with a standard deviation of 4.8 dB. Furthermore, the propagation path loss corresponding to an on-body channel was also measured. The path loss exponent as well as the standard deviation of the shadowing is extracted in Hall and Hao [41]. This is done for several on-body links such as trunk-to-trunk, trunk-to-head, belt-to-chest, and trunk-to-limb in different body posture. A detailed investigation on the channel delay spread and number of MPCs is presented in Hall and Hao [41].

The physical and numerical phantoms can also be used to simulate or measure the interaction of the electromagnetic waves with the biological tissue. They are valuable tools to study radio wave propagation around and inside the human body. The ray tracing method and numerical electromagnetic techniques are also applicable for deterministic propagation prediction in BANs [45–47]. Some numerical analyses and equivalent circuit models for the whole body and arms are presented in Hall and Hao [41].

2.11 METIS Channel Models for 5G Networks

This section summarizes the initial channel models developed in the 7th European framework project METIS (Mobile and wireless communications Enablers for the Twenty-twenty Information Society) [1] for 5G networks. Two different approaches were selected to develop initial channel models, that is, the GSCM and the map-based model. The map based channel model is based on simplified ray tracing and a simple geometrical description of the radio environment by means of geographical maps or 3D indoor models [48]. Within METIS, different channel models per propagation scenario were proposed. If the building/scene models are available, the propagation paths are calculated based on this deterministic environment where some random elements are added on top of it. However, without the building/scene models, the propagation paths must be generated entirely stochastically. Both cases allow applying various antenna models.

2.11.1 Antenna Modeling

The radiation field pattern of an antenna is a vector quantity and is given in the spherical coordinate system by

$$\mathbf{K}(\theta,\phi) = \frac{\mathbf{E}(r,\theta,\phi)e^{jk_0 r}}{\left|\mathbf{E}(r,\theta,\phi)\right|_{max}} \quad \text{for } r \to \infty \quad (2.28)$$

where

$\mathbf{E}$ is the electric field
k_0 is the free space wave number
$\theta \in [0°, 180°]$
$\phi \in [-180°, 180°]$

The radiation field pattern can be written as

$$\mathbf{K}(\theta,\phi) = K_\theta(\theta,\phi)\mathbf{e}_\theta + K_\phi(\theta,\phi)\mathbf{e}_\phi \quad (2.29)$$

where $\mathbf{e}_\theta$ and $\mathbf{e}_\phi$ are the orthonormal basis vectors of the spherical coordinate system.

When an antenna is rotated around its principal axes or mechanically tilted, the radiation field is rotated as well. To calculate the effects of this rotation, a global coordinate system (GCS) with coordinates (θ,ϕ) and a local coordinate system (LCS) with coordinates (θ',ϕ') are presented in the METIS Report [48]. The radiation field pattern in the GCS is defined by

$$\mathbf{K}_{GCS}(\theta,\phi) = \begin{bmatrix} \cos\psi & -\sin\psi \\ \sin\psi & \cos\psi \end{bmatrix} \mathbf{K}_{LCS}(\theta',\phi') \quad (2.30)$$

where

$$\cos\psi = \mathbf{e}_{\theta,GCS}(\theta,\phi)^T \mathbf{R} \, \mathbf{e}_{\theta,LCS}(\theta',\phi') \quad (2.31)$$

$$\sin \psi = \mathbf{e}_{\phi,GCS}\left(\theta,\phi\right)^{\mathrm{T}} \mathbf{R}\, \mathbf{e}_{\theta,LCS}\left(\theta',\phi'\right) \qquad (2.32)$$

where

$\mathbf{R}$ is the arbitrary rotation between the two coordinate systems

$(\cdot)^{\mathrm{T}}$ is the transpose operation

The polarization transfer matrix $\mathbf{P}$ defines the change in the polarization of an electromagnetic wave departing from the transmitting antenna at direction (θ^d, ϕ^d) and arriving at the receiving antenna at direction (θ^a, ϕ^a). In this case, the overall transfer function is described in the METIS Report [48]

$$\mathbf{K}_{GCS} = \mathbf{K}_{GCS,rx}\left(\theta^a,\phi^a\right)^{\mathrm{T}} \mathbf{P}\mathbf{K}_{GCS,tx}\left(\theta^d,\phi^d\right) \qquad (2.33)$$

where $\mathbf{K}_{GCS,tx}(\theta^d,\phi^d)$ and $\mathbf{K}_{GCS,rx}(\theta^a,\phi^a)$ are radiation patterns of the transmitter and receiver antennas, respectively, in their GCSs.

2.11.2 Map-Based Model

The map-based model applies to the following propagation scenarios that provide the necessary geometry data [48]:

- Urban microcell
- Urban macrocell
- Indoor office
- Indoor shopping mall
- Highway
- Open air festival
- Stadium

In this model, a 16-step procedure is proposed to determine the complex impulse response between transmitter and receiver antennas. This procedure includes the detailed schemes for the creation of the propagation environment and transmitter and receiver locations, determination of the propagation paths and channel coefficient matrices, and calculation of the radio channel transfer function. Finally, the impulse response between the receiver antenna element u and the transmitter antenna element v with a true motion of the transceivers is given in the METIS Report [48]

$$H_{u,v}(t,\tau) = \sqrt{\frac{L_{tot}}{C_{u,v}}} \Big(\mathbf{g}_u^{rx}\big(-\mathbf{k}_{u,v}^{rx,\mathrm{LOS}}(t)\big)^\mathrm{T} \mathbf{h}_{u,v}^{\mathrm{LOS}}(t) \mathbf{g}_v^{tx}\big(\mathbf{k}_{u,v}^{tx,\mathrm{LOS}}(t)\big)\delta\big(\tau-\tau_{u,v}^{\mathrm{LOS}}(t)\big)$$

$$+ \sum_{k=1}^{K} \mathbf{g}_u^{rx}\big(-\mathbf{k}_{k,u,v}^{rx,\mathrm{ref}}(t)\big)^\mathrm{T} \mathbf{h}_{k,u,v}^{\mathrm{ref}}(t) \mathbf{g}_v^{tx}\big(\mathbf{k}_{k,u,v}^{tx,\mathrm{ref}}(t)\big)\delta\big(\tau-\tau_{k,u,v}^{\mathrm{ref}}(t)\big)$$

$$+ \sum_{k=1}^{K'} \mathbf{g}_u^{rx}\big(-\mathbf{k}_{k,u,v}^{rx,\mathrm{dif}}(t)\big)^\mathrm{T} \mathbf{h}_{k,u,v}^{\mathrm{dif}}(t) \mathbf{g}_v^{tx}\big(\mathbf{k}_{k,m,n}^{tx,\mathrm{dif}}(t)\big)\delta\big(\tau-\tau_{k,u,v}^{\mathrm{dif}}(t)\big)$$

$$+ \sum_{k=1}^{K''} \mathbf{g}_u^{rx}\big(-\mathbf{k}_{k,u,v}^{rx,\mathrm{sca}}(t)\big)^\mathrm{T} \mathbf{h}_{k,u,v}^{\mathrm{sca}}(t) \mathbf{g}_v^{tx}\big(\mathbf{k}_{k,u,v}^{r,\mathrm{sca}}(t)\big)\delta\big(\tau-\tau_{k,u,v}^{\mathrm{sca}}(t)\big) \Big),$$

$$(2.34)$$

where $\mathbf{g}_u^{rx}$ and $\mathbf{g}_v^{tx}$ are the complex polarimetric antenna pattern vectors corresponding to the receiver antenna element u and transmitter antenna element v, respectively. The superscripts LOS, ref, dif, and sca are referred to the LOS, reflected, diffracted, and scattered paths, respectively, where K, K', and K'' paths result from the reflection, diffraction, and scattering phenomena, respectively. Moreover, $\mathbf{k}$ is the wave vector can be calculated from the geometry and $\mathbf{h}$ is the propagation matrix. The delay of the paths is shown by τ. The path loss is L_{tot} and the normalization factor $C_{m,n}$ is calculated as a sum of the Frobenius norms of matrix coefficients of all paths, that is, LOS, reflected, diffracted, and scattered over all path indices k.

2.11.3 Geometry-Based Stochastic Model

The GSCM proposed in the METIS Report [48] follows the WINNER approach. This model can be applicable to the following METIS propagation scenarios without predefined building structure geometries

- Urban microcell outdoor to outdoor, outdoor to indoor
- Urban macrocell outdoor to outdoor, outdoor to indoor
- Rural macrocell
- Indoor office
- Highway
- Open air festival

The NLOS channel coefficient for each cluster n and each pair of receiver and transmitter elements, u, s is calculated in the METIS Report [48]

$$H_{u,s,n}^{NLOS}(t)$$

$$
= \sqrt{\frac{P_n}{M}} \sum_{m=1}^{M}
\begin{bmatrix} F_{\theta,GCS,rx,u}\left(\theta_{n,m}^a, \phi_{n,m}^a\right) \\ F_{\phi,GCS,rx,u}\left(\theta_{n,m}^a, \phi_{n,m}^a\right) \end{bmatrix}
\begin{bmatrix} \exp\left(j\Phi_{n,m}^{\theta\theta}\right) & \dfrac{\exp\left(j\Phi_{n,m}^{\theta\phi}\right)}{\sqrt{\kappa_{n,m}}} \\ \dfrac{\exp\left(j\Phi_{n,m}^{\phi\theta}\right)}{\sqrt{\kappa_{n,m}}} & \exp\left(j\Phi_{n,m}^{\phi\phi}\right) \end{bmatrix}^{T}
$$

$$
\times
\begin{bmatrix} F_{\theta,GCS,tx,s}\left(\theta_{n,m}^d, \phi_{n,m}^d\right) \\ F_{\phi,GCS,tx,s}\left(\theta_{n,m}^d, \phi_{n,m}^d\right) \end{bmatrix}
\exp\left(j\frac{2\pi}{\lambda_0}\left(\mathbf{e}_r\left(\theta_{n,m}^a, \phi_{n,m}^a\right)^{T} \mathbf{d}_{rx,u}\right)\right)
$$

$$
\times \exp\left(j\frac{2\pi}{\lambda_0}\left(\mathbf{e}_r\left(\theta_{n,m}^d, \phi_{n,m}^d\right)^{T} \mathbf{d}_{tx,s}\right)\right)
$$

$$
\times \exp\left(j\frac{2\pi}{\lambda_0}\left(\mathbf{e}_r\left(\theta_{n,m}^a, \phi_{n,m}^a\right)^{T} \mathbf{v}_{rx} t\right)\right)
\tag{2.35}
$$

where $F_{\theta,GCS,rx,u}$ and $F_{\phi,GCS,rx,u}$ are the radiation patterns along the direction of the spherical coordinate basis vectors, $\mathbf{e}_\theta$ and $\mathbf{e}_\phi$, respectively of the receive antenna element u, while $F_{\theta,GCS,tx,s}$ and $F_{\phi,GCS,tx,s}$ are the radiation patterns in the direction of $\mathbf{e}_\theta$ and $\mathbf{e}_\phi$, respectively of the transmit antenna element s. Furthermore, $\mathbf{d}_{rx,u}$ and $\mathbf{d}_{tx,s}$ are the position vectors corresponding to the receive antenna element u and transmit antenna element s, λ_0 is the wavelength, and $\mathbf{v}_{rx}$ is the velocity vector of the receiver. The parameters, P_n, M, and κ are normalized power of cluster p, number of rays per cluster, and cross polarization ratio, respectively. $\Phi_{n,m}^{\theta\theta}$, $\Phi_{n,m}^{\theta\phi}$, $\Phi_{n,m}^{\phi\theta}$, $\Phi_{n,m}^{\phi\phi}$ are random phases of ray m within each cluster each n, which are uniformly distributed within $[0,2\pi)$.

In the LOS scenario, the channel coefficients are molded by adding an LOS ray and scaling down the rest of the channel coefficients presented in Equation 2.35 as [48]

$$H_{u,s,n}^{\text{LOS}}(t)$$

$$= \sqrt{\frac{1}{K+1}} H_{u,s,n}^{\text{NLOS}}(t) + \begin{bmatrix} F_{\theta,GCS,rx,u}\left(\theta_{\text{LOS}}^a, \phi_{\text{LOS}}^a\right) \\ F_{\phi,GCS,rx,u}\left(\theta_{\text{LOS}}^a, \phi_{\text{LOS}}^a\right) \end{bmatrix}^{\text{T}}$$

$$\times \begin{bmatrix} \exp\left(j\Phi_{\text{LOS}}\right) & 0 \\ 0 & -\exp\left(j\Phi_{\text{LOS}}\right) \end{bmatrix}^{\text{T}}$$

$$\times \delta(p-1)\sqrt{\frac{K}{K+1}} \begin{bmatrix} F_{\theta,GCS,tx,s}\left(\theta_{\text{LOS}}^d, \phi_{\text{LOS}}^d\right) \\ F_{\phi,GCS,tx,s}\left(\theta_{\text{LOS}}^d, \phi_{\text{LOS}}^d\right) \end{bmatrix}$$

$$\times \exp\left(j\frac{2\pi}{\lambda_0}\left(\mathbf{e}_r\left(\theta_{\text{LOS}}^a, \phi_{\text{LOS}}^a\right)^{\text{T}} \mathbf{d}_{rx,u}\right)\right)$$

$$\times \exp\left(j\frac{2\pi}{\lambda_0}\left(\mathbf{e}_r\left(\theta_{\text{LOS}}^d, \phi_{\text{LOS}}^d\right)^{\text{T}} \mathbf{d}_{tx,s}\right)\right)$$

$$\times \exp\left(j\frac{2\pi}{\lambda_0}\left(\mathbf{e}_r\left(\theta_{\text{LOS}}^a, \phi_{\text{LOS}}^a\right)^{\text{T}} \mathbf{v}_{rx}t\right)\right) \tag{2.36}$$

where K shows the Ricean K-factor.

2.12 Conclusions

In this chapter, a summary of the propagation channel modeling is presented. The special aspects of 4G and beyond channels are investigated. As the MIMO technology is an important solution for modern broadband communication, general MIMO channel models that can be applied for different macro, micro, and pico-cell environments are presented. Path loss models and channel impulse response parameters are investigated for outdoor, indoor, and outdoor-to-indoor propagations. Millimeter wave propagation measurements are explored to understand performance of the new generation mobile technology in urban and indoor environments. Channel and propagation modeling in D2D communications and body area networks are also presented as important applications of modern mobile broadband. Furthermore, the METIS channel models proposed for 5G networks are investigated.

2.13 Future Research Directions

Since the new generation of mobile broadband networks may be introduced approximately in the early 2020s, there is still a large extent of debate on 5G solutions, architectures, and standardizations. This opens up a wide research area in all aspects of 5G networks, especially in the field of propagation and channel modeling. A widespread research is needed to understand the detailed radio propagation characteristics for 5G mobile communications systems. One of the most noticeable aspects is radio propagation predictions for millimeter wave frequencies. Another is realistic channel modeling of massive MIMO systems. Furthermore, D2D communication presents one of the most challenging scenarios from the radio propagation viewpoint, especially in the case of V2V communication, where both ends of the link may be moving.

References

1. Mobile and wireless communications Enablers for the Twenty-twenty Information Society (METIS). Final project report, April 2015, https://www.metis2020.com/wp-content/uploads/deliverables/METIS_D8.4_v1.pdf.
2. *NYU Wireless Pulse*, Vol. 3, No. 1, February 2014, NYU Wireless Research Centre, New York University, New York.
3. G. De La Roche, A. A. Glazunov, and B. Allen, *LTE-Advanced and Next Generation Wireless Networks: Channel Modelling and Propagation*, John Wiley & Sons, Chichester, U.K., 2012.
4. T. K. Sarkar, Z. Ji, K. Kim, A. Medour, and M. Salazar-Palma, A survey of various propagation models for mobile communication, *IEEE Antennas and Propagation Magazine*, 45(3), 51–82, June 2003.
5. J. W. McKown and R. L. Hamilton, Jr., Ray tracing as a design tool for radio networks, *IEEE Network*, 5(6), 27–30, November 1991.
6. J. Medbo, K. Börner, K. Haneda, V. Hovinen, T. Imai, J. Järvelainen, T. Jämsä et al., Channel modelling for the fifth generation mobile communications, in *Proceedings of the Eighth European Conference on Antennas and Propagation (EuCAP)*, Hague, the Netherlands, April 2014.
7. C. X. Wang, F. Haider, X. Gao, X. H. You, Y. Yang, D. Yuan, H. M. Aggoune, H. Haas, S. Fletcher, and E. Hepsaydir, Cellular architecture and key technologies for 5G wireless communication networks, *IEEE Communications Magazine*, 52, 122–130, February 2014.

8. K. Zheng, S. Ou, and X. Yin, Massive MIMO channel models: A survey, *International Journal of Antennas and Propagation*, 2014(11), 1–10, 2014.

9. J. Zander, Scientific challenges towards 5G mobile communications, COST IC1004 White paper, December 23, 2013. Available at: http://www.ic1004.org/.

10. J. Meinilä, T. Jämsä, P. Kyösti, and J. Ylitalo, Propagation modeling for evaluation of 4G systems, in *URSI General Assembly*, Chicago, IL, 2008.

11. L. M. Correia, *Wireless Flexible Personalised Communications (COST 259 Final Report)*, John Wiley & Sons, New York, 2001.

12. L. M. Correia, *Mobile Broadband Multimedia Networks Techniques, Models and Tools for 4G*, Academic Press, London, U.K., 2006.

13. WINNER1 WP5, Final report on link level and system level channel models, November 2005, http://www.ist-winner.org/Deliverable Documents/D5.4.pdf.

14. L. Liu, C. Oestges, J. Poutanen, K. Haneda, P. Vainikainen, F. Quitin, F. Tufvesson, and P. D. Doncker, The COST 2100 MIMO channel model, *IEEE Wireless Communications*, 19(6), 92–99, December 2012.

15. P. Kyösti, J. Meinilä, L. Hentilä, X. Zhao, T. Jämsä, C. Schneider, M. Narandzi'c et al., IST-4-027756 WINNER II D1.1.2 V1.2, WINNER II channel models, September 2007.

16. R. Verdone and A. Zanella, *Pervasive Mobile and Ambient Wireless Communications COST Action 2100*, Springer, London, U.K., 2012.

17. M. Steinbauer, A. F. Molisch, and E. Bonek, The double-directional radio channel, *IEEE Antennas and Propagation Magazine*, pp. 51–63, August 2001.

18. J. Poutanen, Geometry-based radio channel modeling: Propagation analysis and concept development, Doctoral dissertation, Aalto University, Espoo, Finland, 2011.

19. F. Quitin, C. Oestges, F. Horlin, and P. De Doncker, Diffuse multipath component characterization for indoor MIMO channel, in *Proceedings of the Fourth European Conference on Antennas and Propagation (EuCAP)*, Barcelona, Spain, April 12–16, 2010.

20. *ITU-R M.2135*, Guidelines for evaluation of radio interface technologies for IMT-advanced, International Telecommunication Union, Technical report, 2009, https://www.itu.int/dms_pub/itu-r/opb/rep/R-REP-M.2135-1-2009-PDF-E.pdf.

21. N. Shabbir, M. T. Sadiq, H. Kashif, and R. Ullah, Comparison of radio propagation models for long term evolution (LTE) network, *International Journal of Next-Generation Networks (IJNGN)*, 3(3), 27–31, September 2011.

22. Cost 231, Digital mobile radio toward future generation systems, Final Report, European Commission, Brussels, Belgium, 1999.

23. Y. Wang, W. J. Lu, and H. B. Zhu, Propagation characteristics of the LTE indoor radio channel with persons at 2.6 GHz, *IEEE Antennas and Wireless Propagation Letters*, 12, 991–994, 2013.

24. G. Salami, S. Burley, O. Durowoju, and C. Kellett, LTE indoor small cell capacity and coverage comparison, in *Proceedings of the 24th IEEE International Symposium on Personal, Indoor and Mobile Radio Communications* (PIMRC), pp. 66–70, London, U.K., September 8–11, 2013.

25. A. H. Zyoud, J. Chebil, M. H. Habaebi, M. R. Islam, and A. M. Zeki, Comparison of empirical indoor propagation models for 4G wireless networks at 2.6 GHz, in *Proceedings of the International Conference on Control, Engineering & Information Technology*, Vol. 3, pp. 7–11, Sousse, Tunisia, June 4–7, 2013.

26. S. G. Glisic, *Advanced Wireless Networks 4G Technologies*, John Wiley & Sons, Chichester, U.K., 2006.

27. C. B. Chae, C. Suh, M. Katz, D. Park, and F. H. P. Fitzeky, Comparative study of radio channel propagation characteristics for 3G/4G communication systems, in *Proceedings of MPRG/SDR Forum Smart Antennas Workshop*, Blacksburg, VA, June 4–6, 2003.

28. S. R. Saunders and A. Aragón-Zavala, *Antennas and Propagation for Wireless Communication Systems*, 2nd edn., John Wiley & Sons, Chichester, U.K., 2007.

29. A. Vilhar, A. Hrovat, I. Ozimek, and T. Javornik, Shooting and bouncing ray approach for 4G radio network planning, *International Journal of Communications*, 6(4), 166–174, 2012.

30. V. Torres, F. Esparza, and F. Falcone, Simulation and analysis of performance of LTE in indoor environments, in *Proceedings of the Mediterranean Microwave Symposium* (*MMS*), Tangiers, Morocco, November 15–17, 2009.

31. D. M. Rose, D. M. T. Jansen, and T. Kurner, Indoor to outdoor propagation—Measuring and modeling of femto cells in LTE networks at 800 and 2600 MHz, in *Proceedings of the IEEE GLOBECOM Workshops*, pp. 203–207, Houston, TX, December 5–9, 2011.

32. A. O. Kaya and D. Calin, Modeling three dimensional channel characteristics in outdoor-to-indoor LTE small cell environments, in *Proceedings of the IEEE Military Communications Conference*, pp. 933–938, San Diego, CA, November 18–20, 2013.

33. G. De La Roche, P. Flipo, Z. Lai, G. Villemaud, J. Zhang, and J. Gorce, Combination of geometric and finite difference models for radio wave propagation in outdoor to indoor scenarios, in *Proceedings of the Fourth European Conference on Antennas and Propagation*, *EUCAP*, Barcelona, Spain, April 12–16, 2010.

34. G. De La Roche, P. Flipo, Z. Lai, G. Villemaud, J. Zhang, and J. Gorce, Combined model for outdoor to indoor radio propagation, in *Proceedings of the 10th COST2100 Management Meeting, TD(10)10045*, Athens, Greece, February 3–5, 2010.

35. G. D. Roche, P. Flipo, Z. Lai, G. Villemaud, J. Zhang, and J. Gorce, Implementation and validation of a new combined model for outdoor to indoor radio coverage predictions, *EURASIP Journal on Wireless Communications and Networking*, 2010(1), 215352, April 2010.

36. D. Umansky, G. D. Roche, Z. Lai, G. Villemaud, J. Gorce, and J. Zhang, A new deterministic hybrid model for indoor-to-outdoor radio coverage prediction, in *Proceedings of the Fifth European Conference on Antennas and Propagation*, *EUCAP*, Rome, Italy, April 11–15, 2011.

37. T. S. Rappaport, S. Sun, R. Mayzus, H. Zhao, Y. Azar, K. Wang, G. N. Wong, J. K. Schulz, M. Samimi, and F. Gutierrez, Millimeter wave mobile communications for 5G cellular: It will work!, *IEEE Access*, 1, 335–349, 2013.

38. G. R. MacCartney, Jr., J. Zhang, S. Nie, and T. S. Rappaport, Path loss models for 5G millimeter wave propagation channels in urban microcells, in *Proceedings of the IEEE Global Communications Conference, Exhibition & Industry Forum*, Atlanta, GA, December 9–13, 2013.

39. V. Nurmela, T. Jämsä, P. Kyösti, V. Hovinen, and J. Medbo, Channel modelling for device-to-device scenarios, *COST IC1004*, Ghent, Belgium, September 26–27, 2013.

40. A. Roivainen, P. Jayasinghe, J. Meinil, V. Hovinen, and M. Latva-Aho, Vehicle-to-vehicle radio channel characterization in urban environment at 2.3 GHz and 5.25 GHz, in *Proceedings of the 25th IEEE International Symposium on Personal, Indoor and Mobile Radio Communications (PIMRC)*, pp. 63–67, Washington, DC, September 2–5, 2014.

41. P. S. Hall and Y. Hao, *Antennas and Propagation for Body-Centric Wireless Communications*, 2nd edn., Artech House, Norwood, MA, 2012.

42. V. Oleshchuk and R. Fensli, Remote patient monitoring within a future 5G infrastructure, *Wireless Personal Communications*, 57(3), 431–439, April 2011.

43. J. Wang and Q. Wang, *Body Area Communications: Channel Modeling, Communication Systems, and EMC*, John Wiley & Sons, Singapore, 2012.

44. Y. Wang, Body-centric radio propagation channels: Characteristics and models, PhD thesis, Aalborg University, Aalborg, Denmark, 2008.

45. J. Q. Wang, M. Komatsu, and O. Fujiwara, Human exposure assessment using a hybrid technique based on ray-tracing and FDTD methods for a cellular base-station antenna, in *Proceedings of the Asia-Pacific Radio Science Conference*, Qingdao, China, August 24–27, 2004.

46. F. J. Meyer, D. B. Davidson, U. Jakobus, and M. A. Stuchly, Human exposure assessment in the near field of GSM base-station antennas using a hybrid finite element/method of moments technique, *IEEE Transactions on Biomedical Engineering*, 50(2), 224–233, February 2003.

47. R. Luebbers and R. Baurle, FDTD predictions of electromagnetic fields in and near human bodies using visible human project anatomical scans, in *Proceedings of the IEEE Antennas and Propagation Society International Symposium, AP-S Digest*, Vol. 3, pp. 1806–180, Baltimore, MD, July 21–26, 1996.

48. METIS report, Initial channel models based on measurements, April 2014, https://www.metis2020.com/wp-content/uploads/deliverables/METIS_D1.2_v1.pdf.

3

LIMITED FEEDBACK FOR 4G AND BEYOND

STEFAN SCHWARZ

Contents

In this chapter, we consider channel state information (CSI) feedback over dedicated limited capacity feedback links, as applied in 3G and 4G cellular networks. Channel state information at the transmitter (CSIT) is useful for achieving the highest performance in multiple antenna wireless communications. Single-user multiple-input multiple-output (MIMO) systems employ CSI to match the spatial signature of the transmit signal to the channel, enabling reliable communication and highest transmission rates over multiple parallel data streams [1–5]. In multiuser MIMO broadcast channels, CSIT is of even higher importance as it allows to spatially separate multiple users and to control the interference caused by parallel transmissions to multiple users [6,7]. Also, most spatial interference coordination and alignment schemes, which are intended to avoid or even exploit the interference caused by parallel transmissions from multiple base stations, rely on accurate CSI at the transmitters to perform reliably [8–12]. Accurate CSIT is therefore an important enabler, facilitating the transition to dense heterogeneous 5G cellular networks. In time division duplex systems, CSIT can be made available by inserting pilot symbols into the uplink transmission and applying well-known channel estimation algorithms at the base stations to determine the CSI; see, for example, [13] and references therein. The majority of long-term evolution (LTE) network deployments, however, is frequency division duplex (FDD) based. In such FDD systems, obtaining CSIT is possible only through dedicated feedback from user equipment (UE).

In this chapter, we introduce and review efficient CSI feedback algorithms that minimize the feedback burden by exploiting time, space, and frequency correlations of the wireless channel.

In the first part of this chapter, we compose a solid overview of CSI feedback concepts that are adopted in Third-Generation Partnership Project (3GPP) LTE and LTE Advanced. We review the basic ideas underlying single- and multiuser MIMO transmission in LTE, utilizing predefined codebooks of precoding matrices and beamformers. We highlight the strengths and weaknesses of such codebook-based precoding for multiuser MIMO transmission and compare its performance to non-codebook-based precoding approaches, which are supported in LTE Advanced (Release 10 upward) through additional reference signals (UE-specific demodulation reference signals), enabling estimation of the precoded transmission channel. We provide a coherent performance comparison of LTE's different single- and multiuser MIMO transmission modes, employing the standard compliant Vienna LTE simulators [14,15].

In the second part of the chapter, we extend our scope to MIMO transmission strategies and corresponding CSI feedback methods that go beyond the capabilities of 4G LTE Advanced. We consider explicit CSI feedback and investigate two competing concepts of CSI quantization: quantization on the Grassmann manifold and quantization on the Stiefel manifold. Both manifolds enable efficient representation of the CSI by reducing the dimensionality of the quantization problem, with slight advantages for Grassmannian quantization. However, the two manifolds are each appropriate for different kinds of transmit strategies. The Grassmannian conveys subspace information and thus enables steering the transmit signals into preferred subspaces of the channel matrices, as is required, for example, for zero-forcing (ZF) beamforming [16], block-diagonalization precoding [17], and interference alignment [8,9]. The Stiefel manifold, on the other hand, not only conveys subspace information but conserves even the orientation of the channel eigenmodes spanning the subspace. In combination with information about the magnitudes of the channel eigenmodes, Stiefel manifold CSI enables the application of signal to noise ratio (SNR)-aware multiuser MIMO and coordinated multipoint transmission strategies, such as regularized ZF beamforming [16], regularized block-diagonalization precoding [18], as well as leakage-based precoding [19,20]. These methods provide advantages in the practically important regime of low to intermediate SNR compared to pure subspace techniques. We review the precoding methods mentioned earlier and identify the corresponding required CSIT. We then provide a thorough overview of existing quantizers for Grassmann and Stiefel manifolds, highlighting important principles of memoryless, differential, and predictive quantization and applying our own proposals [21–23] to achieve a fair comparison of competing precoding methods.

3.1 CSI Feedback Concepts of 3GPP LTE

In this section, we introduce LTE's physical layer (PHY) downlink signal processing chain and briefly review its single-user MIMO transmission modes, that is, codebook-based transmit beamforming, transmit diversity (TxD), and single-user spatial multiplexing employing codebook-based semiunitary (Grassmannian) precoding. We highlight important features of instantaneous and statistical CSI feedback selection algorithms to support these MIMO transmission modes. We then consider multiuser MIMO transmission and compare the performance of implicit and explicit CSI feedback. We finally provide a comprehensive performance comparison of LTE's transmission modes.

3.1.1 LTE's Physical Layer Signal Processing Architecture

The downlink PHY of LTE is based on a combination of MIMO-orthogonal frequency division multiple access (OFDMA) and bit-interleaved coded modulation (BICM), supporting adaptive modulation and coding (AMC) to enable efficient transmission to multiple UEs over time–frequency selective wireless channels [24]. The transmit signal processing chain is shown in Figure 3.1; it consists of an AMC stage, a MIMO transmit processing part, and the OFDMA signal composition and generation:

- With AMC, the channel code rate and modulation order of each user are adapted to the current channel conditions to enable reliable communication over the fading wireless channel. AMC in LTE is achieved with a BICM architecture as

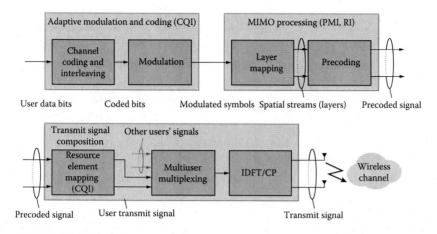

Figure 3.1 MIMO-orthogonal frequency division multiple access transmitter architecture of LTE.

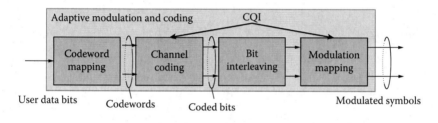

Figure 3.2 Adaptive modulation and coding based on a bit-interleaved coded modulation architecture according to the LTE specifications.

illustrated in Figure 3.2. Block-based transmission is applied in LTE with a transmission time interval (TTI) of 1 ms, corresponding to one LTE subframe. In each TTI, the user data bits are mapped onto codewords. These codewords are independently channel coded and their interleaved bits are mapped onto the applied modulation alphabet (4/16/64 quadrature amplitude modulation [QAM]). For each codeword, a different modulation and coding scheme (MCS) can be employed. To support the base station on deciding for the appropriate MCS, UE provide feedback about the quality of the transmission channel using the channel quality indicator (CQI).

- In the MIMO processing stage, the modulated transmit symbols are mapped onto spatial layers and are then multiplied with the precoding matrix to obtain the precoded transmit signal.

The number of applied spatial layers and the precoding matrix are dependent on the employed transmission mode as well as the rank indicator (RI) and precoding matrix indicator (PMI) feedback from UE. We provide details on this part in Section 3.1.2.

- In the final OFDMA transmit signal composition and generation part, the precoded user signal is mapped onto orthogonal frequency division multiplexing (OFDM) symbols and subcarriers, so-called resource elements (REs), and the signals of all served UE are jointly OFDM processed to obtain the time-domain transmit signal. Here, again the CQI feedback from UE plays an important role as it is required by the scheduler to assign resources to users with good channel conditions, thus enabling exploitation of multiuser diversity during scheduling.

For more details on the LTE PHY, the interested reader is referred to [25,26]. Assuming sufficient cyclic prefix (CP) length and negligible intercarrier interference, the time–frequency selective wireless channel is converted by OFDM into a set of noninterfering subcarriers n and symbols k; the pair $[n, k]$ is denoted as RE index. Alternatively, we also use a single index $[r]$ to indicate REs; however, if not explicitly required, we omit the RE index to simplify notation. The input–output relationship of user u served by base station i is then obtained as

$$\mathbf{y}_{u,i} = \mathbf{H}_{u,i}^{\mathrm{H}} \mathbf{F}_{u,i} \mathbf{x}_{u,i} + \mathbf{H}_{u,i}^{\mathrm{H}} \sum_{\substack{s \in \mathcal{S}_i \\ s \neq u}} \mathbf{F}_{s,i} \mathbf{x}_{s,i} + \tilde{\mathbf{z}}_{u,i} \in \mathbb{C}^{M_{u,i} \times 1}, \qquad (3.1)$$

$$\tilde{\mathbf{z}}_{u,i} = \sum_{\substack{j=1 \\ j \neq i}}^{J} \left(\mathbf{H}_{u,i}^{(j)} \right)^{\mathrm{H}} \sum_{s \in \mathcal{S}_j} \mathbf{F}_{s,j} \mathbf{x}_{s,j} + \mathbf{z}_{u,i}. \qquad (3.2)$$

Here, the transmit signal of user u in cell i is $\mathbf{x}_{u,i} \in \mathbb{C}^{\ell_{u,i} \times 1}$, with $\ell_{u,i}$ being the number of spatial layers of the user. The precoding matrix, which maps the layers onto the N_i transmit antennas, is of dimension $\mathbf{F}_{u,i} \in \mathbb{C}^{N_i \times \ell_{u,i}}$. The set of spatially multiplexed users in cell i is denoted by $\mathcal{S}_i$. The sum of other-cell interference from base stations $j \neq i$ and the Gaussian receiver noise is captured in $\tilde{\mathbf{z}}_{u,i} \in \mathbb{C}^{M_{u,i} \times 1}$, with $M_{u,i}$ denoting the number of receive antennas.

The matrix $\mathbf{H}_{u,i} = \mathbf{H}_{u,i}^{(i)} \in \mathbb{C}^{N_i \times M_{u,i}}$ represents the channel to the own base station, whereas $\mathbf{H}_{u,i}^{(j)} \in \mathbb{C}^{N_j \times M_{u,i}}$ is the channel to base station j. Notice, we employ the conjugate transpose of the channel matrices to simplify later notations. The transmit signal is normalized as

$$\mathbb{E}\left(\mathbf{x}_{u,i}\mathbf{x}_{u,i}^{\mathrm{H}}\right) = \mathbf{I}_{\ell_{u,i}}. \tag{3.3}$$

The allocation of the transmit power P_i among users and layers is considered in the precoders

$$\sum_{s \in \mathcal{S}_i} \mathrm{tr}\left(\mathbf{F}_{s,i}\mathbf{F}_{s,i}^{\mathrm{H}}\right) = P_i. \tag{3.4}$$

3.1.2 LTE's Single-User MIMO Transmission Modes

In single-user MIMO transmission, only a single user is served on a given time–frequency resource, that is, $\mathcal{S}_i = \{u\}$. In this case, multiple antennas at the base station can be exploited to improve the reliability of the transmission using TxD precoding, to enhance the receive SNR through transmit beamforming, and/or to increase the transmission rate by means of spatial multiplexing; see, for example, [27], for an overview of such techniques. Next, we briefly review single-user MIMO transmission in LTE.

3.1.2.1 Transmit Diversity   With TxD transmission, multiple antennas at the base station are utilized to improve the reliability of the transmission by transmitting the symbols of a user over multiple ideally independently fading paths. In LTE, this is achieved by means of Alamouti's space–frequency (or time) block coding scheme [28], as illustrated in Figure 3.3. In case of two transmit antennas, a rate of one information symbol per transmit symbol is obtained at full diversity, whereas with more than two transmit antennas, the Alamouti scheme is applied on subsets of antennas and subcarriers (see Figure 3.3b), thereby sacrificing information rate as well as diversity. Alternative TxD block coding schemes exist that achieve strictly better information rate–diversity trade-offs; see, for example, [29] for a comprehensive overview.

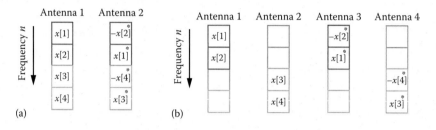

Figure 3.3 Illustration of LTE's transmit diversity scheme: (a) transmit diversity transmission with two transmit antennas and (b) transmit diversity transmission with four transmit antennas.

When TxD transmission is applied, the number of layers $\ell_{u,i}$ is fixed to one and the precoding is determined by the number of transmit antennas. Hence, UE only have to provide CQI feedback for transmission rate adaptation and multiuser scheduling.

3.1.2.2 Open-Loop Spatial Multiplexing LTE's open-loop spatial multiplexing (OLSM) scheme is a combination of two distinct precoding principles that are illustrated in Figure 3.4. The first idea, which is known as cyclic delay diversity (CDD) precoding, is able to transform spatial diversity, as provided by multiple transmit antennas, to frequency diversity. This is achieved by applying a cyclic delay to the transmit signals on different transmit antennas, thereby increasing the richness of the power delay profile of the effective channel [30]. The resulting enhanced frequency selectivity can either be translated to a diversity gain by the forward error correction channel code or be exploited by the scheduler for multiuser diversity. This CDD precoding is combined in LTE with precoder cycling as shown in Figure 3.4b. With precoder cycling, the

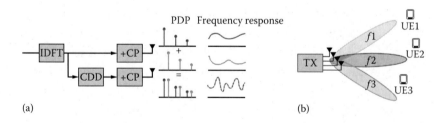

Figure 3.4 Illustration of LTE's open-loop spatial multiplexing scheme: (a) principle of cyclic delay diversity precoding and (b) basic idea of precoder cycling.

precoding matrices applied on consecutive subcarriers are selected cyclically from a predefined set of precoding matrices. In the example illustrated in Figure 3.4b, different beamformers are applied on different frequencies, which are more or less well suited for a given user. This effect can be translated to an opportunistic multiuser gain by the scheduler.

When OLSM is applied, precoding matrices are predetermined by the explained scheme. Hence, UE have to provide CQI feedback for transmission rate adaptation and multiuser scheduling as well as RI feedback to select the number of employed parallel spatial layers, enabling trade-off between beamforming and spatial multiplexing depending on the SNR.

3.1.2.3 Closed-Loop Spatial Multiplexing Closed-loop spatial multiplexing (CLSM) transmission is the most demanding transmission scheme of LTE Release 8 in terms of feedback overhead and signaling. In CLSM, the applied precoders are not predetermined by a specific scheme but are selected by UE from a given quantization codebook of possible precoding matrices. In [31], criteria for the selection of the optimal precoder and corresponding memoryless quantization codebook constructions are proposed for single-carrier systems. With memoryless quantization, the CSI at each time instant is quantized independently, thus neglecting potential correlations of the channel at consecutive time instants. It is shown in [31] that optimal memoryless quantization codebooks for the transmission of ℓ spatial streams are maximally spaced subspace packings on the Grassmann manifold $\mathcal{G}(N_i, \ell)$ of ℓ dimensional subspaces in the N_i dimensional complex Euclidean space. Depending on the considered performance criterion, different distance metrics are employed for the construction of the Grassmannian codebooks, for example, the projection two-norm distance, the Fubini–Study distance, or the chordal distance [31]. The corresponding codebooks can be represented by sets of semiunitary $N_i \times \ell$ dimensional matrices, that is, orthonormal bases, spanning the ℓ dimensional subspaces. Alternative codebook designs that focus more on practical implementation issues are based on discrete Fourier-transform matrices [32], QAM [33], and other concepts [34].

Table 3.1 Two-Transmit-Antenna Precoder Codebook of LTE's Closed-Loop Spatial Multiplexing Scheme

CODEBOOK INDEX	NUMBER OF LAYERS ℓ	
	1	2
0	$\frac{1}{\sqrt{2}}\begin{bmatrix} 1 \\ 1 \end{bmatrix}$	—
1	$\frac{1}{\sqrt{2}}\begin{bmatrix} 1 \\ -1 \end{bmatrix}$	$\frac{1}{2}\begin{bmatrix} 1 & 1 \\ 1 & -1 \end{bmatrix}$
2	$\frac{1}{\sqrt{2}}\begin{bmatrix} 1 \\ j \end{bmatrix}$	$\frac{1}{2}\begin{bmatrix} 1 & 1 \\ j & -j \end{bmatrix}$
3	$\frac{1}{\sqrt{2}}\begin{bmatrix} 1 \\ -j \end{bmatrix}$	—

The design of LTE's CLSM precoder codebooks is strongly influenced by implementation complexity considerations. The individual entries of the precoding matrices fulfill the constant modulus property, which means that precoding can be implemented with simple phase shifters. Additionally, the codebooks for different layer numbers are nested. Consider, for example, the precoder codebook for $N_i = 2$ in Table 3.1; it can be seen that the rank 2 precoders ($\ell = 2$) consist of the columns of the rank 1 precoders. This nested property can be exploited by UE to reuse intermediate results during CQI feedback calculation, reducing the number of required operations; see, for example, [29] for details.

When CLSM is applied, UE have to provide CQI feedback for transmission rate adaptation and multiuser scheduling, RI feedback to select the number of employed parallel spatial layers, as well as PMI feedback to choose the preferred precoding matrices from the given quantization codebook. We denote the precoder codebook for transmission of ℓ streams over N_i transmit antennas as

$$\mathcal{Q}_\ell^{(N_i)} \subset \left\{ \mathbf{Q} \in \mathbb{C}^{N_i \times \ell} \,|\, \mathbf{Q}^H \mathbf{Q} = \mathbf{I}_\ell \right\}. \tag{3.5}$$

Such precoder codebooks are defined for all possible numbers of layers $\ell \in \{1, ..., N_i\}$. Assuming that precoder $\mathbf{Q}_{u,i} \in \mathcal{Q}_{\ell_{u,i}}^{(N_i)}$ is selected as CSI feedback by user u, the precoder applied by the base station is

$$\mathbf{F}_{u,i} = \sqrt{\frac{P_i}{\ell_{u,i}}} \mathbf{Q}_{u,i}, \tag{3.6}$$

equally distributing the available transmit power P_i over the $\ell_{u,i}$ layers.

3.1.2.4 Non-Codebook-Based Precoding In LTE Advanced (LTE Release 10), non-codebook-based precoding is introduced to the standard. With non-codebook-based precoding, precoders are not restricted to predefined schemes or codebooks anymore but can be designed during operation by the base station, providing potential advantages in terms of performance and flexibility; see Section 3.1.5. To exploit these advantages, however, the base station requires explicit and accurate knowledge of the users' channel matrices for precoder calculation, which in general involves significantly increased CSI feedback overhead compared to implicit CSI feedback as employed with CLSM. As benchmark for LTE's codebook-based transmission schemes, we consider maximum ratio transmission (MRT) beamforming and capacity-achieving singular value decomposition (SVD) precoding with water-filling power allocation in our simulations given later. The compact-form SVD of the channel matrix $\mathbf{H}_{u,i}$ is denoted as

$$\mathbf{H}_{u,i} = \mathbf{U}_{u,i} \mathbf{\Sigma}_{u,i} \mathbf{V}_{u,i}^{\mathrm{H}}, \tag{3.7}$$

$$\mathbf{\Sigma}_{u,i} = \mathrm{diag}\left(\sigma_{u,i}^{(1)}, \ldots, \sigma_{u,i}^{(\ell_{\max})}\right), \quad \mathbf{U}_{u,i} \in \mathbb{C}^{N_i \times \ell_{\max}}, \quad \mathbf{V}_{u,i} \in \mathbb{C}^{M_{u,i} \times \ell_{\max}}, \tag{3.8}$$

with $\ell_{\max}$ representing the rank of the channel matrix, typically $\ell_{\max} = \min(N_i, M_{u,i})$, and the singular values in $\mathbf{\Sigma}_{u,i}$ arranged in decreasing order. With SVD-based transmission, the precoder is

$$\mathbf{F}_{u,i} = \mathbf{U}_{u,i} \mathbf{P}_{u,i}^{1/2}, \tag{3.9}$$

where $\mathbf{P}_{u,i} = \mathrm{diag}(p_1, \ldots, p_{\ell_{\max}})$ is a diagonal power-loading matrix as obtained from the water-filling solution [27]. In case of MRT, all power is assigned to the maximum eigenmode of the channel matrix: $p_1 = P_i, p_j = 0 \ \forall j \neq 1$. Notice that LTE's CLSM scheme can also be interpreted as a quantized form of SVD precoding: the water-filling power allocation is replaced with simple on-off switching of the eigenmodes, employing the RI to select the number $\ell_{u,i}$ of activated streams, and the matrix $\mathbf{U}_{u,i}^{(\ell_{u,i})} = [\mathbf{U}_{u,i}]_{:,1:\ell_{u,i}}$, containing the singular vectors corresponding to the $\ell_{u,i}$ largest singular values, is quantized with the Grassmannian codebook $\mathcal{Q}_{\ell_{u,i}}^{(N_i)}$.

3.1.3 Transmission Adaptation and CSI Feedback for Single-User MIMO

In this section, we present methods for the selection of LTE's CSI feedback indicators for single-user transmission according to [5,35]. We focus on CLSM as this mode requires calculation of all three CSI indicators (CQI, RI, PMI); for TxD and OLSM we simply restrict the selection to the appropriate values. We consider instantaneous CSI feedback, aiming at maximizing the instantaneous user throughput, and statistical (or long-term) CSI feedback, adapting the transmission parameters to the statistical behavior of the channel. Naturally, instantaneous CSI feedback is reasonable only if the current channel estimate at the receiver is representative for the time when the feedback is utilized at the transmitter. Hence, the delay in the feedback path must be sufficiently small compared to the coherence time of the channel to ensure similar channel conditions; otherwise, statistical CSI feedback should be employed.

3.1.3.1 Instantaneous CQI Feedback

The selection of the preferred MCS, as signaled by the CQI, is based on maximizing the expected instantaneously achievable throughput. Due to the finite block length of the codewords and other imperfections of the channel code, a vanishing block error ratio (BLER) is in general not achieved. The system is rather designed to operate below a target BLER $P_b^{(t)}$ that is commonly determined by the application driving the data transmission. An investigation of the BLER performance of LTE's 15 different MCSs, obtained through simulations utilizing the standard-compliant Vienna LTE simulators [14,15], is shown in Figure 3.5a for transmission over an additive white Gaussian noise (AWGN) channel of 1.4 MHz bandwidth. The SNR is defined as $\text{SNR} = P_i/\sigma_z^2$, with

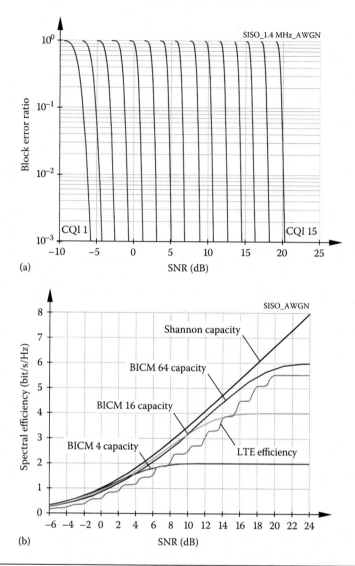

Figure 3.5 Evaluation of LTE's throughput and BLER performance: (a) BLERs of LTE's 15 modulation and coding schemes and (b) throughput versus channel capacity and bit-interleaved coded modulation capacity.

σ_z^2 denoting the power of the Gaussian receiver noise. The corresponding throughput of LTE is compared in Figure 3.5a to Shannon's AWGN channel capacity [36] and to the capacity of BICM [37]. The loss of LTE with respect to the BICM capacity can be attributed to the imperfect operation of the channel code. The BLERs in Figure 3.5a are employed to determine the appropriate highest MCS that achieves the target BLER $P_b^{(t)}$ from the instantaneous signal to interference and noise ratio (SINR) of the νth layer:

$$\text{SINR}_{\nu,u,i} = \frac{\left| \mathbf{g}_{\nu,u,i}^{\text{H}} \mathbf{H}_{u,i}^{\text{H}} \mathbf{f}_{\nu,u,i} \right|^2}{\sum\limits_{\mu=1,\mu\neq\nu}^{\ell_{u,i}} \left| \mathbf{g}_{\nu,u,i}^{\text{H}} \mathbf{H}_{u,i}^{\text{H}} \mathbf{f}_{\mu,u,i} \right|^2 + \tilde{\sigma}_z^2 \left\| \mathbf{g}_{\nu,u,i} \right\|^2}, \quad (3.10)$$

assuming that other-cell interference is treated by the UE as additional Gaussian noise of effective variance $\tilde{\sigma}_z^2$. Here, $\mathbf{f}_{\mu,u,i}$ denotes the μth column of user u's precoding matrix and $\mathbf{g}_{\nu,u,i}$ is the νth column of the linear equalizer $\mathbf{G}_{u,i} \in \mathbb{C}^{M_{u,i} \times \ell_{u,i}}$ applied by the UE. Notice, similar to the channel matrix, we define the equalizer via its conjugate transpose: $\hat{\mathbf{x}}_{u,i} = \mathbf{G}_{u,i}^{\text{H}} \mathbf{y}_{u,i}$. However, in general, the transmission does not take place over AWGN channels but rather over frequency-selective fading channels and thus AWGN-BLER look-up tables are not directly applicable. As the same MCS is employed on all REs assigned to a user, it is necessary to determine an average SNR that represents the performance achievable over the fading channel. This goal can be achieved with effective SINR mapping (ESM); assuming transmission over N_{RE} REs with corresponding SINRs $\{\text{SINR}_{\nu,u,i}[1], \ldots, \text{SINR}_{\nu,u,i}[N_{\text{RE}}]\}$, the effective AWGN-equivalent SNR is defined as

$$\text{SNR}_{\nu,u,i} = \beta_m f_m^{-1} \left(\frac{1}{N_{\text{RE}}} \sum_{r=1}^{N_{\text{RE}}} f_m \left(\frac{\text{SINR}_{\nu,u,i}[r]}{\beta_m} \right) \right). \quad (3.11)$$

Here, the MCS-dependent averaging function $f_m(\cdot)$ is determined by the considered ESM method; its inverse is denoted $f_m^{-1}(\cdot)$. In our simulations, we employ mutual information ESM (MIESM) [38], where the modulation-order dependent BICM capacity is used for $f_m(\cdot)$; see Figure 3.5. The scalar parameters β_m are required for calibration purposes to adapt ESM to the performance of the different MCSs. This AWGN-equivalent SNR can finally be applied to estimate BLERs of different MCSs by means of AWGN channel look-up tables.

3.1.3.2 Statistical CQI Feedback   When the feedback delay is large compared to the channel coherence time, instantaneous CQI feedback is already outdated during transmission, leading to a significant

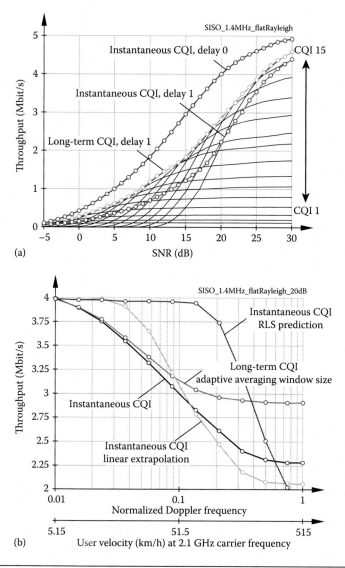

Figure 3.6 Evaluation of channel quality indicator feedback methods for single antenna transmitter and receiver communicating over 1.4 MHz bandwidth: (a) throughput comparison for temporally uncorrelated channels and (b) impact of temporal channel correlation.

throughput reduction as shown in Figure 3.6a. In this figure, we compare the throughput performance of LTE over frequency-flat Rayleigh block-fading channels with a fading block length of 1 subframe (1 ms) and no temporal correlation in-between subframes. We plot the performance of instantaneous CQI feedback with zero feedback delay (instantaneous CQI, delay 0), meaning that the CSIT is available acausally before the transmission, and with a feedback delay of 1 TTI (instantaneous CQI, delay 1). Notice that with 1 TTI feedback delay, the CQI is already completely outdated in the considered temporally uncorrelated situation. We also plot the throughput performance achieved with fixed MCSs corresponding to CQIs 1–15 (thin lines). We observe that rate adaptation with instantaneous but outdated CQI feedback lies substantially below the envelope curve of fixed CQI transmission. The best we can hope for is to achieve the fixed CQI envelope curve in such a situation by determining the long-term average AWGN-equivalent SNR using, for example, an exponential averaging filter:

$$\overline{\text{SNR}}_{v,u,i}[k] = \left(1 - \frac{1}{\alpha}\right)\overline{\text{SNR}}_{v,u,i}[k-1] + \frac{1}{\alpha}\text{SNR}_{v,u,i}, \quad (3.12)$$

with $\text{SNR}_{v,u,i}$ from Equation 3.11. The long-term average $\overline{\text{SNR}}_{v,u,i}[k]$ is then applied to determine the corresponding MCS that achieves the highest transmission rate according to the fixed CQI curves of Figure 3.6a. If the temporal correlation is not zero, the exponential averaging constant α is adapted to match the effective filter length to the channel correlation properties; that is, in case the feedback delay is small compared to the channel coherence time, a small filter length is employed, whereas the filter length is increased with growing feedback delay. The performance of this approach in dependence of the normalized channel Doppler frequency is shown in Figure 3.6b. The temporal channel correlation is determined by the normalized Doppler frequency according to Clarke's model [39]; that is, the autocorrelation function of the channel follows the zeroth-order Bessel function $J_0(2\pi f_d T_s)$ with f_d denoting the maximum channel Doppler frequency, $T_s = 1$ ms being the temporal sampling rate, and $v_d = f_d T_s$ representing the normalized Doppler frequency. In Figure 3.6b, we compare the performance of long-term average CQI feedback to instantaneous feedback using different channel predictors to compensate for the feedback delay of 1 TTI (linear extrapolation and finite impulse response filter-based prediction employing recursive least squares [RLS] filter adaptation [40]). We observe that the feedback delay can partly be compensated by extrapolation and prediction.

However, at a certain Doppler frequency prediction fails and the statistical long-term CQI outperforms the instantaneous rate adaptation.

3.1.3.3 RI and PMI Feedback   In the CQI feedback calculation of the previous sections, it is assumed that the transmission rank $\ell_{u,i}$ and the precoder $\mathbf{Q}_{u,i}$ are already selected. As shown in [5], to optimize the throughput performance of the system, a joint selection of PMI, RI, and CQI is required. Yet this can be computationally very demanding when the precoder codebook size is large and is not likely to be implemented in practice. We thus consider independent selection of PMI, RI, and CQI here. Similar to the CQI, the selection of PMI and RI is also based on maximizing the achievable throughput of the system. As these two indicators are, however, not so crucial for the BLER of the transmission as the CQI, we consider a less accurate throughput estimation in this case to simplify the calculation. To select the preferred precoder and transmission rank for a cluster of N_{RE} REs, we simply maximize the achievable sum-rate over this set

$$\{\ell_{u,i}, \mathbf{Q}_{u,i}\} = \operatorname*{arg\,max}_{\mathbf{Q} \in \mathcal{Q}_{\ell}^{(N_i)}, \ell \in \{1, \ldots, \ell_{\max}\}} \sum_{r=1}^{N_{RE}} \sum_{v=1}^{\ell} f\left(\mathrm{SINR}_{v,u,i}[r]\right), \quad (3.13)$$

where the SINR is determined similar to Equation 3.10 just replacing the precoder $\mathbf{F}_{u,i}$ with the corresponding matrix $\mathbf{Q}$. The function $f(\cdot)$ is employed to map the SINR to spectral efficiency, for example, via the LTE efficiency curve shown in Figure 3.5b. As proposed in [41], the method can be even further simplified by selecting the PMI from the pre-equalization mutual information

$$\mathbf{Q}_{u,i} = \operatorname*{arg\,max}_{\mathbf{Q} \in \mathcal{Q}_{\ell}^{(N_i)}} \sum_{r=1}^{N_{RE}} \log_2 \det\left(\mathbf{I}_{M_{u,i}} + \frac{P_i}{\ell \tilde{\sigma}_z^2} \mathbf{H}_{u,i}[r]^{\mathrm{H}} \mathbf{Q} \mathbf{Q}^{\mathrm{H}} \mathbf{H}_{u,i}[r]\right), \quad (3.14)$$

thus avoiding calculation of the equalizer for each precoder. This function, however, is not appropriate for the selection of the transmission rank as it overestimates the performance of LTE for a given number of layers ℓ; hence, given the precoder selection for each possible layer number $\ell \in \{1, \ldots, \ell_{\max}\}$, the RI $\ell_{u,i}$ should still be determined from the corresponding postequalization SINRs. The PMI and RI

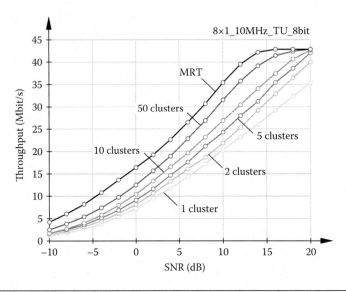

Figure 3.7 Evaluation of precoder feedback granularity with eight transmit antennas and one receive antenna.

selection can account for time–frequency correlation of the wireless channel simply by matching the cluster size N_{RE} to the coherence time-bandwidth of the channel [42]. This is investigated in Figure 3.7 for an $N_i \times M_{u,i} = 8 \times 1$ configuration with 10 MHz bandwidth (600 subcarriers) and a 50% coherence bandwidth (as defined in [43]) of approximately 400 kHz. In this investigation, we reduce the cluster size from 600 subcarriers (1 cluster) to 12 subcarriers (50 clusters), each subcarrier occupying a bandwidth of 15 kHz, which improves the throughput by more than 50% at 10 dB SNR. Yet, the feedback overhead also increases from 8 bits per TTI to $50 \times 8 = 400$ bits per TTI. As a benchmark, we show the performance of MRT assuming perfect CSIT.

In Figure 3.8, we evaluate the performance of transmission rank adaptation utilizing the presented algorithms for the case of spatially uncorrelated and strongly correlated channels. The spatial antenna correlation is determined by a Kronecker correlation model [44]:

$$\mathbf{H}_{u,i} = \mathbf{C}_{TX}^{1/2} \overline{\mathbf{H}}_{u,i} \mathbf{C}_{RX}^{1/2}, \quad \mathbb{E}\left(\mathrm{vec}(\mathbf{H}_{u,i})\mathrm{vec}(\mathbf{H}_{u,i})^{H} \right) = \mathbf{C}_{RX} \otimes \mathbf{C}_{TX}, \quad (3.15)$$

with $\mathbf{C}_{TX}$ and $\mathbf{C}_{RX}$ denoting the transmit- and receive-side correlation matrices and the entries of $\overline{\mathbf{H}}_{u,i}$ being independent and identically

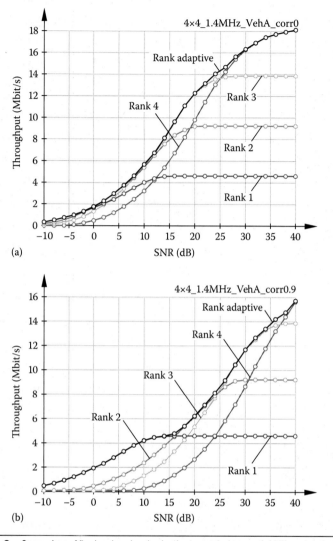

Figure 3.8 Comparison of fixed rank and rank adaptive transmission over 1.4 MHz system bandwidth: (a) performance with no spatial correlation and (b) performance with strong correlation at the receiver.

distributed (iid) Gaussian. For the results in Figure 3.8, we set $\mathbf{C}_{\text{TX}} = \mathbf{I}_{N_i}$ and

$$
\mathbf{C}_{\text{RX}} =
\begin{bmatrix}
1 & \alpha_{\text{corr}}^{1/9} & \alpha_{\text{corr}}^{4/9} & \alpha_{\text{corr}} \\
\alpha_{\text{corr}}^{1/9} & 1 & \alpha_{\text{corr}}^{1/9} & \alpha_{\text{corr}}^{4/9} \\
\alpha_{\text{corr}}^{4/9} & \alpha_{\text{corr}}^{1/9} & 1 & \alpha_{\text{corr}}^{1/9} \\
\alpha_{\text{corr}} & \alpha_{\text{corr}}^{4/9} & \alpha_{\text{corr}}^{1/9} & 1
\end{bmatrix},
\tag{3.16}
$$

according to [45, Appendix B.2.3], with α_{corr} = 0 in Figure 3.8a and α_{corr} = 0.9 in Figure 3.8b. We observe that rank adaptive transmission mostly achieves the envelope curve of fixed rank transmission; only at SNRs where two transmission ranks perform very similar (such as at 25 dB in Figure 3.8a) instantaneous rank, adaptation provides a gain over fixed rank transmission by selecting at each time instant the currently better transmission rank. Thus, rank adaptation based on the long-term average SNR (similar to the presented statistical CQI selection) is sufficient in most cases.

3.1.4 Transmission Adaptation and CSI Feedback for Multiuser MIMO

In multiuser MIMO, several UE are multiplexed in the spatial domain at the same time–frequency resources, exploiting the degrees of freedom (DoF) provided by multiple antennas at the base station. Compared to single-user MIMO, this approach basically provides the advantage that the potential spatial multiplexing gain is not confined by the capabilities of UE, but only by the amount of antennas available at the base station, which is typically much larger than the number of receive antennas for reasons of complexity and available space. Also, multiuser MIMO is less susceptible to spatial antenna correlation at the users, causing a strong singular value disparity of the channel [46] and thus a large SNR gap when increasing the number of layers (see Figure 3.8b), simply by keeping the number of streams per user small; in fact, most proposals in literature consider only single-stream transmission (beamforming) per user. These advantages, however, are only attainable with sufficiently accurate CSI available at the base station. In this respect, LTE is not sufficiently equipped due to the very limited codebook sizes provided for implicit (precoder) or explicit (channel) CSI quantization. Still, multiuser MIMO transmission is foreseen in the standard and we review and evaluate two basic methods later, which can be implemented with LTE's capabilities. Specifically, we consider per user unitary rate control (PU2RC) [47], a multiuser MIMO technique that is based on implicit CSI feedback, and ZF beamforming [16], requiring explicit channel knowledge at the transmitter for precoder calculation. We present more advanced feedback and precoding concepts in Section 3.2.

3.1.4.1 Per User Unitary Rate Control With PU2RC the multiuser MIMO precoders are restricted to be selected from a predefined code-book of precoding matrices, similar to CLSM. In contrast to CLSM, however, each column of the precoding matrix is employed to serve a different user via spatial multiplexing. Similar to CLSM, we write the codebook for transmission of a total number of L streams as (we use L as opposed to ℓ to distinguish between the total number of streams and the number of streams per user)

$$\mathcal{Q}_L^{(N_i)} \subset \left\{ \mathbf{Q} \in \mathbb{C}^{N_i \times L} \,|\, \mathbf{Q}^H \mathbf{Q} = \mathbf{I}_L \right\}. \tag{3.17}$$

Assuming that UE u is served via the single column v of precoder $\mathbf{Q}_i \in \mathcal{Q}_L^{(N_i)}$, the input–output relationship is obtained as

$$\mathbf{y}_{u,i} = \mathbf{H}_{u,i}^H \sqrt{\frac{P_i}{L}} \mathbf{q}_{v,i} x_{u,i} + \mathbf{H}_{u,i}^H \sqrt{\frac{P_i}{L}} \sum_{\substack{\mu=1 \\ \mu \neq v}}^{L} \mathbf{q}_{\mu,i} x_{\mu,i} + \tilde{\mathbf{z}}_{u,i}, \tag{3.18}$$

with $\mathbf{q}_{v,i}$ denoting the vth column of $\mathbf{Q}_i$. As the UE knows the pre-coder codebook, it can design the optimal linear interference-aware minimum mean squared error (MMSE) receiver [48]

$$\mathbf{g}_{u,i} = \left(\tilde{\sigma}_z^2 \mathbf{I}_{M_{u,i}} + \frac{P_i}{L} \mathbf{H}_{u,i}^H \sum_{\substack{\mu=1 \\ \mu \neq v}}^{L} \mathbf{q}_{\mu,i} \mathbf{q}_{\mu,i}^H \mathbf{H}_{u,i} \right)^{-1} \frac{P_i}{L} \mathbf{H}_{u,i}^H \mathbf{q}_{v,i} \tag{3.19}$$

and calculate the corresponding postequalization SINR

$$\text{SINR}_{u,i} = \frac{(P_i/L) \left| \mathbf{g}_{u,i}^H \mathbf{H}_{u,i}^H \mathbf{q}_{v,i} \right|^2}{(P_i/L) \sum_{\mu=1, \mu \neq v}^{L} \left| \mathbf{g}_{u,i}^H \mathbf{H}_{u,i}^H \mathbf{q}_{\mu,i} \right|^2 + \tilde{\sigma}_z^2 \left\| \mathbf{g}_{u,i} \right\|^2}. \tag{3.20}$$

To obtain the CSI feedback, the UE determines the best column v of each precoder $\mathbf{Q}_i \in \mathcal{Q}_L^{(N_i)}$ in terms of SINR. Assuming a total num-ber of n_p precoders in $\mathcal{Q}_L^{(N_i)}$, the UE thus obtains n_p potential precod-ing vectors for CSI feedback. From these n_p vectors, the UE selects the subset of n_l best columns, again in terms of SINR. To signal the n_l selected columns to the base station, a total feedback overhead of $n_l \cdot (\log_2(n_p) + \log_2(L))$ feedback bits is required. In addition to the preferred columns, the UE also feeds back the corresponding SINRs as CQI for multiuser scheduling purposes. The extension to OFDM follows the same approach as explained with CLSM, that is, precoder clustering and MIESM averaging to determine the AWGN-equivalent SNR.

Given the CSI feedback from UE, the base station finds compatible user sets, that is, sets of users that have selected the same precoding matrices $\mathbf{Q}_i$ but different columns $\mathbf{q}_{v,i}$. A greedy scheduler, for example, as proposed in [47], is then applied to determine the best set of compatible users in terms of maximizing the achievable throughput.

One major difficulty of PU2RC is the selection of the feedback parameters n_l, n_p, and L. In principle, it is good if a user feeds back many potential precoding vectors n_l as this increases the chances to be served. However, for a given feedback overhead budget, increasing n_l implies reducing n_p and/or L. Similarly, large n_p is good as it increases the probability that a user finds a precoding matrix that provides high SINR. However, with growing n_p and a limited amount of users in the system, there might not be many users left that select the same precoding matrix, thus reducing the achievable multiplexing gain. Finally, the selection of L impacts the achievable throughput twofold: large L means a large spatial multiplexing gain and hence potentially a large multiuser throughput. However, L also impacts the interference term in the denominator of Equation 3.20, thereby reducing the rate achievable by each user. In our simulations in Section 3.1.5, we determine the best combination of feedback parameters by means of an exhaustive search; see [47] for hints and remarks on how to select these parameters in practice.

3.1.4.2 Zero-Forcing Beamforming   The aim of zero-forcing (ZF) beamforming is to serve multiple users in parallel with each over a single stream (beamforming), while designing the precoding vectors such that the multiuser interference is perfectly eliminated [16]. ZF beamforming was originally proposed for vector channels, that is, $M_{u,i} = 1$. In case $M_{u,i} > 1$, it is sufficient if only a 1D signal subspace of $\mathbf{H}_{u,i}$ is free of interference because this subspace can be filtered out by the antenna combiner $\mathbf{g}_{u,i} \in \mathbb{C}^{M_{u,i} \times 1}$ applied by the UE, thereby separating the intended signal from the interference [48]. Given the antenna combiners $\mathbf{g}_{u,i}$ and a set $\mathcal{S}_i = \{1,\ldots,S_i\}$ of scheduled users satisfying $S_i = |\mathcal{S}_i| \leq N_i$, the base station obtains the precoding vector $\mathbf{f}_{u,i}$ of user $u \in \mathcal{S}_i$ according to [49] from

$$\mathbf{f}_{u,i} = \sqrt{\frac{P_i}{S_i}} \frac{\tilde{\mathbf{f}}_{u,i}}{||\tilde{\mathbf{f}}_{u,i}||_2}, \quad \tilde{\mathbf{f}}_{u,i} = \mathbf{B}_{u,i} \mathbf{B}_{u,i}^{\mathrm{H}} \mathbf{h}_{u,i}^{\mathrm{eff}}, \quad \mathbf{h}_{u,i}^{\mathrm{eff}} = \mathbf{H}_{u,i} \mathbf{g}_{u,i}, \quad (3.21)$$

$$\mathbf{B}_{u,i} \stackrel{\Delta}{=} \mathrm{null}(\overline{\mathbf{H}}_{u,i}), \quad \overline{\mathbf{H}}_{u,i} = \left[\mathbf{h}_{1,i}^{\mathrm{eff}}, \ldots, \mathbf{h}_{u-1,i}^{\mathrm{eff}}, \mathbf{h}_{u+1,i}^{\mathrm{eff}}, \ldots, \mathbf{h}_{S_i,i}^{\mathrm{eff}} \right] \in \mathbb{C}^{N_i \times (S_i - 1)},$$

$$\mathbf{h}_{s,i}^{\mathrm{eff}} = \mathbf{H}_{s,i} \mathbf{g}_{s,i}, \quad \text{for all } s \text{ in } \{1,\ldots,S_i\} \tag{3.22}$$

assuming perfect CSIT. Here, $\mathbf{B}_{u,i} \in \mathbb{C}^{N_i \times (N_i - S_i + 1)}$ denotes an orthonormal basis for the left null space of $\overline{\mathbf{H}}_{u,i}$. The condition $S_i \leq N_i$ assures that this null space is not empty. As $\mathbf{f}_{u,i}$ lies in the null space of all other users' channels, the transmission to user u does not cause any interference to the other users. Similar conditions are fulfilled by all served UEs, thus enabling interference-free multiuser communication.

Conversely, given the precoding vectors $\mathbf{f}_{u,i}$ and the set of scheduled users $\mathcal{S}_i$, the optimal linear interference-aware antenna combiners are obtained according to the MMSE solution given as Equation 3.23. Hence, we end up with a "chicken and egg situation": knowing the antenna combiners, the base station can determine the ZF beamformers from Equation 3.21 and, conversely, knowing the beamformers, UE can calculate the optimal antenna combiners. As pointed out in [49], to optimize the achievable throughput of this precoder and antenna combiner design problem, a joint optimization over the two is required. This, however, can be carried out only by the base station, implying that each UE has to feedback perfectly its channel matrix $\mathbf{H}_{u,i}$. On the other hand, if UE select the effective channels $\mathbf{h}_{u,i}^{\mathrm{eff}}$ beforehand, then only $\mathbf{h}_{u,i}^{\mathrm{eff}}$ must be provided to the base station, reducing the dimensionality of the CSI and giving an advantage in terms of feedback overhead. We follow the second approach here as it appears practically viable.

Notice further that we can replace $\mathbf{h}_{j,i}^{\text{eff}}, \forall j$ in Equations 3.21 and 3.22 with any other scaled vector $\tilde{\mathbf{h}}_{j,i} = c\mathbf{h}_{j,i}^{\text{eff}}, c \in \mathbb{C}$ without affecting the null space basis $\mathbf{B}_{u,i}$ or the precoder $\mathbf{f}_{u,i}$. Hence, all vectors that lie in the 1D subspace spanned by $\mathbf{h}_{j,i}^{\text{eff}}$, that is, $\text{span}(\mathbf{h}_{j,i}^{\text{eff}})$, are equivalent for the precoder calculation. All these vectors can thus be identified with a single point on the Grassmann manifold $\mathcal{G}(N_i, 1)$ of 1D subspaces in the N_i dimensional complex Euclidean space [50]. Therefore, Grassmannian quantization is employed to efficiently provide the CSI to the base station over limited capacity feedback channels. To represent $\text{span}(\mathbf{h}_{u,i}^{\text{eff}})$, the UE employs the normalized vector $\tilde{\mathbf{h}}_{u,i} = \mathbf{h}_{u,i}^{\text{eff}}/(\mathbf{h}_{u,i}^{\text{eff}})_2$ and applies the codebook $\mathcal{Q}_1^{(N_i)}$ for quantization; see Equation 3.5. As quantization metric, we consider the chordal distance:

$$\hat{\mathbf{h}}_{u,i} = \arg\min_{\mathbf{q} \in \mathcal{Q}_1^{(N_i)}} d_c^2\left(\mathbf{h}_{u,i}^{\text{eff}}, \mathbf{q}\right), \quad d_c^2\left(\mathbf{h}_{u,i}^{\text{eff}}, \mathbf{q}\right) = 1 - \text{tr}\left(\mathbf{q}^{\text{H}} \tilde{\mathbf{h}}_{u,i} \tilde{\mathbf{h}}_{u,i}^{\text{H}} \mathbf{q}\right), \quad (3.23)$$

because it has been shown that the expected residual rate loss of ZF with imperfect CSIT due to the quantization error compared to perfect CSIT is determined by $d_c^2(\tilde{\mathbf{h}}_{u,i}, \hat{\mathbf{h}}_{u,i})$ [51]. Here, $\text{tr}(\cdot)$ denotes the trace of a matrix.

Now remains the question of how to select the antenna combiner $\mathbf{g}_{u,i}$, without knowing which precoder is applied by the base station. Different solutions exist [12,48,52–55] from which we summarize a few important ones:

3.1.4.2.1 Maximum Eigenmode Transmission   The goal of maximum eigenmode transmission (MET) antenna combining is to generate a 1D effective channel $\mathbf{h}_{u,i}^{\text{eff}}$ that maximizes the achievable transmission rate of a user in the absence of multiuser interference [52]. The solution to this problem is to set $\mathbf{g}_{u,i}$ equal to that column of the singular vector matrix $\mathbf{U}_{u,i}$ (see Equation 3.7) that corresponds to the maximum singular value. The MET combiner performs well in case the CSI quantization is very accurate because then the assumption of negligible multiuser interference is justified. Specifically, to achieve the same multiplexing gain as with perfect CSIT, the feedback overhead must grow linearly with the logarithmic SNR, with a slope of $(N_i - 1)$ [53].

3.1.4.2.2 Quantization-Based Combining In case $M_{u,i}$ is larger than 1, the required feedback overhead to achieve the same multiplexing gain as with perfect CSIT can be substantially reduced with the quantization-based combining (QBC) method proposed in [53]. Consider a specific $\mathbf{q} \in \mathcal{Q}_1^{(N_i)}$; QBC determines the corresponding antenna combiner such that the chordal distance is minimized: $\min_{\mathbf{g}_{u,i}} d_c^2(\mathbf{h}_{u,i}^{\text{eff}}, \mathbf{q})$. It turns out that

$$\min_{\mathbf{g}_{u,i}} d_c^2\left(\mathbf{h}_{u,i}^{\text{eff}}, \mathbf{q}\right) = d_c^2(\mathbf{H}_{u,i}, \mathbf{q}) = 1 - \text{tr}\left(\mathbf{q}^{\text{H}} \mathbf{U}_{u,i} \mathbf{U}_{u,i}^{\text{H}} \mathbf{q}\right), \quad (3.24)$$

with $\mathbf{U}_{u,i}$ denoting the matrix of left singular vectors of $\mathbf{H}_{u,i}$. Hence, quantization reduces to

$$\hat{\mathbf{h}}_{u,i} = \arg\min_{\mathbf{q} \in \mathcal{Q}_1^{(N_i)}} d_c^2(\mathbf{H}_{u,i}, \mathbf{q}), \quad (3.25)$$

which does not require calculation of the antenna combiner at all and thus can be implemented efficiently. With this approach, the slope of the feedback overhead scaling law reduces to $(N_i - M_{u,i})$ [53].

3.1.4.2.3 Quantization-Based Combining with Dimensionality Adaptation MET and QBC pursue two opposite goals: MET maximizes the gain of the effective user channel, without accounting for the error incurred during quantization of this channel, whereas QBC minimizes the quantization error without considering the gain of the effective channel obtained. Correspondingly, MET is appropriate in a low SNR regime, when the transmission rate is determined by the noise rather than the interference and QBC vice versa [55]. To trade off between these two extrema, QBC can be applied only in the subspace of d dominant eigenmodes of the channel

$$\hat{\mathbf{h}}_{u,i} = \arg\min_{\mathbf{q} \in \mathcal{Q}_1^{(N_i)}} \left(1 - \text{tr}\left(\mathbf{q}^{\text{H}} \mathbf{U}_{u,i}^{(d)} \left(\mathbf{U}_{u,i}^{(d)}\right)^{\text{H}} \mathbf{q}\right)\right), \quad 1 \le d \le \ell_{\max}, \quad (3.26)$$

with $\mathbf{U}_{u,i}^{(d)}$ containing the d columns of $\mathbf{U}_{u,i}$ corresponding to the largest singular values. The preferred dimensionality d of a UE can be determined by estimating the achievable rate of the UE; see [49] for details. We employ this approach in the simulation results presented later as it outperforms the others.

The antenna combiners presented earlier are well suited for CSI feedback calculation without precoder knowledge, as well as for data detection in case UE are not able to accurately estimate the interference caused by the transmission to other users. However, in our simulations presented later, we assume that the users can estimate the interference as soon as the precoders are fixed. Hence, the presented combiners are only applied to determine the CSI feedback; afterward, when the precoders are calculated, UE employ the interference-aware MMSE receiver similar to Equation 3.19 to suppress the residual interference.

In addition to the Grassmannian feedback $\hat{\mathbf{h}}_{u,i}$, UE also have to provide information about the channel quality to the base station for multiuser scheduling. In [52], a lower bound on the expected SINR of a user with ZF beamforming and quantized CSIT is derived, which depends on the CSI quantization error and the gain of the effective user channel. Based on this lower bound, a greedy scheduler is proposed that selects the set of users that is served in parallel. We employ this scheduler and CQI feedback in the simulations later.

3.1.5 Performance Comparison of LTE's MIMO Transmission Modes

In this section, we provide a coherent simulation-based performance comparison of LTE's single- and multiuser transmission modes, employing the Vienna LTE simulators. We start by comparing the uncoded bit error ratio (BER) of 4, 16, and 64 QAM achieved with TxD transmission to single-input single-output (SISO) and single-input multiple-output (SIMO) transmission in Figure 3.9, employing $N_i \in \{1, 2\}$ transmit antennas and $M_{u,i} \in \{1, 2\}$ receive antennas. As expected, the Alamouti TxD scheme improves the robustness of the transmission; more specifically, a diversity order of two and four is achieved with $N_i \times M_{u,i} = 2 \times 1$ and $N_i \times M_{u,i} = 2 \times 2$, respectively. This BER improvement can be translated to a corresponding throughput improvement by means of AMC. In terms of throughput, however, TxD hardly gains over SISO/SIMO transmission because the increased reference symbol overhead, due to using two transmit antennas, consumes most of the potential throughput gains.

Next, we cross-compare the throughput performance of LTE's single-user spatial multiplexing modes, that is, CLSM and OLSM,

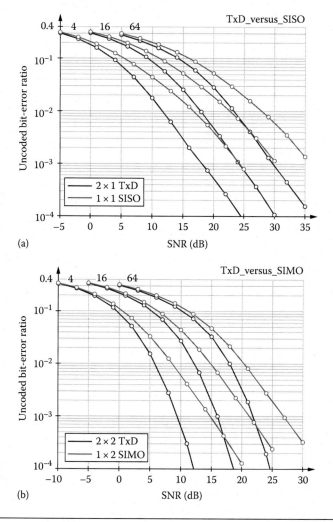

Figure 3.9 Bit error ratio evaluation of transmit diversity compared to single antenna transmission: (a) performance with a single receive antenna and (b) performance with two receive antennas.

in Figure 3.10 for $N_i \times M_{u,i} = 2 \times 2$ and $N_i \times M_{u,i} = 4 \times 4$ and 5 MHz bandwidth. As a benchmark, we also show the performance achieved with SVD precoding and spatial water-filling power allocation assuming perfect CSIT. CLSM achieves a gain of approximately 2–3 dB over OLSM in the case $N_i \times M_{u,i} = 4 \times 4$ over a large SNR range. However, this gain comes at the cost of $25 \times 4 = 100$ bits of PMI feedback overhead per TTI, employing 25 feedback clusters each consisting of $N_{RE} = 12$ subcarriers (i.e., one LTE resource block [RB]). The performance of CLSM is still 2–3 dB worse than SVD precoding,

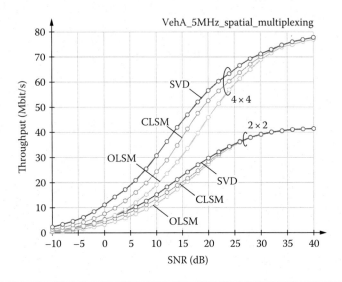

Figure 3.10 Throughput evaluation of single-user spatial multiplexing modes over 5 MHz bandwidth.

which is caused by the mismatch between the quantized precoder and the singular-vector matrix applied with SVD precoding, due to the finite precoder quantization codebook size ($\{2, 4\}$ bits for $N_i = \{2, 4\}$) and due to CSI feedback clustering. Notice that the considered channel model has a 50% coherence bandwidth of approximately 550 kHz, but still the cluster size of 12 subcarriers = 180 kHz is insufficient.

In Figure 3.11, we examine an eight-layer CLSM, as introduced with LTE Advanced, in contrast to the theoretically achievable performance as predicted by Shannon's theory. The left-most curve represents the single-user MIMO capacity of the considered 5 MHz bandwidth channel. If we assume that the transmitter is not aware of the user's channel matrix and thus employs a scaled identity matrix as precoder, we end up with the curve denoted as *mutual information*. It is well known that the scaled identity matrix is the optimal precoder of iid Rayleigh fading channels at high SNR; thus, mutual information converges toward channel capacity with increasing SNR. The *achievable mutual information* curve accounts for the loss caused by the guard bands, reference signals, as well as synchronization signals and CP employed by LTE. For eight-layer transmission, this overhead amounts to approximately 20%. The throughput achieved with LTE assuming perfect CSIT and employing the presented

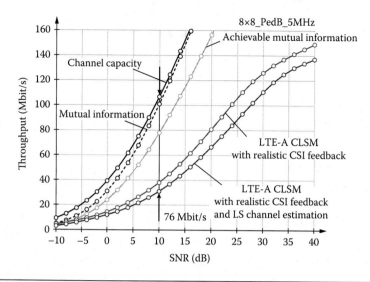

Figure 3.11 Throughput evaluation of closed-loop spatial multiplexing with eight transmit antennas and eight receive antennas over 5 MHz bandwidth.

feedback algorithms is shown by the curve denoted as *LTE-A CLSM with realistic CSI feedback*. From 10 to 30 dB SNR, LTE achieves approximately 40% of channel capacity. Assuming furthermore realistic least-squares channel estimation [13] at the UE, the throughput reduces to approximately 30%–35% of channel capacity. Thus, there is room for improvement of LTE.

In the final simulation of this section, we cross-compare the performance of single-user CLSM to multiuser MIMO based on PU2RC and ZF beamforming. We consider a 1.4 MHz bandwidth system with $N_i = 8$ and $M_{u,i} \in \{1, 4\}$ and a total number of 20 UE. Both multiuser multiplexing schemes are restricted to transmit at most one stream per user, whereas CLSM can transmit up to $M_{u,i}$ layers per user. We employ 8 bits of feedback per user for implicit/ explicit channel quantization in all cases. The results of this investigation are shown in Figure 3.12. We observe that PU2RC outperforms the other schemes in the high SNR regime. Thus, PU2RC can at least partially exploit the eight transmit antennas available at the base station. Notice, however, that the saturation throughput of LTE with $N_i = 8$ lies at approximately 35 Mbit/s in case 8 streams are transmitted in parallel. Hence, the multiplexing capabilities of the base station are by far not exploited. ZF beamforming performs

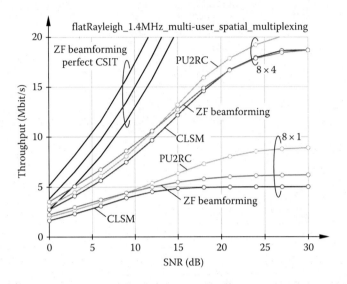

Figure 3.12   Comparison of single- and multiuser MIMO transmission in LTE over 1.4 MHz bandwidth.

similar to CLSM with the coarse CSI quantization achievable with the employed small quantization codebook of LTE. However, with perfect CSIT, ZF beamforming shows a large potential for improvement of the throughput. It is thus worth to consider more advanced quantization codebook constructions and feedback strategies as we do in the following section.

3.2 CSI Feedback Concepts for 5G

In this section, we review several important subspace-based and channel Gramian–based linear precoding techniques for MIMO broadcast and interference channels, identify the required CSIT for precoder calculation, and provide CSI quantization methods to convey the necessary information efficiently to the transmitter. We compare the performance of Grassmannian quantization–based precoding techniques to methods that require information about the orientation and magnitude of the eigenmodes of the channel matrix. This information can be provided by a gain-shape vector quantization approach, combining quantization on the Stiefel manifold with Lloyd quantization of the eigenvalues of the channel Gramian, that is, the singular values of the channel matrix. We consider single-carrier frequency-flat channels in this section; however, the presented results have been extended to frequency-selective OFDM in [49,56], employing feedback clustering and interpolation.

3.2.1 Grassmannian Quantization for Subspace-Based Precoding Techniques

The goal of subspace-based precoding techniques is to steer the transmit signal of UE into preferred subspaces of the channel matrices $\mathbf{H}_{u,i}, \forall u \in \mathcal{S}_i$. We have already encountered two examples of such methods in Section 3.1, namely, single-user CLSM and multiuser ZF beamforming. In CLSM, the single-user precoder $\mathbf{Q}_{u,i} \in \mathbb{C}^{N_i \times \ell_{u,i}}$ ideally aligns the transmit signal with the $\ell_{u,i}$ dimensional subspace of the channel matrix corresponding to the maximum eigenmodes, that is, $\text{span}(\mathbf{U}_{u,i}^{(\ell_{u,i})})$. With ZF beamforming, on the other hand, the multiuser interference is aligned in a subspace of the user's channel matrix $\mathbf{H}_{u,i}$, such that the UE can cancel the interference through appropriate antenna combining. As we have seen, perfect alignment is in general not achieved due to CSI quantization and clustering to restrict the required feedback overhead. Especially multiuser schemes react very sensitively to imperfections in the alignment as this directly impacts the residual interference between users. Next, we briefly review two other subspace-based precoding methods and present Grassmannian quantizers for subspace quantization.

3.2.1.1 Examples of Subspace-Based Precoding Techniques

3.2.1.1.1 Block Diagonalization Precoding
Block diagonalization (BD) precoding is a generalization of ZF beamforming to allow multistream transmission per UE [17]. The precoder design problem is a straightforward extension of Equation 3.21 [49]; to simplify notation, we assume $\ell_{u,i} = \ell, \forall u$:

$$\mathbf{F}_{u,i} = \sqrt{\frac{P_i}{S_i \ell}} \tilde{\mathbf{F}}_{u,i}, \quad \tilde{\mathbf{F}}_{u,i} \overset{\Delta}{=} \text{span}\left(\mathbf{B}_{u,i} \mathbf{B}_{u,i}^H \mathbf{H}_{u,i}^{\text{eff}}\right), \quad \mathbf{H}_{u,i}^{\text{eff}} = \mathbf{H}_{u,i} \mathbf{G}_{u,i}, \quad (3.27)$$

$$\mathbf{B}_{u,i} \overset{\Delta}{=} \text{null}\left(\overline{\mathbf{H}}_{u,i}\right), \quad \overline{\mathbf{H}}_{u,i} = \left[\mathbf{H}_{1,i}^{\text{eff}}, \ldots, \mathbf{H}_{u-1,i}^{\text{eff}}, \mathbf{H}_{u+1,i}^{\text{eff}}, \ldots, \mathbf{H}_{S_i,i}^{\text{eff}}\right] \in \mathbb{C}^{N_i \times (S_i - 1)\ell},$$

$$\mathbf{H}_{s,i}^{\text{eff}} = \mathbf{H}_{s,i} \mathbf{G}_{s,i}, \quad \text{for all } s \text{ in } \{1, \ldots, S_i\} \quad (3.28)$$

with $\widetilde{\mathbf{F}}_{u,i} \in \mathbb{C}^{N_i \times \ell}$ denoting an orthonormal basis for the subspace $\text{span}(\mathbf{B}_{u,i}\mathbf{B}_{u,i}^{H}\mathbf{H}_{u,i}^{\text{eff}})$ and $\mathbf{G}_{u,i} \in \mathbb{C}^{M_{u,i} \times \ell}$ being the antenna combiner of UE u, determining the ℓ dimensional preferred subspace of the user. The feasibility conditions for existence of a solution to this problem are $\ell \le \ell_{\max}$ and $S_i \ell \le N_i$. The MET and QBC antenna combiners presented in Section 3.1.4 are extended to multistream transmission per UE in [55]. Also, the corresponding feedback overhead scaling laws, to achieve the same spatial multiplexing gain as with perfect CSIT, are derived in [55]; that is, the feedback overhead must scale linearly with the logarithmic SNR with a slope of $\ell(N_i - \ell)$ in case of MET and $\ell(N_i - M_{u,i})$ for subspace QBC (SQBC) assuming $M_{u,i} \le N_i$. Similar to ZF beamforming, each matrix $\mathbf{H}_{j,i}^{\text{eff}}$ in Equations 3.27 and 3.28 can equivalently be replaced with any arbitrary other matrix $\widetilde{\mathbf{H}}_{j,i} \in \mathbb{C}^{N_i \times \ell}$ that spans the same subspace:

$$\widetilde{\mathbf{H}}_{j,i} \equiv \mathbf{H}_{j,i}^{\text{eff}} \Leftrightarrow \text{span}(\widetilde{\mathbf{H}}_{j,i}) = \text{span}\left(\mathbf{H}_{j,i}^{\text{eff}}\right), \qquad (3.29)$$

without affecting the precoder solution; specifically, we apply orthonormal bases—$\widetilde{\mathbf{H}}_{j,i}^{H}\widetilde{\mathbf{H}}_{j,i} = \mathbf{I}_{\ell}$. All such matrices can be identified with a single point on the Grassmann manifold $\mathcal{G}(N_i, \ell)$ of ℓ dimensional subspaces in the N_i dimensional complex Euclidean space. Thus, Grassmannian quantization is applicable to represent the CSI feedback for BD precoding. The expected residual rate loss of BD precoding with quantized CSIT compared to the rate with perfect CSIT is determined by the chordal distance quantization error [55,57]. UE thus apply the chordal distance as quantization metric

$$\widehat{\mathbf{H}}_{u,i} = \arg\min_{\mathbf{Q} \in \mathcal{Q}_{\ell}^{(N_i)}} d_c^2\left(\mathbf{H}_{u,i}^{\text{eff}}, \mathbf{Q}\right) = \ell - \text{tr}\left(\mathbf{Q}^{H}\widetilde{\mathbf{H}}_{u,i}\widetilde{\mathbf{H}}_{u,i}^{H}\mathbf{Q}\right),$$
$$(3.30)$$
$$\text{span}\left(\mathbf{H}_{u,i}^{\text{eff}}\right) = \text{span}\left(\widetilde{\mathbf{H}}_{u,i}\right), \quad \widetilde{\mathbf{H}}_{u,i}^{H}\widetilde{\mathbf{H}}_{u,i} = \mathbf{I}_{\ell},$$

with $\mathcal{Q}_{\ell}^{(N_i)}$ as defined in Equation 3.5.

3.2.1.1.2 Interference Alignment Interference alignment (IA) is a precoding method similar to BD precoding, however, not considering the MIMO broadcast channel, that is, the channel between a single base station and multiple UE, but intended for the MIMO interference channel [8,9]. In the original formulation of interference alignment, it is assumed that each of several base stations serves only a single UE, employing a precoder that aligns the interference caused to UE of other base stations in specific subspaces. The goal is to align the interference experienced by a specific UE in a strict subspace of its channel matrix such that the UE can cancel the interference by means of antenna combining. The relevant input–output relationship of user u served by base station i over the MIMO interference channel is

$$\mathbf{y}_{u,i} = \mathbf{H}_{u,i}^{H}\mathbf{F}_i\mathbf{x}_i + \sum_{\substack{j=1 \\ j \neq i}}^{J}\left(\mathbf{H}_{u,i}^{(j)}\right)^{H}\mathbf{F}_j\mathbf{x}_j + \mathbf{z}_{u,i}, \quad \mathbf{x}_i = \mathbf{G}_{u,i}^{H}\mathbf{y}_{u,i} \in \mathbb{C}^{\ell \times 1}. \quad (3.31)$$

As each base station serves only a single user, we omit the UE index in $\mathbf{F}_j$ and $\mathbf{x}_j$ to simplify notation and we also assume equal number of streams $\ell_{u,i} = \ell$ for all served UEs. IA imposes the following constraints to jointly determine the precoders and antenna combiners [58]

$$\mathbf{G}_{u,i}^{H}\left(\mathbf{H}_{u,i}^{(j)}\right)^{H}\mathbf{F}_j = 0, \quad \forall i,j \text{ and } j \neq i, \quad (3.32)$$

$$\text{rank}\left(\mathbf{G}_{u,i}^{H}\mathbf{H}_{u,i}^{H}\mathbf{F}_i\right) = \ell, \quad \forall i. \quad (3.33)$$

The first condition assures that the interference is aligned in an $M_{u,i} - \ell$ dimensional subspace that can be filtered out by the antenna combiner, while the second constraint is required to support the transmission of ℓ streams to user u. Determining feasibility of this precoder and antenna combiner design problem is for itself. In general, a difficult task [59]. However, provided the problem is feasible, it is shown in [60] that the solution can be obtained if each UE feeds back the subspace spanned by all interference channels

$$\widehat{\overline{\mathbf{H}}}_{u,i} = \arg\min_{\mathbf{Q}} d_c^2\left(\overline{\overline{\mathbf{H}}}_{u,i}, \mathbf{Q}\right), \quad \mathbf{Q} \in \mathcal{Q}_{M_{u,i}}^{(\bar{N}_i)}, \quad \bar{N}_i = \sum_{\substack{j=1 \\ j \neq i}}^{J} N_j, \quad (3.34)$$

$$\overline{\overline{\mathbf{H}}}_{u,i} = \left[\left(\mathbf{H}_{u,i}^{(1)}\right)^{H}, \dots, \left(\mathbf{H}_{u,i}^{(i-1)}\right)^{H}, \left(\mathbf{H}_{u,i}^{(i+1)}\right)^{H}, \dots, \left(\mathbf{H}_{u,i}^{(J)}\right)^{H}\right]^{H} \in \mathbb{C}^{\bar{N}_i \times M_{u,i}}. \quad (3.35)$$

Thus, the relevant CSIT is represented as a point on the Grassmann manifold $\mathcal{G}(\bar{N}_i, M_{u,i})$. In [60], the feedback bit scaling law for IA with Grassmannian feedback is derived, showing that the number of feedback bits must scale linearly with the logarithmic SNR with a slope of $M_{u,i}(\bar{N}_i - M_{u,i})$ to achieve the same DoF as with perfect CSIT.

3.2.1.2 Memoryless Grassmannian Quantization Memoryless quantization means that the users quantize the CSI at each TTI independently, utilizing predefined quantization codebooks. If we consider iid Rayleigh fading channels of dimension $N_i \times M_{u,i}$ with $M_{u,i} \leq N_i$, then the subspaces spanned by the channel matrices are uniformly distributed on $\mathcal{G}(N_i, M_{u,i})$ [61]. In this case, the average quantization error, and thus the expected rate loss compared to perfect CSIT, is minimized with quantization codebooks that are maximally spaced subspace packings on $\mathcal{G}(N_i, M_{u,i})$ with respect to the chordal distance [31]. These packings are essentially uniform on the Grassmann manifold in terms of the chordal distance. Unfortunately, finding optimal codebooks is in general hard except for special cases; an algorithm for obtaining good codebooks is provided, for example, in [62]. Instead of maximally spaced subspace packings, we consider random quantization codebook constructions later, which are much simpler to obtain and provide good performance on average [63].

3.2.1.2.1 Isotropic Codebooks In case the elements of the Grassmannian quantization codebook are uniformly distributed on $\mathcal{G}(N_i, M_{u,i})$, the codebook is called isotropic. Such codebooks constructions, also known as random vector quantization (RVQ) [57], can be obtained as

$$\mathcal{Q}_{M_{u,i}}^{(N_i, \text{iso})} = \left\{ \mathbf{Q}_j \mid \mathbf{Q}_j \Sigma \mathbf{V}^H = \mathbf{H} \in \mathbb{C}^{N_i \times M_{u,i}}, [\mathbf{H}]_{m,n} \sim \mathcal{N}_{\mathbb{C}}(0,1) \right\}, \quad (3.36)$$

with $\mathbf{Q}\Sigma\mathbf{V}^H$ representing a compact-form SVD of the iid Gaussian matrix $\mathbf{H}$. RVQ codebooks are shown to perform asymptotically optimal in the codebook size for a number of applications, for example, [64].

3.2.1.2.2 Correlated Codebooks If the channel subspace is not uniformly distributed on $\mathcal{G}(N_i, M_{u,i})$ but correlated in some way, better quantization performance is achievable by matching the distribution of the random quantization codebook to this correlation. For single-user multiple-input single-output beamforming systems, an efficient heuristic codebook construction is proposed in [65], which effectively "colors" a 1D Grassmannian subspace packing according to the channel correlation matrix. This approach is extended in [66] to Grassmannian subspace packings of higher dimension, to obtain random Grassmannian codebooks that are statistically matched to the distribution of the channel subspace. Assuming a Kronecker correlation model as in Equation 3.15, the correlated codebook is

$$
\mathcal{Q}_{M_{u,i}}^{(N_i,\text{corr})} = \left\{ \mathbf{Q}_j \mid \mathbf{Q}_j \Sigma \mathbf{V}^{\mathrm{H}} = \mathbf{C}_{\mathrm{TX}}^{1/2} \mathbf{H} \in \mathbb{C}^{N_i \times M_{u,i}}, [\mathbf{H}]_{m,n} \sim \mathcal{N}_{\mathbb{C}}(0,1) \right\}.
$$
(3.37)

Notice that it is not necessary to consider the receive side correlation in this codebook construction as it has no impact on the distribution of the left singular vectors.

3.2.1.2.3 Codebooks for Quantization of the Effective Channel Subspace The aforementioned codebooks consider quantization of the full $M_{u,i}$ dimensional channel subspace. In case of ZF beamforming and BD precoding, however, we assumed quantization of the $\ell_{u,i}$ dimensional channel subspace obtained by application of an antenna combiner. In this case, the codebook construction can further be tailored to the respective antenna combiner, that is, MET or SQBC [49,67]:

$$
\mathcal{Q}_{\ell_{u,i}}^{(N_i,\text{MET})} = \left\{ \mathbf{Q}_j^{(\text{MET})} = [\mathbf{Q}_j]_{:,\ell_{u,i}} \mid \mathbf{Q}_j \in \mathcal{Q}_{M_{u,i}}^{(N_i)} \right\},
$$
(3.38)

$$
\mathcal{Q}_{\ell_{u,i}}^{(N_i,\text{SQBC})} = \left\{ \begin{array}{l} \mathbf{Q}_j^{(\text{SQBC})} = \mathbf{Q}_j \mathbf{U}_j \mid \mathbf{Q}_j \in \mathcal{Q}_{M_{u,i}}^{(N_i)}, \\ \mathbf{U}_j \in \mathbb{C}^{M_{u,i} \times \ell_{u,i}}, \mathbf{U}_j^{\mathrm{H}} \mathbf{U}_j = \mathbf{I}_{\ell_{u,i}}, \end{array} \right\}
$$
(3.39)

with $[\mathbf{Q}_j]_{:,\ell_{u,i}}$ consisting of the first $\ell_{u,i}$ columns of $\mathbf{Q}_j$, corresponding to the maximum eigenmodes, and $\mathbf{U}_j$ being uniformly distributed on $\mathcal{G}(M_{u,i}, \ell_{u,i})$. Matrix $\mathbf{U}_j$ effectively selects a uniformly distributed subspace within span $(\mathbf{Q}_j)$; this is appropriate for SQBC as this antenna combiner generates such a subspace [55].

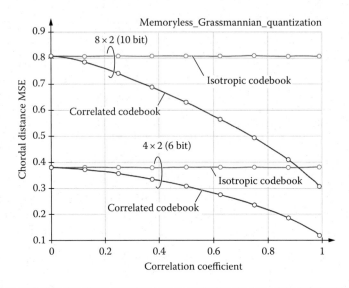

Figure 3.13 Quantization mean squared error of memoryless quantization in dependence of spatial channel correlation.

In Figure 3.13, we compare the performance of isotropic and correlated codebooks for quantization of correlated channel subspaces, in dependence of the channel correlation parameter α_{corr}. We assume that all off-diagonal elements of $\mathbf{C}_{TX}$ are equal to α_{corr}, while the diagonal is equal to one. The performance is evaluated in terms of the average chordal distance distortion. We observe that with increasing spatial channel correlation, the correlated codebook outperforms the isotropic codebook as it is matched to the spatial statistics of the channel. The average distortion achieved with isotropic codebooks is independent of the channel correlation.

3.2.1.3 Differential/Predictive Grassmannian Quantization We have seen in the simulation results of Figure 3.12 that memoryless quantization can mostly not provide sufficient quantization accuracy with a reasonable feedback overhead for multiuser transmission to outperform single-user MIMO. Only if the users are equipped with $M_{u,i} \gg \ell_{u,i}$ excess antennas, memoryless quantization combined with SQBC achieves an acceptable feedback overhead, which, however, comes at the cost of a reduced channel gain compared to MET [55]. As a remedy, we can discard the concept of memoryless quantization and rather try to exploit temporal channel correlation to improve

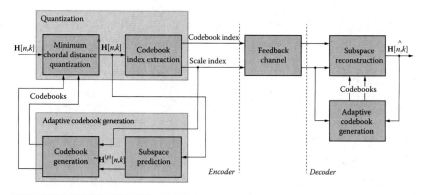

Figure 3.14 Structure of the predictive quantizer visualizing the interplay of its encoder/decoder components.

the quantization accuracy. Commonly, in the source coding literature, temporal correlation is exploited by gathering data over multiple time instants and jointly quantizing/coding this data by means
of vector quantization [68]. This, however, is not an option for CSI
feedback because it implies an encoding delay that cannot be tolerated
as it leads to outdated CSIT. Instead, differential [69,70], predictive
[21,22,71,72], and progressive refinement [73] quantizers have been
proposed. We highlight next the intuition and ideas underlying our
predictive quantizer proposed in [21], yet avoiding complicated mathematical proofs and derivations; for details, the interested reader is
referred to [21,22,56].

The basic idea of differential/predictive quantization on the
Grassmann manifold is to adapt the quantization codebook $\mathcal{Q}_{\ell_{u,i}}^{(N_i)}$
over time, to match the temporal evolution of the channel subspace.
To simplify notation, we denote the quantization codebook at time
instant $[k]$ as $\mathcal{Q}[k]$. The underlying structure of the considered quantizer is illustrated in Figure 3.14. The task and operation of the individual components are as follows.

3.2.1.3.1 Adaptive Codebook Generation This part generates the
codebook $\mathcal{Q}[k]$. It consist of a subspace predictor and a codebook
generator. Based on N_p previously quantized channel subspaces
$\widehat{\mathbf{H}}_{u,i}[k-1],\ldots,\widehat{\mathbf{H}}_{u,i}[k-N_p]$, the subspace predictor calculates a prediction $\widetilde{\mathbf{H}}_{u,i}^{(p)}[k]$ of the current channel subspace $\widetilde{\mathbf{H}}_{u,i}[k] \in \mathcal{G}(N_i,\ell_{u,i})$. With
differential quantization, this prediction simply is $\widetilde{\mathbf{H}}_{u,i}^{(p)}[k] = \widehat{\mathbf{H}}_{u,i}[k-1]$.

However, more sophisticated prediction can further improve the performance of the quantizer as demonstrated later. Specifically, we apply a linear regression in the tangent space of the manifold in the presented simulations; see [22] for details.

The codebook generator takes the prediction $\widetilde{\mathbf{H}}_{u,i}^{(p)}[k]$ as input and calculates two codebooks $\mathcal{Q}_+$ and $\mathcal{Q}_-$, which cover volumes of certain size on the Grassmannian around $\widetilde{\mathbf{H}}_{u,i}^{(p)}[k]$; importantly $\mathcal{Q}_+$ covers a larger volume than $\mathcal{Q}_-$. The size of the volume that needs to be covered by quantization codebooks is basically determined by the accuracy of the prediction. As this is not known beforehand, we apply a simple 1-bit tracking algorithm to adapt the size of the volume over time. Compared to the codebook $\mathcal{Q}[k-1]$ applied at time instant $[k-1]$, the volume covered by $\mathcal{Q}_+$ is larger by a constant factor, whereas the codebook $\mathcal{Q}_-$ is smaller. Both codebooks $\mathcal{Q}_+$ and $\mathcal{Q}_-$ are then provided to the quantization part.

3.2.1.3.2 Quantization   In this part, the channel subspace $\widetilde{\mathbf{H}}_{u,i}[k]$ is first quantized with respect to the two codebooks $\mathcal{Q}_+$ and $\mathcal{Q}_-$. Then, the winning codebook is determined, that is, the one that achieves a smaller quantization error, and this codebook is defined as the employed codebook $\mathcal{Q}[k]$. The scale index signals this decision to the adaptive codebook generation, such that $\mathcal{Q}[k]$ is known by the generator at time instant $[k+1]$.

The scale index and the index of the quantized channel subspace are provided over the feedback channel to the CSI decoder. The decoder duplicates the adaptive codebook generation of the encoder to determine $\mathcal{Q}_+$ and $\mathcal{Q}_-$. By means of the scale index feedback, the subspace reconstruction is informed about the winning codebook, thus enabling successful reconstruction of the quantized channel subspace $\widehat{\mathbf{H}}_{u,i}[k]$.

3.2.1.4 Performance Comparison of Memoryless, Differential, and Predictive Quantization   Let us now compare the performance of memoryless and differential/predictive quantization in dependence of the temporal channel variation. We consider spatially uncorrelated channel matrices of size $N_i \in \{4, 8\}$ and $M_{u,i} = 2$ that are generated with a sum-of-sinusoids channel model as proposed in [74]. From these channel matrices, we determine the corresponding subspaces by applying an

SVD. The amount of temporal channel variation is determined by the normalized Doppler frequency ν_d according to Clarke's model as introduced in Section 3.1.3. We investigate the chordal distance mean squared error (MSE) achieved by different quantization approaches. The results are shown in Figure 3.15. We observe a substantial accuracy improvement of differential/predictive quantization over the

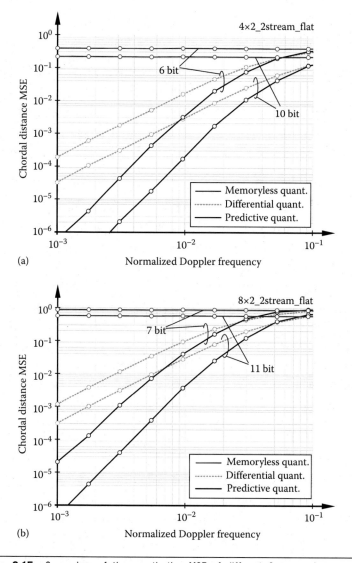

Figure 3.15 Comparison of the quantization MSE of different Grassmannian quantizers: (a) quantization on $\mathcal{G}(4, 2)$ and (b) quantization on $\mathcal{G}(8, 2)$.

memoryless quantizer in case of low user mobility, for example, with $N_i \times M_{u,i} = 4 \times 2$ at $\nu_d = 0.01$ (corresponding to a velocity of 5 km/h at 2 GHz center frequency), the predictive quantizer with 6-bit codebook size improves by more than two orders of magnitude. We can also see that predictive quantization achieves a larger slope of the MSE versus ν_d curve and hence outperforms differential quantization with decreasing Doppler frequency.

3.2.2 Stiefel Manifold Quantization for Gramian-Based Precoding Techniques

Grassmannian (or subspace) precoding techniques, as presented earlier, are appropriate for achieving the optimal multiplexing gain and DoF of multiuser broadcast and interference channels in the regime of high SNR. Yet DoF optimal strategies that are based on subspace information often suffer a substantial loss only with respect to channel capacity in the practically important low to intermediate SNR regime, especially in case of spatially correlated channels [49,75,76]. In this regime, significant throughput gains are possible using precoding methods that account for the SNR experienced on the individual channel eigenmodes, such as SVD precoding along with water-filling power allocation for single-user MIMO (see Section 3.1.2), regularized ZF beamforming/BD precoding for multiuser MIMO [18], and the max-SINR algorithm for IA in the MIMO interference channel [77]. Such methods, however, cannot be applied with subspace information only; the transmitter rather requires knowledge of the directions of the channel eigenmodes and the magnitudes of the corresponding eigenvalues of the channel Gramian. In this section, we review two examples of channel Gramian-based precoding techniques and present quantization strategies to enable efficient operation with a limited amount of CSI feedback overhead.

3.2.2.1 Examples of Gramian-Based Precoding Techniques

3.2.2.1.1 Regularized Zero-Forcing Beamforming and Block-Diagonalization Precoding
Regularized block diagonalization (RBD) precoding can be considered as an extension of BD precoding, not only focusing on the multiuser interference but also considering the received power of the intended signal in the precoder design [18].

The RBD precoder optimally trades off multiuser interference for received signal power in an MSE sense and applies water-filling power allocation to maximize the achievable transmission rate, in contrast to equal power allocation as employed with BD precoding. Similar to our presentation of BD precoding, we also assume here that the precoders are determined from the effective user channels, including the preselected antenna combiners; the antenna combining strategies of Section 3.1.4 are applicable.

The RBD precoder is obtained in two steps: $\mathbf{F}_{u,i} = \mathbf{F}_{u,i}^{(a)}\mathbf{F}_{u,i}^{(b)}$. The partial precoder $\mathbf{F}_{u,i}^{(a)}$ achieves the desired trade-off between residual multiuser interference and intended signal power, while the precoder $\mathbf{F}_{u,i}^{(b)}$ optimizes the transmission over the resulting effective single-user channel $(\mathbf{H}_{u,i}^{\text{eff}})^{\text{H}}\mathbf{F}_{u,i}^{(a)}$, treating residual interference as additive Gaussian noise. For simplicity, we assume $\ell_{u,i} = \ell, \forall u$. Precoder $\mathbf{F}_{u,i}^{(a)}$ is obtained as

$$\mathbf{F}_{u,i}^{(a)} = \frac{\widetilde{\mathbf{F}}_{u,i}^{(a)}}{\left\|\widetilde{\mathbf{F}}_{u,i}^{(a)}\right\|_F}, \quad \widetilde{\mathbf{F}}_{u,i}^{(a)} = \overline{\mathbf{V}}_{u,i}\left(\overline{\Sigma}_{u,i}^{\text{H}}\overline{\Sigma}_{u,i} + \frac{\ell\tilde{\sigma}_z^2}{P_i}\mathbf{I}_{N_i}\right)^{-1/2}, \quad (3.40)$$

$$\overline{\mathbf{H}}_{u,i} = \overline{\mathbf{U}}_{u,i}\overline{\Sigma}_{u,i}\overline{\mathbf{V}}_{u,i}^{\text{H}}, \quad (3.41)$$

with Equation 3.41 denoting a full size SVD of matrix $\overline{\mathbf{H}}_{u,i}$ according to Equation 3.28. Notice that precoder $\mathbf{F}_{u,i}^{(a)}$ is normalized; the actual power allocation is considered in precoder $\mathbf{F}_{u,i}^{(b)}$. After application of precoder $\mathbf{F}_{u,i}^{(a)}$, the resulting effective channel $(\mathbf{H}_{u,i}^{\text{eff}})^{\text{H}}\mathbf{F}_{u,i}^{(a)}$ is assumed as AWGN with the residual interference treated as additional noise. Precoder $\mathbf{F}_{u,i}^{(b)}$ is further separated as $\mathbf{F}_{u,i}^{(b)} = \widetilde{\mathbf{F}}_{u,i}^{(b)}(\mathbf{P}_{u,i}^{(b)})^{1/2}$, with $\mathbf{P}_{u,i}^{(b)}$ denoting a diagonal power allocation matrix. Matrix $\widetilde{\mathbf{F}}_{u,i}^{(b)}$ is calculated as

$$\widetilde{\mathbf{F}}_{u,i}^{(b)} = \mathbf{V}_{u,i}^{(a)}, \quad \left(\mathbf{H}_{u,i}^{\text{eff}}\right)^{\text{H}}\mathbf{F}_{u,i}^{(a)} = \mathbf{U}_{u,i}^{(a)}\Sigma_{u,i}^{(a)}\left(\mathbf{V}_{u,i}^{(a)}\right)^{\text{H}}, \quad (3.42)$$

where we applied a compact-form SVD to the effective single-user channel $(\mathbf{H}_{u,i}^{\text{eff}})^{\text{H}}\mathbf{F}_{u,i}^{(a)}$. Hence, the total transmission channel seen by user u is

$$\left(\mathbf{H}_{u,i}^{\text{eff}}\right)^{\text{H}}\mathbf{F}_{u,i} = \mathbf{U}_{u,i}^{(a)}\Sigma_{u,i}^{(a)}\left(\mathbf{P}_{u,i}^{(b)}\right)^{1/2}, \quad (3.43)$$

which can be diagonalized by the UE through application of $\mathbf{U}_{u,i}^{(a)}$ as receive filter. The diagonal power loading matrices $\mathbf{P}_{u,i}^{(b)}, \forall u$ are jointly calculated to allocate the transmission power P_i over the eigenmodes of the users, applying the water filling algorithm to the diagonal elements of all matrices $\Sigma_{u,i}^{(a)}$ together.

Investigating this precoder design, we observe that the base station requires knowledge of the effective channel matrices $\mathbf{H}_{u,i}^{\text{eff}}, \forall u$. Thus, there does not seem to be hope of reducing the dimensionality of the CSI quantization problem similar to Grassmannian quantization. Yet, this goal can be achieved by applying a change of basis at UE before feedback. Consider an SVD of the product of channel matrix and antenna combiner:

$$\mathbf{H}_{u,i}\mathbf{G}_{u,i} = \widetilde{\mathbf{U}}_{u,i}\widetilde{\Sigma}_{u,i}\widetilde{\mathbf{V}}_{u,i}^{\text{H}} \in \mathbb{C}^{N_i \times \ell} \tag{3.44}$$

The UE can employ the change of basis $\mathbf{r}_{u,i} = \widetilde{\mathbf{V}}_{u,i}^{\text{H}}\mathbf{y}_{u,i}$ without affecting the achievable transmission rate, reducing the effective channel to

$$\mathbf{H}_{u,i}^{\text{eff}} = \mathbf{H}_{u,i}\mathbf{G}_{u,i}\widetilde{\mathbf{V}}_{u,i} = \widetilde{\mathbf{U}}_{u,i}\widetilde{\Sigma}_{u,i}, \quad \widetilde{\mathbf{U}}_{u,i} \in St(N_i, \ell), \tag{3.45}$$

with $St(N_i, \ell)$ denoting the compact Stiefel manifold of all ordered orthonormal Parseval ℓ-frames in the vector space $\mathbb{C}^{N_i}$ ($\ell \leq N_i$)

$$St(N_i, \ell) = \left\{ \mathbf{S} \in \mathbb{C}^{N_i \times \ell} \mid \mathbf{S}^{\text{H}}\mathbf{S} = \mathbf{I}_\ell \right\}. \tag{3.46}$$

Matrix $\widetilde{\Sigma}_{u,i} = \text{diag}(\sigma_{u,i}^{(1)}, \ldots, \sigma_{u,i}^{(\ell)})$ contains the singular values of this effective user channel. The relevant CSIT thus consists of a point on the Stiefel manifold and ℓ positive numbers.

3.2.2.1.2 Leakage-Based Beamforming and Precoding Leakage-based precoding schemes are very popular in literature for optimizing the performance of MIMO interference channels, for example, [20,78,79]. Such methods mostly try to maximize the rate or alternatively the SINR of the users attached to a specific base station, subject to upper bounds on the interference leakage caused to users of other base stations. In this way, game-theoretical Pareto-optimal beamformers can be obtained, provided the leakage upper bounds are selected appropriately [20]; however, finding the right upper bounds is a nontrivial issue, typically requiring complex iterative algorithms.

Let us consider the MIMO interference channel as introduced in Equation 3.31 in the context of interference alignment and assume that each base station serves only a single user. The precoder design problem for optimizing the SINR of the UE subject to interference leakage upper bounds then takes the following form:

$$
\mathbf{F}_i^* = \arg\max_{\mathbf{F}_i \in \mathbb{C}^{N_i \times \ell_i}} \frac{\mathrm{tr}\left(\mathbf{F}_i^{\mathrm{H}} \mathbf{H}_i^{\mathrm{eff}} \left(\mathbf{H}_i^{\mathrm{eff}} \right)^{\mathrm{H}} \mathbf{F}_i \right)}{\ell_i \tilde{\sigma}_z^2 + \sum_{j \neq i} \mathrm{tr}\left(\mathbf{F}_j^{\mathrm{H}} \mathbf{H}_i^{\mathrm{eff},j} \left(\mathbf{H}_i^{\mathrm{eff},j} \right)^{\mathrm{H}} \mathbf{F}_j \right)}, \quad (3.47)
$$

$$
\text{subject to:} \quad \mathrm{tr}\left(\mathbf{F}_i^{\mathrm{H}} \mathbf{F}_i \right) \leq P_i, \quad (3.48)
$$

$$
\mathrm{tr}\left(\mathbf{F}_i^{\mathrm{H}} \mathbf{H}_j^{\mathrm{eff},i} \left(\mathbf{H}_j^{\mathrm{eff},i} \right)^{\mathrm{H}} \mathbf{F}_i \right) \leq \alpha_{ij} \sigma_z^2, \quad \forall j \neq i, \quad (3.49)
$$

where we omit the UE index as only a single user is served per base station. Matrix $\mathbf{H}_i^{\mathrm{eff},j}$ denotes the effective channel of the user attached to base station i with respect to base station j. The first constraint of this optimization problem is a power constraint, while the second constraint upper bounds the interference leakage. We observe that all terms in this optimization problem depend only on channel Gramians $\mathbf{H}_i^{\mathrm{eff},j}(\mathbf{H}_i^{\mathrm{eff},j})^{\mathrm{H}}$. This observation also holds true when we consider multiple users per base station and/or rate optimization instead of the SINR. Also the popular signal to leakage and noise ratio precoder can be calculated from the channel Gramian [12]. Applying an eigendecomposition to these channel Gramians, we obtain, for example,

$$
\mathbf{H}_i^{\mathrm{eff}} \left(\mathbf{H}_i^{\mathrm{eff}} \right)^{\mathrm{H}} = \mathbf{U}_i^{\mathrm{eff}} \left(\Sigma_i^{\mathrm{eff}} \right)^2 \left(\mathbf{U}_i^{\mathrm{eff}} \right)^{\mathrm{H}} \in \mathbb{C}^{N_i \times N_i}, \quad (3.50)
$$

with $\mathbf{U}_i^{\mathrm{eff}} \in St(N_i, \ell_{u,i})$ denoting the matrix of left-singular vectors and diagonal matrix Σ_i^{eff} containing the singular values. We hence end up with similar CSIT requirements as for RBD; that is, UE have to feed back points on the Stiefel manifold and positive numbers.

3.2.2.2 Memoryless Quantization on the Stiefel Manifold The two examples mentioned earlier serve to demonstrate that a broad class of precoder design problems for MIMO broadcast and interference channels can be formulated in terms of channel Gram matrices

$\mathbf{HH}^H = \mathbf{U\Sigma^2U}^H$, thus requiring CSI in the form of directions ($\mathbf{U}$) and magnitudes ($\mathbf{\Sigma}$) of the channel eigenmodes at the transmitter. Basically, this information can be provided to the base station employing limited feedback in two ways: either quantizing the channel Gram/covariance matrix directly and applying an eigendecomposition at the transmitter to determine the eigenmodes [80–82], or, instead, quantizing the directions and magnitudes of the eigenmodes separately [23,83,84]. Memoryless quantization of the channel Gramian is proposed in [80] by independently quantizing the individual entries of the Gram matrix. The efficiency of such an approach is restricted as it does not exploit the geometry associated with channel Gramians. Next, we instead consider separate quantization of the directions and magnitudes of the eigenmodes following the "gain-shape" quantization principle [68], which gives the possibility to vary the feedback bit allocation between these two components. This feature is advantageous in the context of multiuser systems because direction information commonly must be provided with very high precision to avoid multiuser interference, whereas magnitude information is not so crucial as it impacts only power allocation.

For quantization of the matrix of left singular vectors $\mathbf{U}_{u,i}^{(\mathrm{eff})}$ of the effective channel, or equivalently the eigenvector matrix of the Gramian $\mathbf{H}_{u,i}^{\mathrm{eff}}(\mathbf{H}_{u,i}^{\mathrm{eff}})^H$, it is possible to reuse the random codebook constructions proposed in the context of memoryless Grassmannian quantization in Section 3.2.1. This is because orthonormal bases are employed in Section 3.2.1 to represent points on the Grassmannian; that is, each basis represents an entire subspace. Alternatively, any other matrix that spans the same subspace could be used for representation; orthonormal bases are applied as they allow for efficient codebook constructions. When considering quantization of points on the Stiefel manifold as defined in Equation 3.46, however, the task really is to quantize orthonormal bases. It is only necessary to adapt the quantization metric to preserve the directions of the individual eigenmodes. We consider the sum of 1D Grassmannian chordal distances for this purpose, as proposed in [23]:

$$\widehat{\mathbf{U}}_{u,i} = \underset{\mathbf{U}\in\mathcal{Q}_{\ell_{u,i}}^{(N_i)}}{\arg\min}\, \mathrm{d}_{\mathrm{s}}^2\left(\mathbf{U}_{u,i}^{(\mathrm{eff})},\mathbf{U}\right), \qquad (3.51)$$

$$d_s^2\left(\mathbf{U}_{u,i}^{(\mathrm{eff})},\mathbf{U}\right)=\sum_{j=1}^{\ell_{u,i}}d_c^2\left(\mathbf{u}_{u,i}^{(\mathrm{eff},j)},\mathbf{u}^{(j)}\right),$$

$$\mathbf{u}_{u,i}^{(\mathrm{eff},j)}=\left[\mathbf{U}_{u,i}^{(\mathrm{eff})}\right]_{:,j},\quad \mathbf{u}^{(j)}=\left[\mathbf{U}\right]_{:,j}. \tag{3.52}$$

This metric treats all eigenmodes as equally important during quantization. Alternatively, one could employ a weighted sum instead, for example, to weigh the chordal distances of the individual singular vectors in proportion to their corresponding singular values.

In addition to the directions of the eigenmodes, it is also necessary to quantize the associated singular values on the diagonal of $\Sigma_{u,i}^{(\mathrm{eff})}$. As proposed in [84], we apply vector quantization to jointly quantize the $\ell_{u,i}$ positive singular values; the corresponding codebook is denoted as

$$\mathcal{S}=\left\{\mathbf{s}_i\in\mathbb{R}^{\ell_{u,i}\times1}\,\middle|\,\mathbf{s}_i>0\right\}, \tag{3.53}$$

where the inequality is element-wise. The codebook can, for example, be obtained using Lloyd's k-means clustering algorithm [85]. Examples of such codebooks for iid Rayleigh fading channels of dimension $N_i\times M_{u,i}=8\times2$ are shown in Figure 3.16 for the two cases of spatially uncorrelated ($\alpha_{\mathrm{corr}}=0$) and strongly correlated channels ($\alpha_{\mathrm{corr}}=0.9$). As known from theory [46], the singular value disparity increases with growing antenna correlation; correspondingly, the joint distribution of the singular values becomes more concentrated along the larger singular value, facilitating more precise quantization with a given codebook size. For quantization of the singular values, UE apply the Euclidean distance:

$$\hat{\sigma}_u=\arg\min_{s_i\in\mathcal{S}}\left\|\sigma_u-s_i\right\|_2^2,\quad \sigma_u=\left[\left[\Sigma_{u,i}^{(\mathrm{eff})}\right]_{1,1},\ldots,\left[\Sigma_{u,i}^{(\mathrm{eff})}\right]_{\ell_{u,i},\ell_{u,i}}\right]^{\mathsf{T}}. \tag{3.54}$$

3.2.2.3 Differential/Predictive Quantization on the Stiefel Manifold
Similar to Grassmannian differential/predictive quantization, the efficiency of quantization of channel Gramians can also be improved by exploiting temporal correlation. Differential quantization of channel Gram matrices is proposed in [81]. In this work, the authors exploit

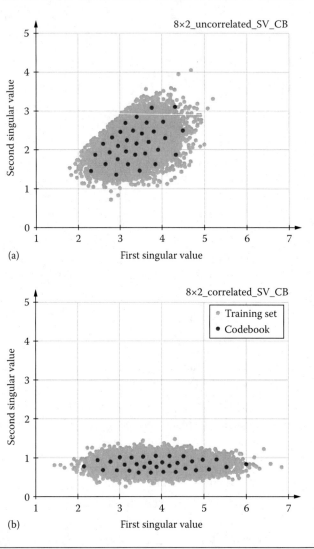

Figure 3.16 Lloyd optimized quantization codebooks for the singular values of an $N_i \times M_{u,i} =$ 8 × 2 channel: (a) singular values of spatially uncorrelated channels and (b) singular values of spatially strongly correlated channels.

the geometry of the convex cone associated with strictly positive definite Gram matrices to enable efficient differential quantization. Despite the success of [81], it has the disadvantage of being applicable only to positive definite Gramians. In the practically important case that the number of receive antennas at UE is less than the number of transmit antennas at the base station, however, the Gramians are only positive semidefinite and hence [81] is not applicable. An alternative predictive quantizer is proposed in [49], which applies separate quantization of the direction and magnitude of the eigenmodes and exploits the geometry of the Stiefel manifold to enable predictive quantization even of semidefinite Gramians. This quantizer applies exactly the same ideas and concepts as explained in the context of predictive Grassmannian quantization; see Section 3.2.1. It is only necessary to replace expressions and calculations that are specific to the Grassmann manifold with the corresponding counterparts of the Stiefel manifold. We do not go into details here; the interested reader is referred to [49].

We evaluate the performance of differential/predictive quantization on the Stiefel manifold in Figure 3.17, considering an antenna configuration of $N_i = 8$ transmit antennas and $M_{u,i} = 2$ receive antennas. We investigate the quantization MSE in terms of Equation 3.51 for the two cases of spatially uncorrelated receive antennas ($\alpha_{corr} = 0$) and strongly correlated antennas ($\alpha_{corr} = 0.9$) in Figure 3.17a and b, respectively. Notice that such a receive-side correlation does not impact the distribution of the subspace spanned by the eigenmodes; it does, however, influence the orientation of the individual eigenmodes within this subspace. Specifically, the variation of the directions of the eigenmodes over time reduces with increasing correlation, explaining why differential and especially predictive quantization performs better with higher receive antenna correlation.

3.2.3 Performance Comparison of Subspace and Gramian-Based Precoding Techniques

In this section, we compare the performance of channel subspace and Gramian-based precoding techniques. We consider a base station that is equipped with $N_i = 16$ transmit antennas and serves a total number of $U = 8$ users, each having $M_{u,i} = 2$ receive antennas. We simulate only a single base station; that is, other-cell interference is equal to zero. We evaluate the performance with perfect and quantized CSIT, employing the CSI quantizers presented in the previous sections.

We consider a low mobility scenario with a normalized Doppler frequency of $\nu_d = 0.01$ and assume receive-side spatial correlation with a correlation parameter of $\alpha_{corr} = 0.9$.

We first investigate the achievable transmission rate of single-user MIMO transmission using CSI feedback on the Grassmann

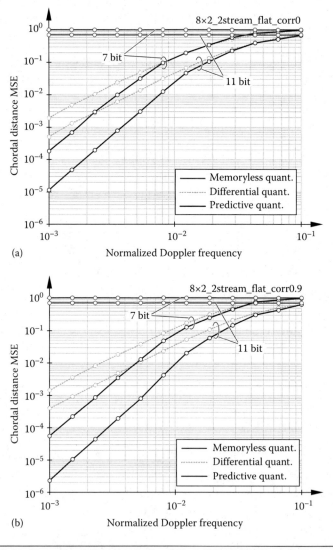

(a)

(b)

Figure 3.17   Comparison of the quantization MSE of different Stiefel manifold quantizers for $N_i \times M_{u,i} = 8 \times 2$: (a) quantization on $\mathcal{S}t(8, 2)$ with no receive-side correlation and (b) quantization on $\mathcal{S}t(8, 2)$ with strong receive-side correlation.

and Stiefel manifold, respectively. The ergodic achievable sum rate is obtained as

$$R_i = \sum_{u=1}^{U} \mathbb{E}\left(\log_{2\det}\left(I_{M_{u,i}} + \frac{1}{\sigma_z^2} H_{u,i}^H F_{u,i} F_{u,i}^H H_{u,i} \right) \right). \quad (3.55)$$

The ergodic rate is estimated by means of Monte Carlo simulations. In case of Grassmannian feedback, we assume that each UE quantizes the first $\ell_{u,i} \in \{1, 2\}$ columns of the left singular vector matrix $U_{u,i}$ (see Equation 3.7) to provide the subspace of $\ell_{u,i}$ dominant eigenmodes to the base station. The chordal distance is applied as a quantization metric:

$$\widehat{H}_{u,i} = \arg\min_{Q \in \mathcal{Q}_{\ell_{u,i}}^{(N_i)}} d_c^2\left(U_{u,i}^{(\ell_{u,i})}, Q \right), \quad U_{u,i}^{(\ell_{u,i})} = [U_{u,i}]_{:,1:\ell_{u,i}}. \quad (3.56)$$

Notice that the chordal distance is unaffected by right multiplication of $U_{u,i}^{(\ell_{u,i})}$ with any unitary matrix. Thus, quantization using the chordal distance as metric does not preserve the directions of the individual eigenmodes of the channel. Correspondingly, power allocation over the eigenmodes is not an option and the base station hence applies equal power allocation according to Equation 3.6. However, UE can determine the achievable rate for different numbers of streams $\ell_{u,i}$ and can thus decide on how many streams to employ at a given time instant. This simple form of power allocation is signaled to the base station via the RI.

With Stiefel manifold feedback, each UE quantizes the matrix $U_{u,i}$ together with the corresponding singular values, employing the methods detailed in Section 3.2.2. Thus, the base station is aware of the quantized directions and magnitudes of the channel eigenmodes and can hence apply SVD-based precoding together with water-filling power allocation as described in Section 3.1.2. Due to CSI quantization errors, however, a mismatch between the channel eigenmodes and the applied precoder is unavoidable. In the considered low mobility scenario, UE provide singular value feedback only every 10 TTIs using 3 bits of feedback; this overhead is thus negligible compared to the {4, 10} bits per TTI employed for Stiefel manifold quantization.

The results of the simulation are shown in Figure 3.18. We observe that precoding based on Stiefel manifold feedback and Grassmannian feedback with transmission rank adaptation performs very similar for single-user MIMO transmission. Hence, in this case no significant gains are obtained from the water-filling power allocation. Yet, with Stiefel manifold feedback, the base station can itself decide on the

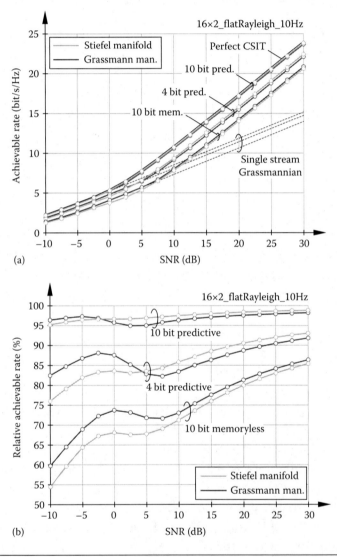

(a)

(b)

Figure 3.18 Comparison of single-user MIMO with Stiefel and Grassmann manifold feedback: (a) performance in terms of absolute achievable rate and (b) performance relative to perfect channel state information at the transmitter.

number of employed spatial streams through power allocation, without having to rely on the decision of the UE. This might be considered as an advantage for network operators because it gives them more control over the transmission. We also observe that predictive feedback significantly outperforms memoryless quantization, providing approximately 5 dB gain in SNR with 10 bits of feedback.

Next, we evaluate the performance of BD precoding and RBD precoding, employing Grassmannian and Stiefel manifold quantization, respectively. We assume transmission of a fixed number of $\ell_{u,i} = 2$ streams per each scheduled user. The set $\mathcal{S}_i$ of users served in parallel is determined by an exhaustive search at each time instant, such as to maximize the achievable sum rate:

$$
R_i = \sum_{u=1}^{U} \mathbb{E}\left[\log_2 \det\left(\mathbf{I}_{M_{u,i}} + \left(\sigma_z^2 \mathbf{I}_{M_{u,i}} + \mathbf{H}_{u,i}^{H} \sum_{s \in \mathcal{S}_i, s \neq u} \mathbf{F}_{s,i} \mathbf{F}_{s,i}^{H} \mathbf{H}_{u,i} \right)^{-1} \mathbf{H}_{u,i}^{H} \mathbf{F}_{u,i} \mathbf{F}_{u,i}^{H} \mathbf{H}_{u,i} \right) \right],
$$

$$(3.57)$$

with $\mathbf{F}_{u,i} = 0$ if a user is not served at a given time instant. UE apply codebooks of size 10 bits for quantization of the channel subspace/Gramian. For quantization of the singular values, 3 bits per 10 TTI are invested. We consider only predictive quantization in this simulation as memoryless quantization has already been evaluated in Section 3.1.5.

The results of this investigation are shown in Figure 3.19. We compare the single- and multiuser throughput achievable with perfect CSIT to the rate obtained with multiuser transmission using quantized CSIT. We observe that multiuser transmission significantly outperforms single-user MIMO because the multiplexing gain of multiuser MIMO is larger; with single-user MIMO, the base station can at most transmit two streams in parallel, while with multiuser MIMO, up to 16 streams are possible. With quantized CSIT, however, we do not achieve this multiplexing gain due to residual multiuser interference caused by the quantization error; at high SNR, a throughput saturation occurs due to this quantization error. This saturation can be mitigated by decreasing the effective number of

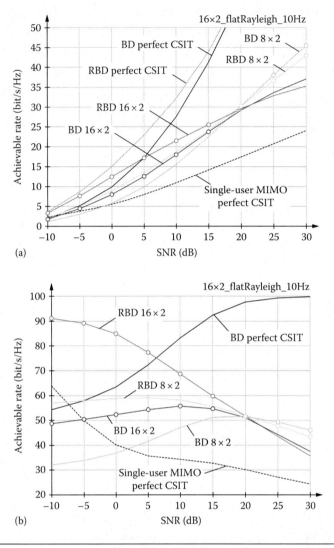

Figure 3.19 Comparison of multiuser MIMO with Stiefel and Grassmann manifold feedback: (a) performance in terms of absolute achievable rate and (b) performance relative to perfect channel state information at the transmitter.

transmit antennas at the base station, as shown in Figure 3.19 with the $N_i \times M_{u,i} = 8 \times 2$ curves, which reduces the dimensionality of the quantization problem and thus leads to more accurate quantization and less residual interference. However, this approach also reduces the maximally attainable multiplexing gain to eight. We achieve this transmit-side dimensionality reduction by applying a random precoder of dimension 16×8 before the actual BD/RBD precoding.

In practice, though, optimized precoding patterns could be designed, for example, to generate several beams pointing into different directions of the network, such as to achieve an additional multiuser diversity, similar to Figure 3.4b.

3.3 Conclusion and Future Work

Over the last decades, feedback has proved to be one of the most successful tools in wireless communications to enable robust and efficient data transmission. With feedback it is possible to adapt transmission parameters to varying channel conditions, thereby exploiting the diversity of the wireless channel and improving the achievable data rate with finite code block length and practical coding schemes [86]. Even more important, the capacity of multi-point channels is substantially increased with feedback [87,88], as it enables transmission of multiple data streams to different users at the same time.

In this chapter, we have reviewed important single- and multiuser transmission schemes that are already implemented and standardized in state-of-the-art 3GPP LTE wireless communication systems, and we have provided a coherent performance comparison of such schemes by means of link-level simulations. These investigations have shown that LTE is already well prepared for high-efficiency single-user spatial multiplexing transmission through its limited feedback-based CLSM transmission mode. Downlink multiuser spatial multiplexing, however, suffers significantly from the low CSI resolution provided by quantization codebooks implemented in the LTE standard. In the second part of the chapter, we have thus focused on improving the multiuser performance of wireless communications by providing more accurate CSI feedback, utilizing high-fidelity adaptive CSI quantizers. To minimize the feedback overhead, we have investigated dimensionality reduction techniques, representing the necessary CSI on appropriate manifolds, that is, Grassmann and Stiefel manifolds, depending on the considered group of precoding strategies. We have furthermore presented predictive CSI quantization, enabling highly efficient limited feedback operation at low to moderate user mobility, and we have evaluated its performance on Monte Carlo simulations, demonstrating close to optimal performance in quasi-static situations.

To enable practical implementations of the presented methods, however, further investigations and research efforts are required. One issue, considering the practical applicability of the considered limited feedback algorithms, is computational complexity. Complex operations, such as matrix inversions and SVDs, are extensively employed in these methods, despite questions over their feasibility in practical realizations. Hence, potentials for savings in terms of algorithmic complexity have to be investigated to enable efficient implementations. Also, the impact of impairments in the feedback channel has to be further investigated. Specifically, feedback errors are problematic in case of predictive quantization because the prediction operation causes error propagation, similar to the effects observed in video codecs. Such errors can be resolved with a simple cyclic redundancy check in combination with a selective repeat automatic repeat request protocol or with a synchronized reset of the quantizers at the encoder and the decoder. Alternatively, a concept similar to the I-frame in video coding [89] can be employed to periodically provide a valid reference. In either case, the additional feedback and signaling overhead must be taken into account in order to realistically gauge the performance of the methods presented here.

List of Abbreviations

3GPP	Third-Generation Partnership Project
AMC	Adaptive modulation and coding
AWGN	Additive white Gaussian noise
BD	Block diagonalization
BER	Bit error ratio
BICM	Bit-interleaved coded modulation
BLER	Block error ratio
CDD	Cyclic delay diversity
CLSM	Closed-loop spatial multiplexing
CP	Cyclic prefix
CQI	Channel quality indicator
CSI	Channel state information
CSIT	Channel state information at the transmitter

DoF	Degrees of freedom
ESM	Effective SINR mapping
FDD	Frequency division duplex
IA	Interference alignment
iid	Independent and identically distributed
LTE	Long-term evolution
MCS	Modulation and coding scheme
MET	Maximum eigenmode transmission
MIESM	Mutual information ESM
MIMO	Multiple-input multiple-output
MMSE	Minimum mean squared error
MRT	Maximum ratio transmission
MSE	Mean squared error
OFDM	Orthogonal frequency division multiplexing
OFDMA	Orthogonal frequency division multiple access
OLSM	Open-loop spatial multiplexing
PHY	Physical layer
PMI	Precoding matrix indicator
PU2RC	Per user unitary rate control
QAM	Quadrature amplitude modulation
QBC	Quantization-based combining
RB	Resource block
RBD	Regularized block diagonalization
RE	Resource element
RI	Rank indicator
RVQ	Random vector quantization
SIMO	Single-input multiple-output
SINR	Signal to interference and noise ratio
SISO	Single-input single-output
SNR	Signal to noise ratio
SQBC	Subspace QBC
SVD	Singular value decomposition
TDD	Time division duplex
TTI	Transmission time interval
TxD	Transmit diversity
UE	User equipment
ZF	Zero forcing

References

1. G. J. Foschini and M. J. Gans, On limits of wireless communications in a fading environment when using multiple antennas, *Wireless Personal Communications*, 6, 311–335, 1998.
2. I. Telatar, Capacity of multi-antenna Gaussian channels, *European Transactions on Telecommunications*, 10(6), 585–595, November 1999.
3. L. Zheng and D. N. C. Tse, Diversity and multiplexing: A fundamental tradeoff in multiple antenna channels, *IEEE Transactions on Information Theory*, 49, 1073–1096, 2002.
4. A. Paulraj, D. Gore, R. Nabar, and H. Bolcskei, An overview of MIMO communications—A key to Gigabit wireless, *Proceedings of the IEEE*, 92(2), 198–218, 2004.
5. S. Schwarz and M. Rupp, Throughput maximizing feedback for MIMO OFDM based wireless communication systems, in: *IEEE 12th International Workshop on Signal Processing Advances in Wireless Communications*, San Francisco, CA, June 2011, pp. 316–320.
6. G. Caire and S. Shamai, On the achievable throughput of a multiantenna Gaussian broadcast channel, *IEEE Transactions on Information Theory*, 49(7), 1691–1706, 2003.
7. D. Gesbert, M. Kountouris, R. Heath, Jr., C. Chae, and T. Sälzer, From single user to multiuser communications: Shifting the MIMO paradigm, *IEEE Signal Processing Magazine*, 24(5), 36, October 2007.
8. V. Cadambe and S. Jafar, Interference alignment and degrees of freedom of the K-user interference channel, *IEEE Transactions on Information Theory*, 54(8), 3425–3441, August 2008.
9. M. Maddah-Ali, A. Motahari, and A. Khandani, Communication over MIMO X channels: Interference alignment, decomposition, and performance analysis, *IEEE Transactions on Information Theory*, 54(8), 3457–3470, August 2008.
10. C.-B. Chae, I. Hwang, R. Heath, Jr., and V. Tarokh, Interference aware-coordinated beamforming in a multi-cell system, *IEEE Transactions on Wireless Communications*, 11(10), 3692–3703, 2012.
11. X. Tao, X. Xu, and Q. Cui, An overview of cooperative communications, *IEEE Communications Magazine*, 50(6), 65–71, 2012.
12. S. Schwarz and M. Rupp, Exploring coordinated multipoint beamforming strategies for 5G cellular, *IEEE Access*, 2, 930–946, 2014.
13. M. Simko, Fast fading channel estimation for UMTS long term evolution, in: *Proceedings of the Junior Scientist Conference 2010*, Vienna, Austria, April 2010, pp. 297–298.
14. C. Mehlführer, J. C. Ikuno, M. Simko, S. Schwarz, and M. Rupp, The Vienna LTE simulators—Enabling reproducibility in wireless communications research, *EURASIP Journal on Advances in Signal Processing (JASP) Special Issue on Reproducible Research*, 2011, 29, 2011.
15. S. Schwarz, J. Ikuno, M. Simko, M. Taranetz, Q. Wang, and M. Rupp, Pushing the limits of LTE: A survey on research enhancing the standard, *IEEE Access*, 1, 51–62, 2013.

16. C. Peel, B. Hochwald, and A. Swindlehurst, A vector-perturbation technique for near-capacity multiantenna multiuser communication— Part I: Channel inversion and regularization, *IEEE Transactions on Communications*, 53(1), 195–202, January 2005.

17. Q. Spencer, A. Swindlehurst, and M. Haardt, Zero-forcing methods for downlink spatial multiplexing in multiuser MIMO channels, *IEEE Transactions on Signal Processing*, 52(2), 461–471, February 2004.

18. V. Stankovic and M. Haardt, Generalized design of multi-user MIMO precoding matrices, *IEEE Transactions on Wireless Communications*, 7(3), 953–961, 2008.

19. R. Zhang and S. Cui, Cooperative interference management with MISO beamforming, *IEEE Transactions on Signal Processing*, 58(10), 5450–5458, October 2010.

20. J. Park, G. Lee, Y. Sung, and M. Yukawa, Coordinated beamforming with relaxed zero forcing: The sequential orthogonal projection combining method and rate control, *IEEE Transactions on Signal Processing*, 61(12), 3100–3112, June 2013.

21. S. Schwarz, R. Heath, Jr., and M. Rupp, Adaptive quantization on a Grassmann-manifold for limited feedback beamforming systems, *IEEE Transactions on Signal Processing*, 61(18), 4450–4462, 2013.

22. S. Schwarz, R. Heath, Jr., and M. Rupp, Adaptive quantization on the Grassmann-manifold for limited feedback multi-user MIMO systems, in: *Thirty-Eighth International Conference on Acoustics, Speech and Signal Processing*, Vancouver, British Columbia, Canada, May 2013.

23. S. Schwarz and M. Rupp, Predictive quantization on the Stiefel manifold, *IEEE Signal Processing Letters*, 22(2), 234–238, 2014.

24. 3GPP, Technical specification group radio access network; Evolved Universal Terrestrial Radio Access (E-UTRA); Physical channels and modulation (Release 8), September 2009 [Online], http://www.3gorg/ ftp/Specs/html-info/36211.htm (accessed December, 2015).

25. E. Dahlman, S. Parkvall, and J. Sköld, *4G LTE/LTE-Advanced for Mobile Broadband*. Oxford, UK: Elsevier Academic Press, 2011.

26. S. Caban, C. Mehlführer, M. Rupp, and M. Wrulich, *Evaluation of HSDPA and LTE: From Testbed Measurements to System Level Performance*. John Wiley & Sons, Chichester, U.K., 2012.

27. D. Tse and P. Viswanath, *Fundamentals of Wireless Communications*. Cambridge, UK: Cambridge University Press, 2008.

28. S. Alamouti, A simple transmit diversity technique for wireless communications, *IEEE Journal on Selected Areas in Communications*, 16(8), 1451–1458, October 1998.

29. B. Clerckx and C. Oestges, *MIMO Wireless Networks: Channels, Techniques and Standards for Multi-Antenna, Multi-User and Multi-Cell Systems*, Oxford, UK: Academic Press, Elsevier, 2013.

30. A. Dammann and S. Kaiser, Standard conformable antenna diversity techniques for OFDM and its application to the DVB-T system, in: *IEEE Global Telecommunications Conference*, Vol. 5, San Antonio, TX, 2001, pp. 3100–3105.

31. D. Love and R. Heath, Jr., Limited feedback unitary precoding for spatial multiplexing systems, *IEEE Transactions on Information Theory*, 51(8), 2967–2976, 2005.

32. D. Yang, L.-L. Yang, and L. Hanzo, DFT-based beamforming weight-vector codebook design for spatially correlated channels in the unitary precoding aided multiuser downlink, in: *IEEE International Conference on Communications*, Cape Town, South Africa, 2010, pp. 1–5.

33. D. Ryan, I. Vaughan, L. Clarkson, I. Collings, D. Guo, and M. Honig, QAM codebooks for low-complexity limited feedback MIMO beamforming, in: *IEEE International Conference on Communications*, Glasgow, Scotland, 2007, pp. 4162–4167.

34. T. Inoue and R. Heath, Jr., Kerdock codes for limited feedback MIMO systems, in: *IEEE International Conference on Acoustics, Speech and Signal Processing*, Las Vegas, NV, 2008, pp. 3113–3116.

35. S. Schwarz, C. Mehlführer, and M. Rupp, Calculation of the spatial preprocessing and link adaption feedback for 3GPP UMTS/LTE, in: *2010 Sixth Conference on Wireless Advanced (WiAD)*, London, U.K., June 2010.

36. C. E. Shannon, A mathematical theory of communication, *The Bell System Technical Journal*, 27, 379–423, July 1948.

37. G. Caire, G. Taricco, and E. Biglieri, Capacity of bit-interleaved channels, *Electronics Letters*, 32(12), 1060–1061, June 1996.

38. L. Wan, S. Tsai, and M. Almgren, A fading-insensitive performance metric for a unified link quality model, in: *Proceedings of the IEEE Wireless Communications & Networking Conference WCNC*, Vol. 4, Las Vegas, NV, 2006, pp. 2110–2114.

39. R. H. Clarke, A statistical theory of mobile radio reception, *Bell Systems Technical Journal*, 47, 957–1000, 1968.

40. T. K. Moon and W. C. Stirling, *Mathematical Methods and Algorithms for Signal Processing*, Upper Saddle River, NJ: Prentice Hall, 2000.

41. S. Schwarz, M. Wrulich, and M. Rupp, Mutual information based calculation of the precoding matrix indicator for 3GPP UMTS/LTE, in: *International ITG Workshop on Smart Antennas*, Bremen, Germany, February 2010, pp. 52–58.

42. T. Pande, D. Love, and J. Krogmeier, Reduced feedback MIMO-OFDM precoding and antenna selection, *IEEE Transactions on Signal Processing*, 55(5), 2284–2293, 2007.

43. T. Rappaport, *Wireless Communications: Principles and Practice*, Upper Saddle River, NJ: Prentice Hall Communications Engineering and Emerging Technologies Series. Dorling Kindersley, 2009.

44. C. Oestges, Validity of the Kronecker model for MIMO correlated channels, in: *IEEE 63rd Vehicular Technology Conference*, Melbourne, Victoria, Australia, May 2006, Vol. 6, pp. 2818–2822.

45. 3GPP, Technical specification group radio access network; Evolved Universal Terrestrial Radio Access (E-UTRA); User Equipment (UE) radio transmission and reception, December 2010, http://www.3gorg/ftp/Specs/html-info/36101.htm (accessed December, 2015).

46. A. Zanella, M. Chiani, and M. Win, On the marginal distribution of the eigenvalues of Wishart matrices, *IEEE Transactions on Communications*, 57(4), 1050–1060, April 2009.

47. J. S. Kim, H. Kim, C. S. Park, and K. B. Lee, On the performance of multiuser MIMO systems in WCDMA/HSDPA: Beamforming, feedback and user diversity, *IEICE Transactions*, 89-B(8), 2161–2169, 2006.

48. S. Schwarz and M. Rupp, Antenna combiners for block-diagonalization based multi-user MIMO with limited feedback, in: *IEEE International Conference on Communications, WS: Beyond LTE-A*, Budapest, Hungary, June 2013.

49. S. Schwarz and M. Rupp, Evaluation of distributed multi-user MIMO-OFDM with limited feedback, *IEEE Transactions on Wireless Communications*, 13(11), 6081–6094, 2014.

50. A. Edelman, T. A. Arias, and S. T. Smith, The geometry of algorithms with orthogonality constraints, *SIAM Journal on Matrix Analysis and Applications*, 20(2), 303–353, 1998.

51. N. Jindal, MIMO broadcast channels with finite-rate feedback, *IEEE Transactions on Information Theory*, 52(11), 5, November 2006.

52. M. Trivellato, F. Boccardi, and F. Tosate, User selection schemes for MIMO broadcast channels with limited feedback, in: *Sixty-Fifth IEEE Vehicular Technology Conference*, Spring 2007, Dublin, Ireland, April 2007.

53. N. Jindal, Antenna combining for the MIMO downlink channel, *IEEE Transactions on Wireless Communications*, 7(10), 3834–3844, October 2008.

54. E. Bjornson, M. Kountouris, M. Bengtsson, and B. Ottersten, Receive combining vs. multi-stream multiplexing in downlink systems with multi-antenna users, *IEEE Transactions on Signal Processing*, 61(13), 3431–3446, July 2013.

55. S. Schwarz and M. Rupp, Subspace quantization based combining for limited feedback block-diagonalization, *IEEE Transactions on Wireless Communications*, 12(11), 5868–5879, 2013.

56. S. Schwarz, Limited feedback transceiver design for downlink MIMO OFDM cellular networks, PhD dissertation, Vienna University of Technology, Vienna, Austria, 2013.

57. N. Ravindran and N. Jindal, Limited feedback-based block diagonalization for the MIMO broadcast channel, *IEEE Journal on Selected Areas in Communications*, 26(8), 1473–1482, October 2008.

58. C. Yetis, T. Gou, S. Jafar, and A. Kayran, Feasibility conditions for interference alignment, in: *IEEE Global Telecommunications Conference*, Honolulu, HI, November 2009, pp. 1–6.

59. O. Gonzalez, C. Beltran, and I. Santamaria, A feasibility test for linear interference alignment in MIMO channels with constant coefficients, *IEEE Transactions on Information Theory*, 60(3), 1840–1856, March 2014.

60. M. Rezaeekheirabadi and M. Guillaud, Limited feedback for interference alignment in the k-user MIMO interference channel, in: *Proceedings of the Information Theory Workshop*, Lausanne, Switzerland, September 2012, pp. 1–5.

61. B. Hassibi, Random matrices, integrals and space-time systems, in: *DIMACS Workshop on Algebraic Coding Theory and Information Theory*, Piscataway, NJ, December 2003.

62. I. S. Dhillon, R. Heath, Jr., T. Strohmer, and J. A. Tropp, Constructing packings in Grassmannian manifolds via alternating projection, arXiv e-prints, September 2007.

63. W. Dai, Y. Liu, and B. Rider, Quantization bounds on Grassmann manifolds and applications to MIMO communications, *IEEE Transactions on Information Theory*, 54(3), 1108–1123, March 2008.

64. W. Santipach and M. Honig, Capacity of a multiple-antenna fading channel with a quantized precoding matrix, *IEEE Transactions on Information Theory*, 55(3), 1218–1234, 2009.

65. D. Love and R. Heath, Jr., Limited feedback diversity techniques for correlated channels, *IEEE Transactions on Vehicular Technology*, 55(2), 718–722, March 2006.

66. S. Schwarz, R. Heath, Jr., and M. Rupp, Multiuser MIMO in distributed antenna systems with limited feedback, in: *IEEE Fourth International Workshop on Heterogeneous and Small Cell Networks, GLOBECOM*, Anaheim, CA, December 2012.

67. S. Schwarz, R. Heath, Jr., and M. Rupp, Single-user MIMO versus multi-user MIMO in distributed antenna systems with limited feedback, *EURASIP Journal on Advances in Signal Processing*, 2013(1), 54, 2013.

68. A. Gersho and R. Gray, *Vector Quantization and Signal Compression*, The Kluwer International Series in Engineering and Computer Science: Communications and Information Theory. Dordrecht, the Netherlands Publishers, 1992.

69. T. Kim, D. Love, and B. Clerckx, MIMO systems with limited rate differential feedback in slowly varying channels, *IEEE Transactions on Communications*, 59(4), 1175–1189, April 2011.

70. O. El Ayach and R. Heath, Jr., Grassmannian differential limited feedback for interference alignment, *IEEE Transactions on Signal Processing*, 60(12), 6481–6494, December 2012.

71. D. Zhu, Y. Zhang, G. Wang, and M. Lei, Grassmannian subspace prediction for precoded spatial multiplexing MIMO with delayed feedback, *IEEE Signal Processing Letters*, 18(10), 555–558, October 2011.

72. T. Inoue and R. Heath, Jr., Grassmannian predictive coding for limited feedback multiuser MIMO systems, in: *IEEE International Conference on Acoustics, Speech and Signal Processing*, Prague, Czech Republic, May 2011, pp. 3076–3079.

73. R. Heath, Jr., T. Wu, and A. Soong, Progressive refinement of beamforming vectors for high-resolution limited feedback, *EURASIP Journal on Advances in Signal Processing*, 2009(1), 743–747, 2009.

74. W. C. Jakes and D. C. Cox, *Microwave Mobile Communications*. New York: Wiley-IEEE Press, 1994.

75. J. Lee and N. Jindal, High SNR analysis for MIMO broadcast channels: Dirty paper coding versus linear precoding, *IEEE Transactions on Information Theory*, 53(12), 4787–4792, December 2007.

76. A. Tenenbaum and R. Adve, Linear processing and sum through-put in the multiuser MIMO downlink, *IEEE Transactions on Wireless Communications*, 8(5), 2652–2661, May 2009.
77. K. Gomadam, V. Cadambe, and S. Jafar, Approaching the capacity of wireless networks through distributed interference alignment, in: *IEEE Global Telecommunications Conference*, New Orleans, LA, December 2008, Vol. 30(4), pp. 1–6.
78. H. Du and P.-J. Chung, A probabilistic approach for robust leakage-based MU-MIMO downlink beamforming with imperfect channel state information, *IEEE Transactions on Wireless Communications*, 11(3), 1239–1247, March 2012.
79. T. Rüegg, A. Amah, and A. Wittneben, On the trade-off between transmit and leakage power for rate optimal MIMO precoding, in: *IEEE Workshop on Signal Processing Advances for Wireless Communications (SPAWC)*, Toronto, Canada, June 2014.
80. C.-B. Chae, D. Mazzarese, N. Jindal, and R. Heath, Jr., Coordinated beamforming with limited feedback in the MIMO broadcast channel, *IEEE Journal on Selected Areas in Communications*, 26(8), 1505–1515, October 2008.
81. D. Sacristan-Murga, M. Payaro, and A. Pascual-Iserte, Transceiver design framework for multiuser MIMO-OFDM broadcast systems with channel Gram matrix feedback, *IEEE Transactions on Wireless Communications*, 11(5), 1774–1787, May 2012.
82. R. Krishnamachari and M. Varanasi, On the geometry and quantization of manifolds of positive semi-definite matrices, *IEEE Transactions on Signal Processing*, 61(18), 4587–4599, September 2013.
83. Y. Li, S. Zhu, H. Tong, and M. Xu, Enhanced limited rate implicit CSI feedback and its usage in covariance matrix based MU-MIMO, in: *Wireless Communications and Networking Conference*, Shanghai, China, April 2013, pp. 3067–3071.
84. S. Schwarz and M. Rupp, Subspace versus eigenmode quantization for limited feedback block-diagonalization, in: *Sixth International Symposium on Communications, Control and Signal Processing*, Athens, Greece, May 2014, pp. 1–4.
85. S. Lloyd, Least squares quantization in PCM, *IEEE Transactions on Information Theory*, 28(2), 129–137, 1982.
86. Y. Polyanskiy, H. Poor, and S. Verdu, Feedback in the non-asymptotic regime, *IEEE Transactions on Information Theory*, 57(8), 4903–4925, 2011.
87. N. Gaarder and J. Wolf, The capacity region of a multiple-access discrete memoryless channel can increase with feedback (corresp.), *IEEE Transactions on Information Theory*, 21(1), 100–102, 1975.
88. L. Ozarow and S. Leung-Yan-Cheong, An achievable region and outer bound for the Gaussian broadcast channel with feedback (corresp.), *IEEE Transactions on Information Theory*, 30(4), 667–671, 1984.
89. T. Wiegand, G. Sullivan, G. Bjontegaard, and A. Luthra, Overview of the H.264/AVC video coding standard, *IEEE Transactions on Circuits and Systems for Video Technology*, 13(7), 560–576, 2003.

4

GAME THEORY APPLICATION FOR POWER CONTROL AND SUBCARRIER ALLOCATION IN OFDMA CELLULAR NETWORKS

EMMANUEL DUARTE-REYNOSO,
DOMINGO LARA-RODRÍGUEZ, AND
FERNANDO RAMÍREZ-MIRELES

Contents

This chapter provides an in-depth study of the problem of dynamic power control and subcarrier allocation in 4G cellular networks and beyond based on orthogonal frequency division multiple access (OFDMA) technology through a game theoretic perspective. The challenge of spectrum channel allocation and power control in relation with 4G and 5G main design components is discussed. The resource (power and subcarriers) allocation schemes based on game theory presented in the literature are discussed and classified according to their optimization objectives, including a novel scheme called PRISMAP. The PRISMAP allocation scheme is conceived to improve fairness and capacity while maintaining energy efficiency and minimizing no-transmission probability. We take this approach as in 5G networks it will be necessary to implement distributed power control schemes [1]. With the introduction of the preferred reuse concept and a new utility function, PRISMAP overcomes the limitations of other compared schemes. The mathematical approach is formal and a theorem to guarantee Nash equilibrium and some propositions (for PRISMAP) are demonstrated. The chapter ends with a section on future research directions.

4.1 4G and Beyond Networks: A Challenge for Dynamic Spectrum Allocation and Power Control

As envisioned by the International Telecommunications Union Radiocommunication Sector (ITU-R) [2], some of the key features and objectives in 4G communication systems are high data rates, efficient spectrum utilization, adaptive radio interface, and long battery life of user equipment. The materialization of these objectives is accomplished through different novel techniques and design components. Among the main design components in 4G and 5G are [3,4]

- Very low power consumption of both access points and terminals
- Efficient support of device-to-device communication
- Flexibility in spectrum usage
- Self-optimized ultradense deployment of access points
- Massive multiple-input-multiple-output (MIMO) techniques
- Millimetric wave communications
- Support of relay communications between base stations (self-backhauling)
- Simple and low-cost design

Despite the integration of multiple radio access technologies in user equipment such as WiFi, long-term evolution (LTE), and millimetric wave communications, it is anticipated that all devices should maintain compatibility with currently deployed LTE [5]. Indeed, to achieve the expected high data rates (1 Gbps) it will be necessary to implement massive MIMO techniques that enable both robustness and multiplexed transmissions [6].

Finally, the structure of the network will suffer important changes motivated by spatial densification [6] and the deployment of different-size cells (macro, micro, pico, femto), evolving to the concept of heterogeneous networks (HetNets) that are groups of different size cells, each size called a tier, that support aggressive spectrum reuse.

On the other side, current 4G technologies are based on OFDMA [7], which makes it possible to achieve high data rates through rate adaptation and an efficient utilization of the available spectrum, and even if 5G networks may have a new physical layer, this would part from OFDMA to include new concepts [8]. However, as it will be seen, in order to exploit OFDMA at its maximum and make HetNets a reality, it is necessary to perform an advanced interference management [5], which is strongly impacted by the allocation of subcarriers, transmit power, and modulation rates to the users in the system. Thus, taking into account that very low power consumption and flexibility in spectrum usage are part of the design components in

4G and 5G networks, in this chapter we focus on dynamic resource allocation and energy efficiency through the game theoretic approach.

In this section, an overview of OFDMA basics is presented in order to understand its potential and the need of dynamic resource allocation including assignation of subcarriers to the users and power control over those subcarriers. Finally, we will present the motivation and justification for the use of game theory in the study of power control and subcarrier allocation in OFDMA cellular networks.

4.1.1 OFDMA Overview

OFDMA is a multiple access scheme that extends orthogonal frequency division multiplexing (OFDM) in such a way that enables the data transmission of multiple users simultaneously. Despite that OFDM requires tight synchronization and signaling overhead for channel estimation, it provides some important features that enable systems to comply with 4G objectives:

- OFDM systems are robust to multipath phenomena thanks to the longer duration of transmitted symbols and the cyclic prefix application.
- OFDM has a high spectral efficiency due to the overlap of subcarriers.
- OFDM implementation in the digital domain has low complexity.

In addition to these features, OFDMA allows the partition of time and frequency resources in OFDM symbols and OFDM subcarriers [9]. This situation is depicted in Figure 4.1.

In this way, any subcarrier could be assigned to a different user every symbol time interval. Furthermore, taking into account that symbol detection in every subcarrier can be done independently, each user can allocate a different power level and modulation scheme for every allocated subcarrier. However, in practical systems, subcarriers are grouped into blocks (called subchannels) and these blocks are allocated to the users in the system, making easier channel estimation due to the correlation between channel responses in adjacent subcarriers.

For these reasons, several multiple access schemes have been developed, such as single-carrier frequency division multiple access

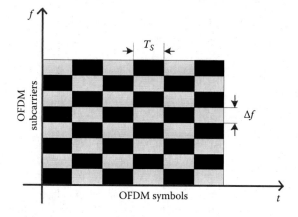

Figure 4.1   Resource partition in orthogonal frequency division multiple access.

(SC-FDMA). SC-FDMA can be considered as a precoded OFDMA that achieves a low peak to average power ratio [10]. Thus, this scheme would be more suitable for the uplink where the transmitter has limited amplification and processing capabilities. However, because of the similarities between SC-FDMA and OFDMA, the presented approach is focused on OFDMA.

4.1.2 Dynamic Resource Allocation: Taking Advantage of Multiuser Diversity

In multiuser OFDMA systems, each user sees a different channel quality, which has to be taken into account in allocating subcarriers to users because even if a subcarrier is in deep fading for a given user, this will be different for the rest of the users. This is called multiuser diversity and it is possible to take advantage of this feature.

To take advantage of this multiuser diversity, a dynamic resource allocation that considers three aspects is required [11]:

1. *Subcarrier allocation*: In 4G cellular networks, universal reuse is imposed to achieve maximum spectral efficiency. However, two users in the same cell/sector cannot use a given subcarrier simultaneously. Furthermore, it is desirable that the allocated subcarriers to a user are those in better channel conditions. Furthermore, the subcarrier allocation has to be done aiming to provide users with an acceptable performance while considering the intracell interference constraint.

2. *Adaptive modulation*: High data rates and adaptive modulation come together. As stated before, each user sees different channel conditions and may have different rate requirements. For this reason, it is necessary to select a specific modulation scheme.
3. *Power control*: Typically, cellular systems have been limited by co-channel interference from adjacent cells. When considering universal frequency reuse, interference becomes a critical issue. Consequently, the transmitted power has to be controlled in order to meet or maximize transmission rates while maintaining interference to other users in an acceptable level.

Research in this topic has been intensified in the last years. Nevertheless, there is not an accepted unified method to address dynamic resource allocation because the problem becomes more complex when considering the complete scenario, that is, a multicell and multiuser system with discrete subcarrier allocation and interdependence between cells through co-channel interference. In addition, fairness is another issue that is present in this kind of environments as some approaches results in an excellent performance for cell-center users and poor performance for cell-edge users.

In 4G and beyond networks, a more efficient use of current spectrum will not be sufficient to keep pace with the mobile usage increase. This could only be realized with much more spectrum [5]. However, efficient and dynamic resource allocation are still needed [12], especially when considering HetNets, which will need further cooperation among tiers [1].

In conclusion, when addressing dynamic resource allocation in 4G and beyond cellular networks based in OFDMA, it is imperative to develop schemes that jointly consider fairness, efficiency, complexity, and capacity. As will be shown later, the game theoretic approach is appropriate to accomplish this challenge.

4.1.3 Game Theoretic Approach

4.1.3.1 What Is Game Theory?   Game theory is a mathematical discipline focused on modeling the interactions between greedy agents that take decision in circumstances such that the actions of each agent affects the payoff (utility) that other agents obtain in the interaction [13].

Basic underlying assumptions of game theory are that the agents or decision-makers have well-defined objectives (they are rational) and that they are able to infer others' behavior from the available information (they think in terms of strategies) [14]. Each agent's objective is represented by a utility function that each agent tries to maximize, and this function depends on both own agent's actions and others' actions. This results in potentially conflictive situations, inducing the competition or cooperation between the various agents.

There are many game models and each one has different components. However, all game models have three basic components [15]: (1) a set of players $I = \{1,...,N\}$, (2) a set of strategies for each player S_i with elements s_i, and (3) a set of preference relations that are represented through utility functions $\{u_i(s_i, s_{-i})\}_{i \in I}$, where s_{-i} represents the vector including the strategies from all players but player i. A strategy profile is denoted as $s = (s_i, s_{-i})$ and the set of possible strategy profiles is S.

The players are the agents or decision-makers in the modeled scenario. The strategies are the available alternatives to each player, that is, they are the options from which a player can make a decision. Finally, the preference relation of each player represents its evaluation of the possible game outcomes. Preference relation is represented by a utility function $u_i(s)$ that assigns a number to every possible resulting strategy profile $s \in S$.

The fundamental question that game theory is intended to answer is which strategy profile is more likely to occur as the outcome of the game. The basic solution concept is the Nash equilibrium, which is defined in the following.

Definition 3.1: Strategy profile s^* is a Nash equilibrium if, for every player i,

$$u_i\left(s_i^*, s_{-i}^*\right) \geq u_i\left(s_i, s_{-i}^*\right) \forall s_i \in S_i. \tag{4.1}$$

In other words, Nash equilibrium produces consistent predictions of how the game will be played, in the sense that if all players predict that a particular Nash equilibrium will occur, then no player has an incentive to play in a different way [16]. Alternatively, Nash equilibrium is a strategy profile in which each player's strategy is the best response to other players' strategies.

Indeed, best response dynamic is a method to obtain or to approximate the outcome to a Nash equilibrium in which each turn, players update their strategies to maximize their utility function as they would do it if the strategies of the other players remain the same in the next iteration.   ■

Basic concepts of game theory have been presented up to this point. In the following section, the application of game theory in OFDMA cellular networks is justified. For readers willing to learn more about game theory concepts, please refer to [11].

4.1.3.2 Game Theory in 4G and Beyond Cellular Networks   Unlike classic application areas of game theory in which basic assumptions, that is, rationality and the capacity to infer other players' actions are only partially true, in cellular networks in which the players are programmed agents, it is more reasonable to assume that the device will maximize the expected value of a utility function [17].

In the context of communications networks, players are common nodes in the network; and their strategies could include the selection of a modulation scheme, coding rate, protocol, flow control parameter, transmit power, etc.; and the preference relations are utility functions related to bit error rate, signal-to-noise ratio, data rate, or power consumption.

Given these features, game theory is a powerful tool to analyze wireless communication problems such as resource allocation, power control, random access, cooperative spectrum sensing, admission control, synchronization, security, network selection, rate selection, multihop routing, and cooperative transmission.

However, specific and universally accepted game theory algorithms do not exist as the algorithms obtained with game theory depend on the different system models and objectives. In spite of this, important solution concepts do exist such as Nash equilibrium, best response dynamics, Stackelberg equilibrium, learning, pricing, Nash bargaining solution (NBS), and auction game.

Three important applications of game theory in 4G and 5G wireless networks are synchronization, scheduling (planned bandwidth allocation), and power control. The synchronization problem has been addressed for both code division multiple access (CDMA) and OFDMA networks. In [18], the CDMA synchronization problem is examined through a noncooperative game in which its transmitter–receiver pair seeks to maximize the ratio of the probability of code alignment detection to the transmitted energy per acquisition, achieving a good trade-off between synchronization and energy savings. Another example is the distributed iterative algorithm derived in [19], which performs uplink power control in the initial synchronization procedure of OFDMA networks letting the mobile terminals maximize their own energy efficiency in terms of average synchronization time.

Scheduling in wireless networks is the process of bandwidth resource request and distribution over time through a given mechanism. This problem has been studied using game theory in [20–22]. In [20], scheduling is modeled as a competition game between competing traffic types that distribute the resources to increase the aggregated utility of the whole system and provide fairness to individual traffic types at the same time. Besides, in [21] the problem of scheduling between relay stations is addressed. As the centralized scheduling implies a severe overhead for channel reporting, a distributed scheduling algorithm is proposed, exploiting frequency reuse and providing quality of service (QoS) to mobile stations. Another example, [22], takes some game theoretic solutions for resource allocation and uses them as schedulers, comparing their performance. The compared solutions are NBS, proportional fairness, and Kalai–Smorodinsky approach.

Indeed, scheduling is closely related to dynamic resource allocation in wireless networks, which in a more general sense can include power control as well as adaptive modulation. The specific problem of resource allocation in 4G multicell OFDMA networks has two important issues. On one side, it is required to develop dynamic allocation schemes that consider at the same time fairness, system efficiency, complexity, etc. On the other side, the implementation of these schemes with a centralized control implicates signaling overhead and an excessive computation time. Because of this, it becomes necessary to implement distributed schemes.

Through a general analysis of the conventional resource allocation schemes, it can be said that these proposals consider some approximations and simplifications that are not always close to a real system. Such is the case of the scheme presented in [23], where some interference sources are neglected, or the proposal in [24] in which the described advantages of OFDMA are not exploited. At the same time, most of these schemes do not consider the interdependence between cells and users as they consider the downlink of a single-cell system [25,26].

All of this is due to the difficulty in capturing those effects and including them in conventional optimization methods. Indeed, the multicell OFDMA environment raises interesting competitive and cooperative situations among system users [11] because wireless channel characteristics and co-channel interference among adjacent cells provoke that the actions taken by each user (player) impact the performance (utility) of other users.

In fact, in [27], it is demonstrated that even if power control is a convex problem that is tractable, subcarrier allocation is a strongly NP-hard problem, which means that it is not solvable in a polynomial-defined time. Taking this into account, it is necessary to use alternative tools that permit the inclusion of a more complete system model, considering the interactions among the network users (nodes) that participate in a multicell multiuser 4G network.

Indeed, it has been foreseen that in order to achieve the dramatic improvements in capacity for 4G and 5G without significant cost, new tools for jointly optimizing network spectrum efficiency, network efficiency, and quality of experience will be essential [28]. And it is predicted that these new tools will consider smarter devices, capable of taking a greater role in sharing info and making decisions in the network. One of these tools widely studied in the literature [29–34] is game theory.

Given that game theory focuses on distributed solutions to agent interaction problems [17], it makes possible to abstract the resource allocation problem with a different approach and to apply fundamental results of the theory to simplify its analysis. Therefore, game theory can be viewed as a mathematical tool for the design and analysis of dynamic resource allocation schemes in multicell multiuser OFDMA 4G/5G networks.

4.2 Basic Power Control Algorithms and Capacity Maximization

There are two basic power control algorithms discussed in the literature that typically have been applied to achieve maximum capacity in Gaussian vector channels: water-filling and successive bit allocation. As OFDM transmission is a case of transmission over Gaussian vector channels, many dynamic resource allocation schemes in OFDMA systems make use of these basic power control algorithms. Therefore, this section studies both water-filling and successive bit allocation algorithms.

4.2.1 Iterative Water-Filling

The simple version of this method considers a system with a single transmitter and n-independent parallel Gaussian channels that share a common power constraint. The objective is to distribute the total power P_{tot} among the channels to maximize capacity, which is given in every channel by Shannon's formula:

$$C_i = W \log_2 \left(1 + \frac{P_i}{\sigma_i^2} \right). \tag{4.2}$$

Then, capacity maximization problem can be expressed as

$$\underset{\{P_i\}, i=1,\dots,n}{\arg\max} \sum_i C_i, \tag{4.3}$$

$$\text{s.t.} \quad \sum_i P_i = P_{tot}. \tag{4.4}$$

This is a standard optimization problem, and following the procedure in [35], it can be solved using Lagrange multipliers. Writing the functional as

$$J(P_1,\dots,P_n) = \sum_i W \log_2 \left(1 + \frac{P_i}{\sigma_i^2} \right) + \lambda \left(\sum_i P_i \right), \tag{4.5}$$

and differentiating with respect to P_i, yields

$$\frac{1}{P_i + \sigma_i^2} + \lambda = 0, \tag{4.6}$$

where P_i is obtained as

$$P_i = v - \sigma_i^2. \tag{4.7}$$

However, as values for P_i should be nonnegative, it will not always be possible to find a solution in this form. In the later case, using Karusch–Kuhn–Tucker (KKT) conditions [35], it is verified that the solution

$$P_i = \left(v - \sigma_i^2\right)^+ \tag{4.8}$$

is the power allocation that maximizes the capacity, where v is chosen such that

$$\sum_i \left(v - \sigma_i^2\right)^+ = P. \tag{4.9}$$

This solution is depicted in Figure 4.2. The vertical levels indicate noise levels through the different channels. If the power is increased,

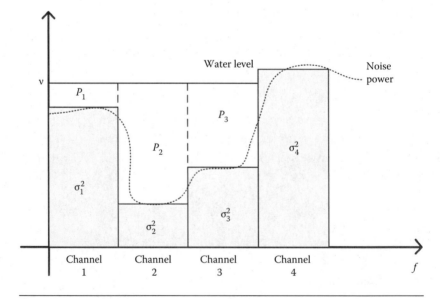

Figure 4.2   Water-filling method.

more power is allocated to the channels with lower noise. If the available power is higher, some power is allocated in more noisy channels. The distribution process of the power among the different channels is identical to the way in which water distributes in a recipient; for this reason, the method is called water-filling.

This method can be directly applied in point-to-point OFDM systems or in the downlink of a single-cell OFDMA system. Furthermore, the problem can be easily reformulated for the case in which the objective is to minimize transmitted power and constraint shown in Equation 4.4 is substituted by a required rate constraint.

In [36], it is demonstrated that this method can be generalized for the case of multiple access channels that allows its application in OFDMA systems for both uplink and downlink. To make use of this generalization, it is considered that the power at the receiver corresponding to interfering users is Gaussian distributed; thus, it can be treated as an additional noise source that gives the capacity for a specific user and channel i as

$$C_i = W \log_2 \left(1 + \frac{P_i}{\sigma_i^2 + I} \right). \tag{4.10}$$

From this result, an algorithm known as iterative water-filling is proposed. At each iteration, the users perform the water-filling method considering $\sigma_i^2 + I$ instead of σ_i^2 as the total noise power. This is repeated until the transmitted powers converge. In [36], convergence of iterative water-filling is demonstrated. As will be seen later, this algorithm is widely used in dynamic resource allocation schemes.

4.2.2 Successive Bit Allocation

The water-filling method considers that the transmission rate of a user can take continuous values; however, in practical systems with adaptive modulation and codification (AMC), these values are discrete. Consequently, the power control problem becomes a constrained discrete optimization problem. Fortunately, there exists in the literature a well-known optimal solution for this kind of problem. This solution is explained in the following as presented in [37].

In general, the discrete optimization problem can be expressed as

$$\max_{x}\{\phi(x) : x \in S, C(x) \le M\}, \tag{4.11}$$

where

the allocation vector S is the set of n-tuples of nonnegative integers $(x_1, \ldots, x_n)$

M is the number of available resources

$C(x)$ is the cost function given by

$$C(x) = \sum_{j=1}^{n} c_j x_j \tag{4.12}$$

$\phi(x)$ is the function subject to optimization and it can be decomposed as

$$\phi(x) = \sum_{j=1}^{n} \phi_j(x_j) \tag{4.13}$$

The method to find the optimal solution is given by Algorithm 4.1.

Algorithm 4.1

- Start with the allocation
- $k = 1$
- $x^k = x^{k-1} + e_i$, where e_i is the ith unitary vector and i is any index s.t.

$$\frac{\left[\phi_j\left(x_j^{k-1} + 1\right) - \phi_j\left(x_j^{k-1}\right)\right]}{c_j}$$

- If $C(x^k) > M$ stop; otherwise $k \leftarrow k + 1$ and return to step 3.

In [37], it is shown that if $\phi_i(x_i)$ is a strictly increasing concave function, the difference between the cost $C(x)$ obtained by Algorithm 4.1 and the optimum cost is lower than the maximum cost among the elements, that is, $\max(c_j)$. Furthermore, it is established that the stop condition can be substituted, that is, instead of stop when the cost reaches a maximum resource value, it may stop when the objective function $\phi(x)$ reaches a target value, which corresponds to the problem

$$\max_{x}\{\phi(x) : x \in S, \phi(x) = R\}. \tag{4.14}$$

In the case of a point-to-point communication system with n-independent Gaussian channels, the objective function $\phi(x)$ corresponds to the bit per symbol sum along the n channels, where x_j is the number of bits per symbol in channel j; while the cost function $c_j(x_j)$ would be defined as the required power to transmit x_j bits per symbol in channel j to achieve a specified bit error rate (BER). Taking this into account, in [38] it is shown that if the required BER is the same for all channels then the result obtained through Algorithm 4.1 corresponds to the optimal solution.

4.3 Game Theoretic Dynamic Resource Allocation

Many game theoretic dynamic resource allocation schemes have been presented in the literature. Each scheme is designed to accomplish a different objective and make different assumptions related to the system model. Thus, a convenient classification for these schemes is based in their specific objective. For instance, we take into account the following: (1) capacity maximization, (2) transmit power minimization, (3) system fairness, and (4) energy efficiency. However, capacity maximization is well treated in conventional resource allocation methods [26]. Thus, in this section, we make an overview of schemes that can be classified into categories 2 and 3. Taking into account the trend toward green communications [7,39] and that 4G and beyond systems need to make an efficient use of battery power, we take energy efficiency as a case of study, which is presented in the next section.

4.3.1 Transmit Power Minimization

Among various transmit power minimization resource allocation schemes based on game theory, two have been selected to be studied here: a noncooperative scheme with a virtual referee and a potential game-based scheme.

4.3.1.1 Virtual Referee   A common situation in 4G systems is that every user in the system has a different rate requirement, so a design objective could be to satisfy each user's rate requirement using the minimum amount of transmit power. A game theoretic scheme widely cited in the literature (including patents) is the one presented in [40]. This scheme's model considers the uplink of a multicell OFDMA system in which there is only one user per cell. In this way, the subcarrier allocation is done at cell level instead of at user level.

Under these assumptions, a noncooperative game in which the users are the players and the best response dynamic is given by the iterative water-filling is proposed. Nevertheless, it will not be always possible to satisfy all players' rates. Indeed, through an example with two users, two base stations, and two subcarriers, it is found that the Nash equilibrium is unique and optimal for the whole system. While the rate requirements increase, some local optima appear and the global optimum occurs when some user does not transmit over a specific subcarrier. If the rate requirements increase even more, then there will not be a feasibility region.

Based on these observations, a virtual referee that prevents some cells to transmit over certain subcarriers is proposed. This is done to make that the game equilibrium gets closer to the global optimum. The way in which the virtual referee operates is explained in the following.

All the users begin playing the noncooperative game to minimize their transmit power through the iterative water-filling algorithm. Users that transmit power reach the maximum and switch to a dual game in which they are intended to maximize their rate using their available power. After convergence, if a user switches to the dual game, it is considered that the equilibrium point coincides with a local optimum; in consequence, a user is not permitted to transmit in one subcarrier at each iteration until no user can be not permitted, system performance cannot be improved, or the desired equilibrium is reached. However, to remove a user, some conditions have to be satisfied: (1) Every user must have at least one subcarrier to transmit, (2) at least one user must transmit in every subcarrier, and (3) a user who has switched to the dual game cannot be removed.

Through numerical simulations of a seven-cell system, the average achievable rate and the transmit power are obtained as a function of the reuse distance. The results show that the virtual referee scheme achieves a higher rate and a lower-power transmission than the water-filling method alone, especially for small reuse distances, while for larger distances this scheme is equivalent to the water-filling method.

The presented analysis for the design of this scheme establishes some important results about the feasibility region at the time that it demonstrates that subcarrier allocation impacts in a significant way the performance of the system. Even if it considers a single user per cell, this could be extended to the multiuser case if there is a previous subchannel assignment among them.

The disadvantage of the virtual referee is that it has to be implemented in a centralized manner, requiring a continuous information transmission from the base stations to the central controller.

4.3.1.2 Potential Game   An alternative approach to minimize transmit power is through interference minimization when all users in the system have a common signal to interference plus noise ratio (SINR) objective. Based on this approach, a resource allocation scheme using potential games is presented in [41].

This scheme considers the downlink of a multicell OFDMA system, in which every user is assigned to a single subcarrier (block of subcarriers). The set of base stations corresponds to the set of players and their strategy space is the set of feasible allocations (allocations in which a subcarrier is exclusively assigned to one user in each base station). Each player tries to maximize its own utility subject to the maximum transmit power constraint. Utility function U_k for every player (base station) k is defined as the generalized interference sum, which is given by

$$U_k = -(\text{interference generated by user } k)$$

$$+ (\text{interference experimented by user } k). \qquad (4.15)$$

In [41], it is demonstrated that the proposed game guarantees equilibrium convergence through the best response dynamic as the proposed game is an exact potential game with potential function

$$F = \frac{1}{2} \sum_k U_k. \qquad (4.16)$$

Based on this result, the authors propose an iterative algorithm in which each base station starts with a random feasible subcarrier allocation. In their corresponding turn, each player estimates his/her own utilities for the different strategies, choosing the best response among these strategies. The process is repeated until the convergence is reached. To estimate the corresponding utility, it is necessary to establish the transmit power in each subcarrier. In this proposal, transmit power is computed through a suboptimum method that consists in assigning a power p_n^{min} such that the target SINR ($\gamma_{n(dB)}^*$) is guaranteed considering a fading margin $\mu_{n(dB)}$ in each subcarrier n:

$$p_{n(dBm)}^{min} = \sigma_{(dBm)}^2 + \mu_{n(dB)} + \gamma_{n(dB)}^*. \tag{4.17}$$

This method reaches the convergence in approximately 10 iterations. In general, the SINR of all users is more than the initial SINR after the algorithm converges, improving the capacity while reducing the transmit power. Nevertheless, an important limitation is associated with this scheme: subcarrier allocation implies a centralized search among all the feasible allocations that introduce a considerable processing load. Besides, the use of the interference sum as the utility function requires each base to know the whole system information, implying an important signaling overload.

4.3.2 Fairness

In this section, allocation schemes with an objective to improve the system fairness, that is, to achieve an acceptable trade-off between system capacity and cell-edge users' performance, are presented. Proportional fairness, max–min fairness, and NBS are studied in the following.

4.3.2.1 Proportional Fairness   A resource allocation scheme that addresses power and subcarrier allocation in the uplink of a single-cell OFDMA system is presented in [42]. It considers the proportional fairness as the concept to achieve system fairness. In this case, the utility function of the user (player) i is dependent of the user overall rate R_i and is given by

$$U_i(R_i) = \ln(R_i). \tag{4.18}$$

Imposing the KKT conditions, it is shown that in order to maximize the utility function, each subcarrier n has to be assigned to the user that fulfills the condition

$$i_n = \arg\max_i \left\{ U_i'(R_i) \ln\left(1 + p_{i,n} \frac{h_{i,n}^2}{\sigma^2} \right) \right\}, \tag{4.19}$$

where

$p_{i,n}$ is the power transmitted by user i over subcarrier n

$h_{i,n}^2$ is the corresponding channel coefficient

σ^2 is the additive white Gaussian noise (AWGN) noise power

Taking this into account and that the optimal power allocation for every user would be given by the water-filling method, the authors propose a method in which the users perform the water-filling method over their allocated subcarriers and the unassigned subcarriers. Then, the factor $U_i'(R_i)\ln(1 + p_{i,n}(h_{i,n}^2/\sigma^2))$ is calculated for each pair (i, n) restricting n to the unassigned subcarriers; finally, the pair that maximizes this factor is assigned. When all subcarriers have been allocated, the users perform once again the water-filling method over their assigned subcarriers.

In [42], the authors demonstrate that the obtained result by means of this algorithm cannot be improved by any combination in which each user changes, up to one subcarrier. Furthermore, they obtain a low-complexity implementation algorithm using a binary tree.

The presented results indicate that the proposed algorithm is close to the upper performance limit established by the optimal solution. The drawback of the presented results resides in the fact that they are presented as a function of the number of subcarriers in the system that usually cannot be changed within a given system.

The game model is similar to an auction because in every iteration the players (users) make a bid (a transmit power) based on which the optimality condition is calculated. The base station (auctioneer) allocates the subcarriers considering the bids. Finally, the water-filling method is performed to maximize the capacity. In spite of the mathematical analysis being complete and some important results are derived, the implementation of this scheme in a multicell scenario is not as simple because the base station in every cell would have to allocate subcarriers considering both power bids and the interdependence with other cells.

4.3.2.2 Max–Min Fairness   Another option is to consider a fairness criterion in which there is a maximum achievable rate constraint per user $R_k^{\max}$ in such a way that the users with the best channel conditions could not monopolize the resources. Instead, the user only can use the necessary resources to achieve $R_k^{\max}$. This approach is presented in [43], where a subcarrier and transmit power allocation scheme is proposed for the uplink of a multicell system.

In the subcarrier allocation stage, the objective is to improve system fairness by allocating each subcarrier to the user that values it most. The valuation is established as a function of the normalized channel gain, which is defined for the user k in cell m over subcarrier n as

$$q_{m,k}(n) = \frac{h_{m,k}^2(n)}{\sum_{l \in N} h_{m,k}^2(j)}, \tag{4.20}$$

where $h_{m,k}^2(n)$ is the power channel gain between the user k and base m over subcarrier n.

On the other hand, power allocation has as its main objective that every user maximizes its rate but reducing as much as possible its transmit power in such a way that the produced interference can be supported by all the users. Taking into account this objective, the users must minimize a function composed by a utility function and a cost function given by

$$J_k = U_k + C_k, \tag{4.21}$$

where U_k is the utility function of user k defined as

$$U_k = \left(R_k^{\max} - \sum_{n \in \psi_k} R_{k,n} \right)^{1/2}, \tag{4.22}$$

where $R_{k,n}$ is the transmission rate of user k over subcarrier n, and is the set of allocated user to user k. Besides, C_k represents the cost function given by

$$C_k = a_k \sum_{n \in \psi_k} h_{m,k}^2(n) p_{k,n}, \tag{4.23}$$

where a_k is the pricing factor of user k.

In the original article, it is demonstrated that J_i satisfies the conditions of the Nash theorem, which guarantees the equilibrium existence. In addition, using Lagrange multipliers establishes a best response dynamic in which each user minimizes J_i. In the reported results, it is observed that when adjustable cost factors are considered as a function of the distance, this scheme maintains a rate close to the no interference case for the cell-edge users, indicating the fairness of this scheme.

However, this scheme has some open issues: each subcarrier is treated in an independent way, the maximum transmit power constraint is not explicitly considered, and it is not specified how to obtain the maximum rate value.

4.3.2.3 Nash Bargaining Solution   An alternative fairness criterion that has its origins in game theory is the NBS. The intuitive idea behind this solution is that after the minimum rate requirements of each user R_{min}^i have been satisfied, the remaining resources are assigned proportionally to the users, according to their achievable rate R_i.

In [44], a dynamic resource allocation scheme is presented that has as its objective to maximize the capacity sum while the minimum rate requirements of the different users are satisfied. The system model considers the uplink of a single-cell OFDMA system. It is assumed that (1) the users can negotiate through the base station, cooperating in taking the decisions related to the subcarriers' utilization, and (2) once the subcarrier allocation has taken place, every user performs the water-filling method to maximize its own capacity. Thus, we focus on the subcarrier allocation.

NBS is obtained when the global utility function is maximized. This function is given by

$$U = \prod_i \left(R_i - R_{min}^i \right). \tag{4.24}$$

NBS is proposed for two reasons. First, it guarantees a fair allocation in the sense that the NBS is a generalized form of proportional fairness (proportional fairness is the case of NBS when the minimum rate requirement is $R_{min}^i = 0 \; \forall i$). Given that in practice it is required that all users satisfy a minimum rate different of zero, it is more feasible to use NBS. The second reason is because cooperative game theory demonstrates that it exists a unique and efficient solution for this problem.

Based on this approach, the solution for the case in which there are only two users is directly derived and presented in [44]. Basically, both users interchange subcarriers until NBS is reached. Using this algorithm, a two-step iterative scheme is proposed. First, the users are grouped in pairs called coalitions. Then, at every coalition the bargaining algorithm for two users is applied. Finally, the users form new coalitions until the convergence is reached.

However, an optimal coalition formation method is necessary to facilitate the cooperation among users. The classical option in this case is known as the Hungarian method [45], which always finds the optimal coalition pairs. Finally, the convergence of the method can be demonstrated based on that function U in Equation 4.24 is a nondecreasing function and that the optimal solution is upper bounded.

The reported results show that performance loss of the NBS scheme compared to a capacity maximization scheme is low, which implies that NBS achieves a good trade-off between fairness and general system performance. Furthermore, this difference is reduced when the number of users is higher because there are more negotiation possibilities.

This scheme shows that when considering the resource allocation problem as a cooperative game, it is possible to form coalitions and provide system fairness without degrading the capacity. Nevertheless, it requires a centralized control and a significant amount of signaling. Also, each iteration could last longer than the channel variation time, which would be more severe in a multicell environment.

4.4 Case of Study: Energy Efficiency

In the previous section, an overview of some game theoretic schemes that aim to minimize the transmit power or improve system fairness was given. This section has a different approach: it is dedicated to energy-efficient resource allocation schemes and an in-depth study of this problem is accomplished. This is done because as stated before energy efficiency is a relevant issue in 4G and beyond networks particularly in the uplink where the available energy is limited.

Due to the complexity of the multicell OFDMA scenario, and that energy efficiency is a function of transmit power and data rate, since the seminal work of Goodman et al. [46,47], the energy efficiency problem has been treated using game theory and many game theoretic resource allocation schemes have been presented in the literature [32,34,48,49]. However, some of these schemes make assumptions that are not feasible or consider a too simplified system model. For this reason, only two schemes from the literature along with a novel scheme called PRISMAP are presented and compared here. The comparison is done through numerical simulations under a common system model. So first, the common system model is presented; then, the three mentioned schemes are described. Finally, numerical results are presented.

4.4.1 System Model

Let us consider the uplink channel of a multicell OFDMA network with universal frequency reuse. Let $M = \{m, m = 1,...,|M|\}$ denote the set of cells in the network and $K = \{k, k = 1,...,|K_m||M|\}$, the set of active users (mobile stations) with $K_m = \{k|m_k = m\}$ the set of active users in cell m where m_k is the base station to which user k is assigned.

The available spectrum is divided into N physical resource blocks (PRBs) such that each PRB is formed by N_b subcarriers with frequency spacing BW. Each cell allocates the complete set of PRBs among its active users through the allocation matrix $A = \{a_{m,n}|m \in M, n \in N\}$, where element $a_{m,n}$ designates the user to which PRB n has been allocated. PRB allocation is repeated at each transmission time interval (TTI) that consists of D_s signaling symbols and D_d user information symbols.

It is assumed that the receivers and transmitters have perfect channel state information (CSI) and that it is possible to use a single coefficient $h_{m,k}^2(n)$ to represent the channel power gain between user k and base m over PRB n. The transmit power vector of user k is denoted by $p_k = \{p_{k,n}|p_{k,n} \geq 0\}$ where element $p_{k,n}$ indicates the transmit power of the user for every subcarrier in PRB n. This vector has the restrictions:

$$p_{k,n} = 0, \quad a_{m_k,n} \neq k, \tag{4.25}$$

$$\sum_{n \in N} p_{k,n} \leq \frac{P_{max}}{N_b},$$ (4.26)

where Equation 4.25 prevents the existence of intracell interference and Equation 4.26 stands for the power limitation due to the available power P_{max} at the mobile station.

The effect of AWGN with power σ^2 at the receiver is considered; thus, the SINR at base m over PRB n is computed as

$$\gamma_{m,n} = \frac{p_{k,n} h_{m,k}^2(n)}{\sigma^2 + I_{m,n}},$$ (4.27)

where
k is the user such that $a_{m,n} = k$
$I_{m,n}$ is the interference received at base m over PRB n

$$I_{m,n} = \sum_{j \in K \mid j \neq k} p_{j,n} \cdot h_{m,j}^2(n)$$ (4.28)

Energy efficiency in a specific subcarrier for user k that transmits over PRB n is computed as

$$e_{k,n} = \frac{r_{k,n}}{p_{k,n}},$$ (4.29)

where $r_{k,n}$ is the error-free transmission rate of user k over the subcarriers conforming PRB n. It is established that $r_{k,n} = 0$ if the SINR $\gamma_{m,n}$ corresponding to the user k with allocated PRB n in base m is below a minimum SINR threshold S_{min}.

Finally, a metric that is not defined in other references but gives important information about the performance of the resource allocation is the no-transmission probability, which is defined here as the probability that in a given TTI, the total transmitted power of a user k is equal to zero, that is,

$$\sum_{n \in N} p_{k,n} = 0.$$ (4.30)

This situation occurs in one of two conditions: (1) the user has no allocated subcarriers and (2) channel conditions are such that the user cannot achieve the minimum SINR threshold.

Once the system model has been defined, the mentioned schemes are presented in the context of this model.

4.4.2 Potential Game-Based Scheme

A potential game is guaranteed to converge to the Nash equilibrium point through the best (or better) response dynamic. Taking this into account, a noncooperative potential game is proposed in [34]. It is considered a strategy space in which each user chooses an equal number of subcarriers L as well as its transmit power aiming to maximize the system energy efficiency. The potential function is defined as the natural logarithm of the efficiencies of each user over each subcarrier:

$$V = \ln \left(\prod_{k=1}^{K} \prod_{\substack{n \in N \\ a_{m,n}=k}} \frac{e(\gamma_{m,n})}{p_{k,n}} \right), \tag{4.31}$$

where $e(\gamma_{m,n})$ is a modified efficiency function of the fixed quadrature phase shift keying (QPSK) modulation and is defined as

$$e(\gamma_{m,n}) = (e^{-\beta/\gamma_{m,n}})^{D_d}, \tag{4.32}$$

where $\beta = \gamma^*/M$ and γ^* is the optimum SINR derived in [46] and given by

$$\frac{f(\gamma^*)}{\gamma^*} = f'(\gamma)\big|_{\gamma=\gamma^*}, \tag{4.33}$$

where $f(\gamma)$ is the modulation efficiency function such that the error-free transmission rate is given by

$$r_{k,n} = BW \times f(\gamma). \tag{4.34}$$

In [34], it is shown that V is indeed the potential function for the case where the user's utility is given by

$$u_k(a_{m_k,n}, p_{k,n}) = -\beta D_d \sum_{\substack{n \in N \\ a_{m,n}=k}} \frac{\sigma^2 + \sum_{j=1, j \neq k}^{K} p_{j,n} h^2_{m_k,j}(n)}{p_{k,n} h^2_{m_k,k}(n)}$$

$$-\beta D_d \sum_{\substack{i=1, i \neq k}}^{K} \sum_{\substack{n \in N \\ a_{m_i,n}=i}} \left(\frac{p_{k,n} h^2_{m_i,k}(n)}{p_{i,n} h^2_{m_i,i}(n)} \right) - \sum_{\substack{n \in N \\ a_{m,n}=k}} \ln(p_{k,n}).$$

$$(4.35)$$

This indicates that the best response dynamic, in which every user updates its subcarriers (trying all possible combinations) and transmit power to maximize Equation 4.35 will converge to a Nash equilibrium point. However, in order to reduce the computational complexity of this scheme, which comes from the combinatorial search performed for each user, a suboptimal implementation in which users make combinations with their $N_1 < N$ best subcarriers.

Once this scheme has been described, some aspects related to the model presented in Section 4.4.1 have to be pointed out. As it was established before, all the PRBs have to be allocated; thus, the value of L becomes a function of the number of active users in each cell. The authors did not impose any constraint to prevent the frequency reuse within the cell, so the best dynamic response could lead to this situation. If so, the next rule can be applied: if more than one user is allocated a given subcarrier the base station takes a decision, allocating the subcarrier to the user that achieves the highest efficiency.

Due to the need for base station intervention and the form of the potential function, which require CSI from all users, this scheme can be considered as a centralized scheme. In addition, the allocation of PRBs instead of subcarriers and the minimum SINR threshold $S_{\min}$ have to be considered.

4.4.3 Repeated Game-Based Scheme

Due to the difficulty of optimizing the nonconvex function of the sum of energy efficiency, the authors in [32] claim that it is more feasible to maximize it in a time-averaged manner instead taking an averaging window of size w.

The proposed scheme in [32] considers that the base stations are the players that form $|N|$-independent noncooperative games, one game per subcarrier. In each game, at TTI t, the base stations will attempt to maximize the average efficiency in the corresponding subcarrier, allocating it to the user with the greatest ratio between optimal achievable rate $r_{m,n}^{k^*}$ and its average transmit power at $t-1$, $P_{k,n}(t-1)$. It is considered that users are capable of selecting its modulation scheme adaptively from a set Q so that the optimal achievable rate could be not feasible; thus, it is necessary to define an efficiency gap:

$$\Delta g_{m,n}^k(t) = \min_{q \in Q} \left| \tilde{r}_q - r_{m,n}^{k^*}(t) \right|, \qquad (4.36)$$

where $\tilde{r}_q$ is the rate corresponding to modulation scheme $q \in Q$. Hence, each subcarrier is assigned to the user that maximize

$$\eta_{m,n}^k(t) = \frac{\hat{r}_{m,n}^{k^*}(t) - \beta \Delta g_{m,n}^k(t)}{P_{k,n}(t-1)}, \qquad (4.37)$$

where β is a design parameter that takes values between 0 and 1. After this, each user adopts the modulation scheme that minimizes its efficiency gap and the transmit power required to achieve the SINR threshold corresponding to the adopted modulation scheme.

It is worth noting that in the proposed game, a change in the strategy of a base station might change the assigned user that usually results in different interfering power to other cells, so other cells will continue the best response dynamic until an equilibrium point is reached. However, the game does not guarantee to converge to an equilibrium point so it is necessary to limit the maximum iteration time to prevent infinite loops.

Some refinements are necessary for the implementation of this scheme considering the model proposed in Section 4.4.1. First, the modulation schemes and its block error rate (BLER) curves reported in [50] are adopted; those curves give the information required to compute the error-free transmission rate $r_{k,n}$. Second, this scheme does not consider the case where a user is not capable of achieving the minimum SINR threshold, so it can be established that in this case, the user sets the transmit power to zero. Finally, as we stated for the potential game scheme, we consider the allocation of PRBs instead of subcarriers.

4.4.4 PRISMAP

As can be seen, any of the presented schemes considers (or quantifies) the possibility that a user cannot transmit in a certain TTI. Thus, a novel resource allocation scheme is proposed. The PRISMAP allocation scheme is conceived to improve fairness and capacity while maintaining energy efficiency and minimizing no-transmission probability.

This problem can be decoupled into a PRB allocation problem at the base station level and a power control and modulation adaptation problem at the user level. In the following, both problems are analyzed individually, and later, the joint implementation of the proposed solutions is described.

4.4.4.1 PRB Allocation   As in some other schemes, the subcarrier allocation can be used to improve system fairness. In order to provide an acceptable performance to cell-edge users while allowing universal frequency reuse, we propose the concept of preferred reuse 3, which is a dynamic version of the partial isolation fractional reuse presented in [51] and can be related to prioritized power control [1]. The way of operation with preferred reuse 3 is as follows.

Available bandwidth is divided into three bands. Each base station is assigned one of the bands according to a reuse 3 pattern; nevertheless, every base station must assign all the available subcarriers among its active users; thus, every base station forms two subcarrier blocks: S_{m3} block, corresponding to the assigned reuse 3 band, and S_{m1} block, corresponding to the remaining available spectrum.

Then, for every user k the vector $F_k = \{n_1,...,n_{|N|}\}$ is created; this vector contains the complete set of PRBs ordered according to the user's preferences. These preferences will be a function of the distance of the user to the center of the cell, that is, the users in the center of the cell will prefer PRBs in block S_{m1} than PRBs in the block S_{m3} whereas cell-edge users will have a higher preference for PRBs in S_{m3}.

Based on the vectors $\{F_k\}$, each base station m orders its users according to their channel conditions as follows:

$$K_{index} = \left\{k_1,...,k_{K_m} \middle| \max\left(F_{k_{i+1}}\right) > \max\left(F_{k_i}\right)\right\}. \qquad (4.38)$$

Following the order established by K_{index}, the base station assigns to each user its most preferred available subcarrier denoted by n_a

and computes the user achievable rate r_k using Shannon's formula considering all the assigned PRBs and neglecting the interference from other cells. Then, if there are unassigned PRBs, the average achievable rate r_{avg} between users is calculated and the unassigned PRBs continue to be assigned to the users with an achievable rate under the average until $r_k > r_{avg}$ for all $k \in K_m$. While there are unassigned PRBs, r_{avg} is updated and the process is repeated. The complete procedure executed by each base station $m \in M$ is detailed in Algorithm 4.2.

Algorithm 4.2

1. *Initialization*
   a. Obtain $F_k \; \forall \, k \in K_m$ and K_{index}
   b. $r_k = 0 \; \forall \, k \in K_m$
   c. $r_{avg} = 0$
   d. $K_l = K_{index}$
   e. $a_{m,n} = 0 \; \forall \, n \in N$

2. *Assignment*
   a. For all $k \in K_l$, if there are PRBs s.t. $a_{m,n} = 0$:
      1. $n_a = \underset{n:\, a_{m,n}=0}{\arg\max}\, h_{m,k}^2(n)$
      2. $a_{m,n_a} = k$
      3. Update r_k including all assigned PRBs
      4. If $r_k > r_{avg}$, then $K_l = K_l - \{k\}$
      5. If $K_l \ne \{\varnothing\}$, go back to step 2.

3. *Update*
   a. If there are PRBs s.t. $a_{m,n} = 0$:
      1. Update r_{avg}
      2. $K_l = K_{index} - \{k | r_k > r_{avg}\}$
      3. Go back to step 2.
   b. Otherwise, go to step 4.

Output: $\{a_{m,n}\}$ Stop.

4.4.4.2 Power Control with Adaptive Modulation   Now, assuming that PRB allocation has taken place, it is possible to find the transmit power vectors $p_k \; \forall \, k \in K$, and the modulation schemes vectors

q_k $\forall\, k \in K$ through an energy-efficient game theoretic approach. It is assumed that users are capable of adapting its transmission rate over each PRB by choosing a modulation-codification scheme q from the set of available schemes Q. Let r_q be the effective transmission rate of scheme q, and let Q be a ordered set in such a way that $r_{q+1} > r_q$. In order to transmit using the scheme q, a user should satisfy the corresponding SINR threshold S_q.

To address this problem, it is convenient to analyze first the case of a user with a single allocated PRB. Next, based on the results for this simple case, the complete problem in which the user has multiple PRBs allocated is studied.

4.4.4.2.1 Simple Case: One Block per User   This case is focused on the power control of the user k in cell m that has been allocated the PRB n. Assuming a constant interference power level $I_{m,n}$ at base m, the achievable data rate for a specified BER using M-QAM modulation is approximated by [52]

$$r_{k,n} = BW \log_2(1 + c_3 \gamma_{m,n}), \tag{4.39}$$

where c_3 is a constant depending on the desired BER, which corresponds to the inverse of the SINR gap defined in [53]. Therefore, the energy efficiency defined in Equation 4.29 can be expressed as

$$e_{k,n} = \frac{r_{k,n}}{p_{k,n}} = \frac{BW \log_2\left(1 + c_3\left(\left(p_{k,n} b_{m,k}^2(n)\right)\big/\left(\sigma^2 + I_{m,n}\right)\right)\right)}{p_{k,n}}. \tag{4.40}$$

From this definition, we aim to find the transmit power that maximizes a utility function that should be related to energy efficiency and good put over the specific PRB. In various schemes presented in the literature [17,34,47,49] in which a single modulation scheme is considered, error-rate transmission rate is established as in Equation 4.34.

In [47], it is shown that under this approach, user utility function (energy efficiency) is an optimizable function in terms of the transmit power $p_{k,n}$. However, when adaptive modulation and codification is considered, effective transmission rate becomes a logarithmic function of the transmit power and energy efficiency behaves as a continuously decreasing function that cannot be optimized in terms of a nontrivial solution. This implies that the definition of the energy efficiency as the utility function is not convenient as the users could be willing to obtain a higher transmission rate even if they have to increase their transmit power, reducing its efficiency in a controlled way. In order to represent these preferences, that is, to apply context awareness [5], here a new utility function is proposed:

$$u_{k,n} = \begin{cases} 0, & p_{k,n} = 0, \\ \dfrac{(r_{k,n})^{1+\lambda_k}}{p_{k,n}}, & p_{k,n} \neq 0, \end{cases} \tag{4.41}$$

where $r_{k,n}$ is defined in Equation 4.9, whereas $0 < \lambda_k < 1$ is a parameter that allows us to model the users' desire to achieve a higher rate at the cost of increasing their transmit powers; in general, central users will have values of λ_k closer to 1 than cell-edge users. To simplify its analysis, the proposed utility function is rewritten in terms of the SINR $\gamma_{m,n}$ as

$$u_{k,n} = BW^{1+\lambda_k} \frac{h_{m,k}^2(n)}{\sigma^2 + I_{m,n}} \frac{(\log_2(1+c_3\gamma_{m,n}))^{1+\lambda_k}}{\gamma_{m,n}}, \quad p_{k,n} > 0. \tag{4.42}$$

Normalizing the term $BW^{1+\lambda_k}((h_{m,k}^2(n))/(\sigma^2 + I_{m,n}))$ the plot in Figure 4.3 is obtained, where we can observe that this function is continuous and has a maximum for a value of SINR $\gamma_{m,n} > 0$. These observations are formally stated in the following proposition.

Proposition 4.1: The utility function defined in Equation 4.41 has a unique maximum point, and this maximum point is located at the SINR value γ^* that satisfies

$$(1+\lambda_k)\frac{c_3\gamma^*}{1+c_3\gamma^*} = \ln\left(1+c_3\gamma^*\right), \quad \gamma^* \geq 0. \tag{4.43}$$

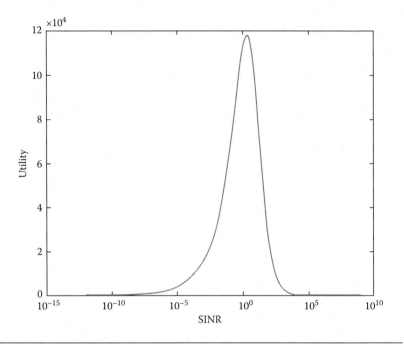

Figure 4.3   Proposed utility function.

The complete proof is omitted for the sake of brevity; however, it can be directly derived obtaining $(du/d\gamma)|_{\gamma=\gamma^*} = 0$, then showing that γ^* is unique when $\lambda_k > 0$ and that this value is a maximum through the second derivative criterion $d^2u/d\gamma^2|_{\gamma=\gamma^*} < 0$.

According to this proposition, the optimum SINR value $\gamma^*_{m,n}$ that maximizes $u_{k,n}$ can be found by evaluating Equation 4.43. Thus, the best response of a user k in base m over PRB n is given by

$$p^*_{k,n} = \min\left(\frac{\gamma^*_{m,n}\left(\sigma^2 + I_{m,n}\right)}{h^2_{m,k}(n)}, \frac{P_{\max}}{N_b}\right). \tag{4.44}$$

Using this result, the single block power control with continuous rate adaptation noncooperative game for PRB n is defined as $G^n = [K^n, \{P_{k,n}\}, \{u_{k,n}\}]$, where $K_n = \{k \in K \mid a_{m_k,n} = k\}$ is the set of players (the set of users to which PRB n has been assigned); $P_{k,n} = [0, P_{\max}]$ is the set of strategies of player k (set of feasible transmit powers); and $u_{k,n}$ defined in Equation 4.41 is the utility function of player k over PRB n. Considering this game, the following result is derived. ∎

Theorem 4.1: **The single block power control noncooperative game with continuous rate adaptation G^n admits a unique Nash equilibrium point (in pure strategies) that is achieved through the best response dynamics.**

The outline of the proof is explained in the following. According to Milgrom and Roberts' theorem [54], a supermodular game has at least one Nash equilibrium point and, if this equilibrium is unique, the best response dynamic converges to this equilibrium point. Thus, it suffices to show that the game G^n is a supermodular game and that the Nash equilibrium is unique. Supermodularity is demonstrated using the fact that the strategy spaces $P_{k,n}$ are compact subsets of real numbers and that the utility function is continuous and has no decreasing differences in $\{P_{k,n}, P_{-k,n}\}$. Then, knowing that the best response function in Equation 4.44 has only one value, the uniqueness of the equilibrium point is demonstrated.

It is important to notice that under the assumptions made, the users transmitting over the same PRB can reach an equilibrium point by simply updating their transmit power according to Equation 4.44.

This result has been obtained assuming that users can select their transmission rate from a continuous set; however, the proposed system model has a finite number of modulation and codification schemes, hence, applying a similar approach to that presented in [32], the optimum SINR value $\gamma^*_{m,n}$ is quantized to choose the scheme $\hat{q}^*_{k,n}$ with the closest SINR threshold to $\gamma^*_{m,n}$:

$$\hat{q}^*_{k,n} = \arg\min_{q \in Q} \left| S_q - \gamma^*_{m,n} \right|. \tag{4.45}$$

Then, the user update its transmit power to achieve the corresponding SINR threshold $S_{\hat{q}^*_{k,n}}$ yielding

$$\hat{p}^*_{k,n} = \frac{S_{\hat{q}^*_{k,n}} (\sigma^2 + I_{m,n})}{h^2_{m,k}(n)}, \tag{4.46}$$

then, considering the power restriction in Equation 4.26, the transmit power of user k over subcarrier n is given by

$$p_{k,n} = \min\left(p_{k,n}^*, \frac{P_{max}}{N_b} \right). \tag{4.47}$$

As the quantization of the optimum SINR only implies a change in λ_k, the properties of existence, uniqueness, and convergence to the equilibrium still hold, that is, the best response dynamics between the users converge to the Nash equilibrium point even if the set of modulation schemes is finite. ∎

4.4.4.2.2 Multiple Blocks per User In a practical scenario, each user is allocated more than one PRB. On the other hand, from Equations 4.43 and 4.45 it is clear that the value of $\hat{q}_{k,n}^*$ is independent of the PRB n as it only depends on λ_k. Hence, it is possible to drop the PRB subindex, and it follows that each user k will try to achieve the same SINR $S_{\hat{q}_k}$ in all its assigned subcarriers to maximize its utility, given in this case by the sum of the energy efficiency over the available PRBs:

$$u_k = \sum_{n \in N} u_{k,n}, \tag{4.48}$$

where $u_{k,n}$ as in the previous case is defined by Equation 4.41. Though, due to the power constraint in Equation 4.26, a user may not be able to achieve the desired SINR $S_{\hat{q}_k}$ in all its allocated PRBs.

Thus, a method to distribute the available power is required. Based on the successive bit allocation method described in Section 4.2, a similar algorithm is proposed. At each iteration, the user k selects a PRB among the set of PRBs that have not reached the desired SINR $S_{\hat{q}_k}$ to achieve the SINR threshold corresponding to the next modulation scheme in Q. The user k selects the PRB with the highest utility gap, which is given by

$$\Delta u_{k,n} = \frac{\left(r_{q_n+1} \right)^{1+\lambda_k}}{p_{k,n}(q_n+1)} - \frac{\left(r_{q_n} \right)^{1+\lambda_k}}{p_{k,n}(q_n)}. \tag{4.49}$$

This operation is repeated until the desired SINR is reached over all the assigned PRBs or the transmit power of the user is equal to P_{max}. The complete algorithm to allocate the transmit power over the assigned PRBs is described next.

Algorithm 4.3

1. *Initialization*
   a. $P_{un} = P_{max}/N_b$
   b. $p_{k,n} = 0 \ \forall \, n \in N$
   c. $q_{k,n} = 0 \ \forall \, n \in N$
   d. $N_{av} = \{n \, | \, a_{m,n} = k\}$
2. *If* $N_{av} \neq \{\varnothing\}$ *and* $P_{un} > 0$
   a. For all $n \in N$, obtain

$$p_{req}(n) = p_{k,n}(q_{k,n} + 1) - p_{k,n} \quad \text{and} \quad \Delta u_{k,n}$$

   b. Select n^* s.t. $\Delta u_{k,n^*} = \max_l \Delta u_{k,l}$
   c. If $P_{un} \geq p_{req}(n^*)$:
      1. $p_{k,n^*} = p_{k,n^*} + p_{req}(n^*)$
      2. $P_{un} = P_{un} - p_{req}(n^*)$
      3. $q_{k,n} = q_{k,n} + 1$
      4. If $q_{k,n} = q_{k,n}^*$, then $N_{av} = N_{av} - \{n^*\}$
         Otherwise, $P_{un} = 0$
   d. Go back to step 2
3. *Output*: $\{p_{k,n}\}, \{q_{k,n}\}$ Stop.

From this algorithm, another important result is derived.

Proposition 4.2: Once that PRB allocation has taken place, Algorithm 4.3 constitutes the best response of a user whose utility function is given by Equation 4.48.

The proof of Proposition 4.2 is outlined here. It can be deduced that to maximize the sum in u_k, the user should adjust its power in every subcarrier to achieve the corresponding SINR threshold $S_{q_k^*}$ in all subcarriers that is equivalent to set a rate constraint $R_{tar} = \sum_{n | a_{m,n} = k} \frac{r_{q_k^*}^*}{q_k^*} \cdot$ Considering that the required power to achieve modulation scheme $\hat{q}_k^* + 1$ tends to an infinite value, the problem adopts a well-known form, of which the optimal solution is the successive bit allocation modifying the allocation criterion; that is, instead of assigning a bit to the subcarrier that requires the least additional power, it is assigned to the subcarrier that has a greater increase in energy efficiency as indicated in Algorithm 4.3.

Based on Proposition 4.2, the multiple subcarrier game for power control and modulation adaptation is directly formulated as $G = [K,\{P_k\},\{u_k\}]$, where K is the set of players (set of active users in the system); $P_k = \{p_{k,n}, \forall\ k \in K\ \forall\ n \in N\}$ subject to Equations 4.25 and 4.26 is the set of strategies of player k (set of possible transmit power vectors); and u_k defined in Equation 4.48 is the utility function of player k. Within this context, it seems reasonable to follow a best response dynamic, given by Algorithm 4.3 to establish the transmit powers of all active users in the system. Although game G cannot be decoupled in $|N|$-independent subgames, it can be demonstrated [34] that it inherits the existence and convergence to equilibrium properties of the single-block game G^n defined in the previous section, given the nondecreasing differences property of the utility function $u_{k,n}$. ∎

4.4.4.2.3 Implementation   The fair and energy-efficient resource allocation problem has been addressed next decoupling PRB and power/modulation allocation. In this section, it is described the implementation of the novel proposed joint allocation method, called PRISMAP.

The direct implementation of a joint allocation method would be as follows: Before each TTI, base stations perform the PRB allocation executing Algorithm 4.2; later, during the signaling symbols D_s, the users interact through the best response dynamic established in Algorithm 4.3, trying to approximate to the equilibrium point; these operations are repeated each TTI.

However, a practical implementation requires some refinements related to the system efficiency since some users may experiment deep fading conditions, making it not possible for them to reach even the minimum SINR threshold over any PRB. In this case, it would be better not to allocate PRBs to these users or reallocate the PRBs in deep fading conditions to other users with better channel conditions. Therefore, in the stage of PRB allocation the set of active users K_m is restricted to the set of users that satisfy

$$d_v(k) \leq d_{vm}, \tag{4.50}$$

where d_v denotes the *virtual distance*, a novel concept defined as the distance at which the user would be located if the channel coefficients were affected only by path loss, that is, it is a way to estimate the distance of the user to the base station based on its channel coefficients. It is computed as

$$d_v(k) = L^{-1} \left\{ \max_n h_{m_k,k}^2(n) \right\}, \qquad (4.51)$$

where $L^{-1}\{\cdot\}$ corresponds to the inverse function of the path loss. The term d_{vm} represents the maximum virtual distance a user may have in order to allocate subcarriers to it; that is, if a user k has a virtual distance $d_v(k)$ higher than d_{vm}, the base station will not allocate PRBs to that users in that given TTI.

Furthermore, as some users may not transmit in an allocated PRB during the power control stage, a mechanism of reallocation is considered. When this occurs, the base station will reassign that given PRB to the user k' with the highest channel gain obtained as

$$k' = \arg\max_{j \in K_m} \left(h_{m_k,j}^2(n) \right). \qquad (4.52)$$

The final implementation of the proposed PRISMAP scheme includes the modifications explained earlier, and its flow chart is depicted in Figure 4.4. As can be seen, this scheme is suitable for implementation in practical 4G cellular systems.

4.4.5 Numerical Evaluation

In this section, the presented schemes for the case of study (energy efficiency) are compared through numerical simulations. First, the simulation environment is described and then the results are reported and commented upon.

4.4.5.1 Simulation Environment In order to obtain results from a realistic scenario, the simulation environment adopts recommendations in [55]. A 19-cell grid, that is, a center cell and the first two tiers of surrounding cells with a cell radius $R = 1000$ m, is considered. To mitigate the border effect, the wraparound technique [56] is applied.

The number of PRBs available in the system is $|N| = 50$, with $N_b = 12$ subcarriers and a subcarrier spacing $BW = 15$ kHz. Location

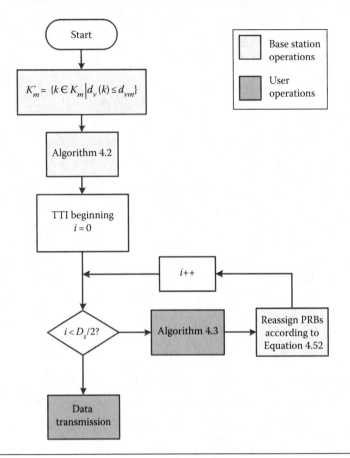

Figure 4.4   Operation of the proposed PRISMAP scheme.

of active users follows a uniform distribution restricted to distances larger than 35 m from the base station. The center area of a cell is equal to $2A_{cell}/3$ where A_{cell} is the total cell area.

AMC schemes and their efficiency are given by the BLER curves reported in [50]. Maximum transmission power per user P_{max} is set to 21 dBm and noise power spectral density at base stations is $N_0 = -170$ dBm/Hz. Channel coefficients are computed as

$$h_{m,k}^2(n) = \mu_{m,k} \hat{h}_{m,k}^2(n), \tag{4.53}$$

where

$\hat{h}_{m,k}^2(n)$ models the fast fading and is computed from ITU pedestrian-A channel profile [57]

$\mu_{m,k}$ is a frequency-independent factor that includes shadowing and path-loss effects:

$$\mu_{m,k} = 10^{((-L_{m,k}+X_\sigma)/10)}, \qquad (4.54)$$

where

X_σ is a Gaussian random variable with standard deviation of 10 dB
$L_{m,k}$ represents the path loss that is computed as

$$L_{m,k} = 128.1 + 37.6\log_{10}\left(\frac{d_{m,k}}{d_0}\right), \qquad (4.55)$$

where

$d_{m,k}$ is the distance between user k and base m
d_0 is the reference distance equal to 1 km

We consider a TTI of 1 ms, having $D_s = 4$ signaling symbols and $D_d = 10$ information symbols; hence, the effective data transmission rate $\hat{r}_{k,n}$ are given by

$$\hat{r}_{k,n} = \frac{N_b}{1.4} r_{k,n}. \qquad (4.56)$$

The parameters concerning each compared scheme are detailed in the following. According to the results reported in [34], the value of the threshold λ in the potential game-based scheme of Section 4.4.2 is set to –20 dB. For the repeated game-based scheme of Section 4.4.3, the averaging window size is set as $w = 10$ and the parameter as $\beta = 0.7$. QPSK is considered as the modulation for both schemes, whereas the minimum SINR S_{min} was set to –1.43 dB.

In the case of PRISMAP, virtual distance d_v is computed using Equation 4.51 and the maximum virtual distance d_{vm} is set to 1250 m. As stated before, the value of λ_k captures the preferences of each user according to its distance to the base; thus, central users will have higher values of λ_k than cell-edge users. To establish this value, the first and the last values are fixed, in particular $\lambda_k = 0.9$ for users with a virtual distance $35 < d_v(k) \le 250$ m and $\lambda_k = 0.2$ for users with a virtual distance $750 < d_v(k) \le 1000$ m; then through an interpolation process with respect to the path loss, the rest of the values can be computed, laying $\lambda_k = 0.3383$ and 0.5403 for users with virtual distances $500 < d_v(k) \leq 750$ m and $250 < d_v(k) \leq 500$ m, respectively.

4.4.5.2 Results   Having the simulation environment setup, it is possible to obtain various metrics considering overall cell performance or individual user average performance as a function of the distance to the base station. Here, the second approach is taken as it leads to a deep understanding of the operation of the different schemes as well as evaluate qualitatively their fairness, comparing cell-edge users and center users' performance. Figures 4.5 through 4.9 present these results when considering a number of users per cell $|K_m| = 20$.

Figure 4.5 shows the goodput per user, where it can be observed that PRISMAP has the highest goodput for all the distances. Furthermore, it is demonstrated that PRISMAP improves system fairness as the difference between center users with respect to cell-edge users is the smallest among the compared schemes. This is due to both the fair PRB allocation of PRISMAP and the control of the interference levels for users at cell edge.

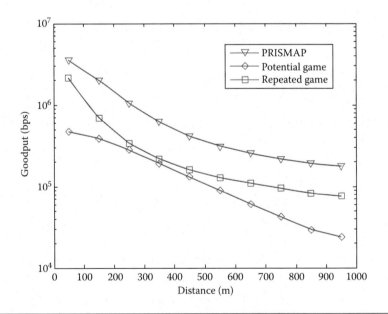

Figure 4.5   Average goodput of a user as a function of its distance to BS.

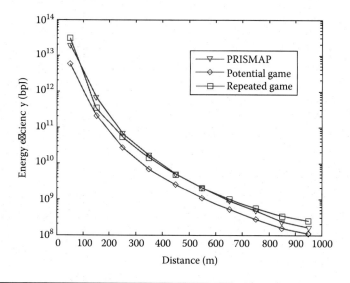

Figure 4.6   Average energy efficiency of a user as a function of its distance to BS.

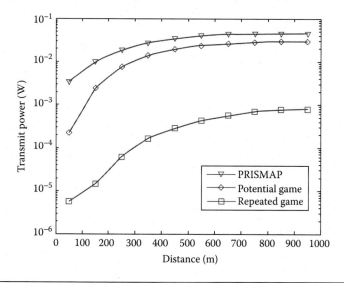

Figure 4.7   Average transmitted power of a user as a function of its distance to BS.

In Figure 4.6, the energy efficiency per user is shown. As can be seen, in the region between 50 and 150 m, repeated game-based scheme achieves the highest energy efficiency, whereas from 250 to 450 m PRISMAP has a better performance; and for distances above 850 m, repeated game's efficiency is slightly higher. Nevertheless, PRISMAP

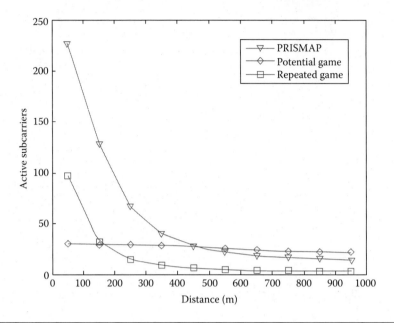

Figure 4.8 Average number of active subcarriers of a user as a function of its distance to BS.

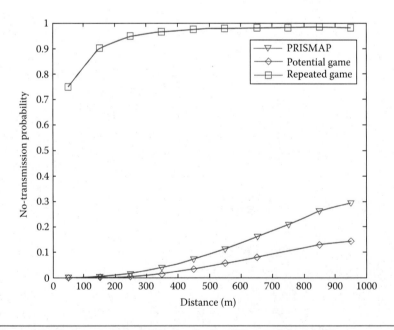

Figure 4.9 No-transmission probability as a function of the user distance to BS.

is always close to highest efficiency. Additionally, we can note that in general, energy efficiency decreases with the distance.

In Figure 4.7, transmit power per user is plotted. The lowest transmit power corresponds to a repeated game, whereas PRISMAP always transmits with the highest power. Since PRISMAP has the highest goodput and adequate energy efficiency, it is considered that it establishes an acceptable trade-off between goodput, energy efficiency, and transmitted power. Besides, a potential game has a transmit power close to that of PRISMAP. We claim that in comparison, with a slight increase in the transmit power, PRISMAP achieves a significantly higher goodput than the potential game.

Figure 4.8 presents the number of active subcarriers per user. From this plot, it can be observed that in PRISMAP as well as in the repeated game, users near the base station have much more active subcarriers. However, PRISMAP keeps a number of active subcarriers for the farther users close to that of the potential game that allocates the same number of PRBs for all users; though, with the increase on the distance, some users are not capable of transmitting in all the allocated PRBs. This is a fundamental difference from PRISMAP in that despite it allocating less PRBs to farther users, it has acceptable channel conditions because of the preferred reuse.

In Figure 4.9, the no-transmission probability is shown as a function of the user distance. Notice that the repeated game has a high no-transmission probability reaching levels above 0.9 for distances larger than 700 m. This phenomenon is directly related to the way that PRBs are allocated as farther users are allocated with few PRBs and their interference conditions are not controlled, which implies that these users cannot achieve the minimum SINR. Besides, PRISMAP and the potential game have similar performances: they keep the no-transmission probability below 0.3. Comparing them, scheme 3 has a no-transmission probability up to 15% lower. As stated before, this difference is due to the even allocation performed by the potential game and that its operation is more likely centralized. Nevertheless, PRISMAP achieves a higher throughput and energy efficiency than this scheme.

The comparison between PRISMAP and the selected schemes shows that in an overall perspective, PRISMAP outperforms those schemes since it achieves a no-transmission probability much lower than the first schemes, its energy efficiency is near the highest among those schemes, and it has the highest goodput in all the simulated cases. Derived from this, some final remarks can be pointed out:

- Preferred reuse, based on the fractional frequency reuse improves system fairness not only with regard to the number of allocated PRBs but to the efficient mitigation of the interference at cell edge.
- The proposed utility function models in a more accurate way users' preferences, making it possible for the center users to achieve higher rates and for the cell-edge users to maintain transmission of information.
- The formulation of power control as a supermodular game guarantees equilibrium convergence through best response dynamics that can be performed by users during the signaling symbols in a TTI.
- Successive bit allocation algorithm allows users to optimize power control and to include in an explicit way the modulation adaptation according to channel conditions.
- PRISMAP is a robust scheme that can be implemented in practical scenarios and executed in a distributed fashion.

4.5 Conclusions

OFDMA is a multiple access technique that can be used to exploit frequency and multiuser diversity. Furthermore, it is robust to channel effects and achieves a high spectral efficiency. As a consequence, OFDMA cellular networks have been widely studied because of their potential to achieve 4G and beyond objectives.

In order to take the maximum advantage of OFDMA features, requirements include an adequate and dynamic allocation of the available resources: subcarriers, rate, and transmit power. Dynamic resource allocation is directly affected by the interdependence between users and base stations, and the variability in interference levels. While there are optimal solutions to the power control alone as the water-filling method and successive bit allocation, this does not apply to subcarrier allocation problem due to the NP-hard nature of this problem.

Under these circumstances, game theory is presented as a useful tool for the analysis and solution of the dynamic resource allocation problem in a distributed way given that this theory studies the interaction among independent agents that influence each other. Using game theory, it is possible to address the problem with a different approach applying well-known theorems and results of game theory to find a solution.

In the literature, several game theoretic resource allocation schemes have been proposed, having a special interest in those dedicated to maximize energy efficiency. Two of these schemes were studied and compared to a novel proposed scheme called PRISMAP. This novel scheme introduces the concept of preferred reuse and proposes a new utility function considering user preference according to the distance to the base station. It makes use of supermodular game theory and the successive bit algorithm.

The comparison performed under a common system model and a realistic simulation environment shows that the mechanisms implemented in PRISMAP make it outperform other schemes as they achieve an acceptable no-transmission probability while maintaining energy efficiency and improving data rates.

4.6 Future Research Directions

Taking into account the evolution of wireless networks toward 5G, we envision the following research directions for game theory in wireless networks that remain as open issues:

- Femtocell resource allocation
- QoS-aware resource allocation
- Cooperative transmission in relay networks
- HetNets deployment and planning
- Algorithms considering incomplete information (Bayesian games)

List of Abbreviations

AMC	Adaptive modulation and coding
AWGN	Additive white Gaussian noise
BER	Bit error rate
BLER	Block error rate

CDMA	Code division multiple access
D2D	Device-to-device
HetNet	Heterogeneous network
ITU-R	International Telecommunications Union Radiocommunication Sector
KKT	Karusch–Kuhn–Tucker
LTE	Long-term evolution
MIMO	Multiple-input-multiple-output
NBS	Nash bargaining solution
OFDM	Orthogonal frequency division multiplexing
OFDMA	Orthogonal frequency division multiple access
PAPR	Peak to average ratio
PRB	Physical resource block
QAM	Quadrature amplitude modulation
QoE	Quality of experience
QoS	Quality of service
QPSK	Quadrature phase shift keying
RS	Relay station
SC-FDMA	Single-carrier frequency division multiplexing
SINR	Signal to interference plus noise ratio
SNR	Signal-to-noise ratio
TTI	Transmission time interval

References

1. E. Hossain, M. Rasti, H. Tabassum, and A. Abdelnasser, Evolution toward 5G multi-tier cellular wireless networks: An interference management perspective, *IEEE Wireless Communications*, 21(3), 118–127, 2014.
2. ITU-R M.1645, Framework and overall objectives of the future development of IMT-2000 and systems beyond IMT-2000, 2010.
3. P. Mogensen et al., 5G small cell optimized radio design, in: *GLOBECOM*, Atlanta, GA, 2013.
4. F. Boccardi, R. W. Heath, A. Lozano, T. Marzetta, and P. Popovski, Five disruptive technology directions for 5G, *IEEE Communications Magazine*, 52(2), 74–80, 2014.
5. B. Bangerter, S. Talwar, R. Arefi, and K. Stewart, Networks and devices for the 5G era, *IEEE Communications Magazine*, 52(2), 90–96, 2014.
6. N. Bhushan et al., Network densification: The dominant theme for wireless evolution into 5G, *IEEE Communications Magazine*, 52(2), 82–89, 2014.

7. C.-X. Wang et al., Cellular architecture and key technologies for 5G wireless communication networks, *IEEE Communications Magazine*, 52(2), 122–130, 2014.

8. G. Fettweis and S. Alamouti, 5G: Personal mobile internet beyond what cellular did to telephony, *IEEE Communications Magazine*, 52(2), 140–145, 2014.

9. S. Pietrzyk, *OFDMA for Broadband Wireless Access*. Boston/London: Artech House, 2006.

10. T. Hwang, C. Yang, G. Wu, S. Li, and G. Y. Li, OFDM and its wireless applications: A survey, *IEEE Transactions on Vehicular Technology*, 58(4), 1673–1694, 2009.

11. Z. Han, D. Niyato, W. Saad, T. Basar, and A. Hjorungnes, *Game Theory in Wireless and Communication Networks*. Cambridge, UK: Cambridge University Press, 2012.

12. R. Q. Hu and Y. Qian, An energy efficient and spectrum efficient wireless heterogeneous network framework for 5G systems, *IEEE Communications Magazine*, 52(5), 94–101, 2014.

13. M. Wooldridge, Does game theory work? *IEEE Intelligent Systems*, 27(6), 76–80, 2012.

14. M. J. Osborne and A. Rubinstein, *A Course in Game Theory*. MIT Press, Cambridge, MA, 1994.

15. L. A. DaSilva, H. Bogucka, and A. B. MacKenzie, Game theory in wireless networks, *IEEE Communications Magazine*, 49(8), 110–111, 2011.

16. D. Fudenberg and J. Tirole, *Game Theory*. MIT Press, Cambridge, MA, 1991.

17. A. B. MacKenzie and S. B. Wicker, Game Theory in communications: Motivation, explanation, and application to power control, in: *IEEE Global Telecommunications Conference*, San Antonio, TX, 2001.

18. G. Bacci and M. Luise, A game-theoretic perspective on code synchronization for CDMA wireless systems, *IEEE Journal on Selected Areas in Communications*, 30(1), 107–118, 2012.

19. G. Bacci, L. Sanguinetti, M. Luise, and V. Poor, Energy-efficient contention-based synchronization in OFDMA systems with discrete powers and limited feedback, in: *WCNC*, Shangai, China, 2013.

20. R. Jayaparvathy and S. Geetha, Resource allocation and game theoretic scheduling with dynamic weight assignment in IEEE 802.16 fixed broadband wireless access systems, in: *SPECTS*, Edinburgh, UK, 2008.

21. J. Ae Han and W. Sook Jeon, Game-theoretic approach to distributed scheduling for relay-aided OFDMA systems, in: *VTC Spring*, Yokohama, Japan, 2011.

22. A. Ibing and H. Boche, Fairness vs. efficiency: Comparison of game theoretic criteria for OFDMA scheduling, in: *ACSSC*, Pacific Grove, CA, 2007.

23. Y. Yu, E. Dutkiewicz, X. Huang, and M. Mueck, Downlink resource allocation for next generation wireless networks with inter-cell interference, *IEEE Transactions on Wireless Communications*, 12(4), 1783–1793, 2013.

24. S. V. Hanly, L. L. H. Andrew, and T. Thanabalasingham, Dynamic allocation of subcarriers and transmit powers in an OFDMA cellular network, *IEEE Transactions on Information Theory*, 55(12), 5445–5462, 2009.

25. M.-L. Tham, C.-O. Chow, K. Utsu, and H. Ishii, BER-driven resource allocation in OFDMA systems, in: *IEEE International Symposium on Personal, Indoor and Mobile Radio Communications*, London, UK, 2013.

26. C. Y. Wong, R. S. Cheng, K. B. Letaief, and R. D. Murch, Multiuser OFDM with adaptive subcarrier, bit, and power allocation, *IEEE Journal on Selected Areas in Communications*, 17(10), 1747–1758, 1999.

27. Y.-F. Liu and Y.-H. Dai, On the complexity of joint subcarrier and power allocation for multi-user OFDMA systems, *IEEE Transactions on Signal Processing*, 62(3), 583–596, 2014.

28. I. Chih-Lin et al., Toward green and soft: A 5G perspective, *IEEE Communications Magazine*, 52(2), 66–73, 2014.

29. A. F. Al Rawi, S. Aissa, and C. C. Tsimendis, Game theoretic framework for future generation networks modelling and optimization, *IEEE Transactions on Wireless Communications*, 13(3), 1153–1163, 2014.

30. Z. Chang, T. Ristaniemi, and N. Zhisheng, Radio resource allocation for collaborative OFDMA relay networks with imperfect channel state information, *IEEE Transactions on Wireless Communications*, 13(5), 2824–2835, 2014.

31. H. Liu, Z. Wei, H. Zhang, Z. Zhicai, and W. Xiangming, An iterative two-step algorithm for energy efficient resource allocation in multi-cell OFDMA networks, in: *IEEE Wireless Communications and Networking Conference*, Shangai, China, 2013.

32. C. Yuan-Ho and H. Ching-Yao, Non-cooperative multi-cell resource allocation and modulation adaptation for maximizing energy efficiency in uplink OFDMA cellular networks, *IEEE Wireless Communications Letters*, 1(5), 420–423, 2012.

33. J. Zheng et al., Optimal power allocation and user scheduling in multi-cell networks: Base station cooperation using a game-theoretic approach, *IEEE Transactions on Wireless Communications*, 12(13), 6928–6942, 2014.

34. S. Buzzi, G. Colavolpe, D. Saturnino, and A. Zappone, Potential games for energy-efficient power control and subcarrier allocation in uplink multicell OFDMA systems, *IEEE Journal on Selected Topics in Signal Processing*, 6(2), 89–103, 2012.

35. T. M. Cover and J. Thomas, *Elements of Information Theory*. New York: John Wiley & Sons, 1991.

36. W. Yu, W. Rhee, S. Boyd, and J. M. Cioffi, Iterative water-filling for Gaussian vector multiple-access channels, *IEEE Transactions on Information Theory*, 50(1), 145–152, 2004.

37. B. Fox, Discrete optimization via marginal analysis, *Management Science*, 13(3), 210–216, 1966.

38. J. Camppello, Optimal discrete bitloading for multicarrier systems, in: *International Symposium in Information Theory*, Cambridge, MA, 1998.

39. J. B. Rao and A. Fapojuwo, A survey of energy efficient resource management techniques for multicell cellular networks, *Communications Surveys and Tutorials*, 16(1), 154–180, 2014.

40. Z. Han, Z. Ji, and K. J. Ray-Liu, Non cooperative resource competition game by virtual referee in multicell OFDMA networks, *IEEE Journal on Selected Areas in Communications*, 25(6), 1079–1090, 2007.

41. Q. Duy La, Y. H. Chew, and B.-H. Soong, Subcarrier assignment in multi-cell OFDMA systems via interference minimization game, in: *IEEE Wireless Communications and Networking Conference*, Shangai, China, 2012.

42. C. Y. Ng and C. W. Sung, Low complexity subcarrier and power allocation for utility maximization in uplink OFDMA systems, *IEEE Transactions on Wireless Communications*, 7(5), 1667–1675, 2008.

43. Z. Hou, D. Wu, and Y. Cai, Subcarrier and power allocation in uplink multicell OFDMA systems based on game theory, in: *International Conference on Communication Technology*, Nanjing, China, 2010.

44. Z. Han and K. J. R. L. Z. Ji, Fair multiuser channel allocation for OFDMA networks using Nash bargaining solutions and coalitions, *IEEE Transactions on Communications*, 53(8), 1366–1376, 2006.

45. S. H. Kuhn, The Hungarian method for the assignment problem, *Naval Research Logistic Quarterly*, 2, 83–97, 1955.

46. D. Goodman and N. Mandayam, Power control for wireless data, *IEEE Personal Communications*, 7(2), 48–54, 2000.

47. C. U. Saraydar, N. B. Mandayam, and D. J. Goodman, Pricing and power control in a multicell wireless data network, *IEEE Journal on Selected Areas in Communications*, 19(10), 1883–1892, 2001.

48. S. M. Alavi, C. Zhou, and W. W. Gen, Efficient and fair resource allocation scheme for OFDMA networks based on auction game, in: *IEEE Vehicular Technology Conference*, Quebec City, Quebec, Canada, 2012.

49. G. Bacci, A. Bulzomato, and M. Luise, Uplink power control and subcarrier assignment for an OFDMA multicellular network based on game theory, in: *VALUETOOLS*, Brussels, Belgium, 2011.

50. 3GPP TR 25.892, Feasibility study for OFDM for UTRAN enhancement (release 6), 2004.

51. S.-E. Elayoubi, O. B. Haddada, and B. Fourestié, Performance evaluation of frequency planning schemes in OFDMA-based networks, *IEEE Transactions on Wireless Communications*, 7(5), 1623–1633, 2008.

52. X. Qiu and K. Chawla, On the performance of adaptive modulation in cellular systems, *IEEE Transactions on Communications*, 47(6), 884–895, 1999.

53. J.M. Cioffi, A multicarrier primer, ANSI T1E1.4/91–157, Committee Contribution, Technical Report, 1991.

54. P. Milgrom and J. Roberts, Rationalizability, learning and equilibrium in games with strategic complementarities, *Econometrica*, 58, 1255–1277, 1990.

55. 3GPP TR 25.814, Physical layer aspects for evolved UTRA (release 7), 2006.

56. Y.-B. Lin and V. W. Mak, Eliminating the boundary effect of a large-scale personal communication service network simulation, *ACM Transactions on Modeling and Computer Simulation*, 4(2), 165–190, 1994.

57. ITU-R M.1225, Guidelines for evaluation of radio transmission technologies for IMT-2000, 1997.

5

MILLIMETER-WAVE TECHNOLOGY FOR 5G NETWORKS

NARGES NOORI

Contents

5.1 Introduction

The electromagnetic spectrum contains a range of possible radiation frequencies. Although commercial wireless systems cover bands from a few MHz up to more than 100 GHz [1], most of the existing mobile communication systems operate at a frequency range of 300 MHz to 3 GHz with a few MHz bandwidth. However, users ranging from enterprise data centers to smart phone users require wider bandwidths. The huge demand for more spectrum bands as well as bandwidth scarcity has encouraged exploration of the millimeter-wave (mmWave) spectrum for future fifth-generation (5G) broadband mobile communications. This spectrum would allow significant expansion of the channel bandwidth and increase in data capacity [2]. The availability of a bandwidth, which is about several times that of the present cellular networks, offers an opportunity for developing multigigabit broadband wireless networks.

5.2 Millimeter-Wave Technology Issues and Applications

A millimeter wave represents frequencies between 30 and 300 GHz or a wavelength from 1 to 10 mm. But, in the context of communication networking and equipment, this term mostly refers to a few bands of radio frequencies around 38, 60, and 94 GHz. Recently, a frequency spectrum of 70–90 GHz, previously assigned for wireless communication in the public domain, was also considered. With about 5 GHz of spectrum available in each of these bands, the total spectral bandwidth available exceeds that of all of the allocated bands in the microwave frequency.

The 60 GHz frequency band is applicable for data and voice communications as well as broadband Internet access. This frequency offers key benefits, such as highly secure unlicensed interference-free operation, a high level of frequency reuse (due to short transmission distances as a result of oxygen absorption), mature technology, and high availability (up to 99.999%) [3]. Compared to microwave frequencies, millimeter waves have higher free space loss and atmospheric attenuation. They are obstructed by building walls and experience foliage attenuation. Additionally, rain fade and absorption due to humidity pose serious problems. However, this allows for smaller frequency reuse distances in mmWave frequencies. The narrower beams of millimeter-wave antennas permit utilization of several independent links in the vicinity. These characteristics of millimeter waves make them good candidates for personal area network applications in densely populated areas.

Although the mmWave spectrum was principally applied in military and radio astronomy [4–6], it is currently used in a wider range of applications. mmWave radars can be used for short-range fire control. Furthermore, satellite-based remote sensing around 60 GHz can estimate temperature in the upper atmosphere. This band is used for atmospheric monitoring and climate sensing applications as well. mmWave frequencies also have a wide range of applications in wireless communication. Different frequency bands in the mmWave spectrum are applicable for cellular backhauling, intervehicle communication, connection redundancy and failure recovery, campus and enterprise facility networks, licensed high-speed microwave data links, unlicensed short range data communication, and point-to-point (P2P) as well as point-to-multipoint (P2MP) transmissions [7,8]. The mmWave imaging technology is applied in security systems and medicine. Moreover, mmWave therapy is a new and emerging form of treatment in health care [9].

5.3 Millimeter-Wave Regulation and Standardization

5.3.1 Millimeter-Wave Regulation

Regulatory authorities in various countries are responsible for the management, coordination, and licensing of the national radio frequency spectrum. In the United States, the 59–64 GHz band (known as 60 GHz band or the V-band) is governed by the Federal Communication Commission (FCC) for unlicensed operations [7]. Furthermore, in many countries worldwide, at least 5 GHz of continuous spectrum can be used at this band [10]. The significant attenuation of 60 GHz signals due to oxygen absorption makes this band suited for short-range P2P and P2MP applications. The 92–95 GHz band (known as 94 GHz band or the W-band) is similarly assigned by the FCC for indoor unlicensed operations and outdoor licensed P2P applications. The band between 70 and 90 GHz (known as the E-band) has been considered open for public use in some countries like the United States and the European Union with precise rules for their use [7]. In 2003, the FCC opened the E-band spectrum for licensed operations like broadband P2P wireless local area networks, high-speed Internet access, and other innovative products and services. Many other countries in the Americas, the Middle East, Africa, and Asia have already taken steps to open E-band spectrum for licensed applications [11].

5.3.2 Millimeter-Wave Standardization

5.3.2.1 IEEE 802.15.3c The IEEE 802.15.3c Task Group (TG3c) was formally created in March 2005 to provide an alternative physical (PHY) layer in mmWave spectrum for wireless personal area networks (WPANs). mmWave WPANs operate in the 60 GHz unlicensed band according to the FCC regulation. The PHY of

IEEE 802.15.3c is intended to support at least 2 Gbps data rate (with achievable data rates of more than 3 Gbps) over a few meters range. The task group decided to reuse an existing medium access control (MAC) layer with essential extensions and adjustments. In September 2009, IEEE 802.15.3c-2009 standard was approved [12]. This standard is the first to address multigigabit short-range wireless systems [13]. Three different PHY modes have been developed: single carrier (SC), high-speed interface (HSI), and orthogonal frequency division multiplexing (OFDM) audio/video (A/V). The SC PHY is used for low-power, low-cost mobile applications. The HSI PHY is applied for high-speed, low-latency bidirectional transmission. The AV PHY is designed for A/V-specific applications. All three PHY modes can support beamforming. The standard also determines an unequal error protection support that can be realized in PHY and MAC layers. Common mode signaling (CMS) is a low-data-rate SC PHY modulation and coding scheme (MCS) designed to prevent interference between different PHY modes.

5.3.2.2 ECMA-387 The first edition of ECMA-387 was approved by the ECMA Technical Committee (TC48) in December 2008 for subsequent submission to ISO/IEC JTC 1 fast-track procedure. The standard was published in 2009 as ISO/IEC 13156. The second edition of this standard was approved in December 2010 [14] and published in 2011 after submission to the JTC 1 Fast-Track Procedure [15]. This standard provides a PHY along with distributed MAC, and a high-definition multimedia interface (HDMI) protocol adaptation layer (PAL) for 60 GHz heterogeneous networks. The heterogeneous network consists of two types of devices (Types A and B). Both A and B device types can fully interoperate with other devices. Moreover, every device type can operate individually. By employing adaptive antenna arrays, the standard supports a variety of implementations for including simple and low-power, short-range applications to high-rate, longer distance multimedia streaming. The Type A device is a high-end, high-performance device and delivers numerous features to support high data rate and longer range transmission, immunity against multipath, adaptive arrays, and multilevel quality of service (QoS). The Type B high data rate device is simpler with low power consumption and low cost. The A and B device types can support up to 6.350 and 3.175 Gbps data rates in a single channel, respectively. Four standard frequency channels are defined with a separation of 2.160 GHz. These bands may be bonded together to improve data rates by a factor of 2, 3, or 4. The ECMA-387 standard describes one decentralized MAC protocol for Type A and Type B devices. This MAC protocol delivers interoperability and coexistence for both types of devices and improves QoS, bandwidth efficiency, and spatial reuse capability [14]. Several PALs can exist on the top of the MAC layer and can communicate with this layer via a multiplexing sublayer.

5.3.2.3 IEEE 802.11ad   IEEE 802.11ad was created by Task Group "ad" (TGad) of IEEE 802.11 working group in January 2009 as an amendment to both the IEEE 802.11 PHYs and MAC layers to allow operation at around 60 GHz and provide very high throughput with a minimum rate of 1 Gbps. The final IEEE 802.11ad-2012 specification was published in December 2012 [16]. The MAC and PHY protocols can support short range data communication between devices over ad hoc networks with a data rate up to 7 Gbps. This standard also supports fast session transfer among 2.4, 5, and 60 GHz frequencies. The standard beamforming can be supported through all PHY modes. Beamforming is accomplished by using a bidirectional training sequence at each transmission.

5.3.2.4 Wireless Gigabit Alliance   In May 2009, the Wireless Gigabit Alliance (WiGig) was created for short-range wireless transmission at the unlicensed 60 GHz band. The PHY and MAC layers of this technology contributed to the IEEE 802.11ad standardization procedure. WiGig can be used for audio, high-definition video, and multimedia content transmission at a speed of 7 Gbps over a distance of 10–12 m. This standard enables devices to switch between operating bands of 2.4, 5, and 60 GHz of 802.11 networks. It supports beamforming, advanced security, and high-performance wireless implementations of HDMI, and enables wireless data transfer between devices such as tablets, PCs, display ports, and USBs [17].

5.3.2.5 WirelessHD The WirelessHD® Consortium was formed in 2006 by joining together of different companies to define 60 GHz multigigabit mmWave communication for both mobile and stationary applications. WirelessHD is the first interface to operate in the unlicensed 60 GHz spectrum [18]. It can support data transmission rates at 10–28 Gbps; 3D video formats; low-, medium-, and high-rate PHY modes; smart antenna; beamforming; and data privacy.

5.4 Millimeter-Wave Systems

A mmWave system is constructed predominantly on PHY and MAC layers. The MAC functions comprise radio resource management, multiple access, optimization of transmission parameters, rate adaptation, and QoS [19]. Additional functions such as probing, link setup, and maintenance that need support from MAC layer in the system with antenna array are also employed. The PHY layer includes a baseband back end and an RF front end.

5.4.1 Modulation for mmWave Communications

Digital modulation schemes are used to transform digital signals into mmWave signals. Some modulation schemes, such as phase shift keying (PSK) and frequency shift keying (FSK), use a constant amplitude carrier with variations in frequency or phase. Other schemes, such as quadrature amplitude modulation (QAM) and amplitude shift keying (ASK), are based on carrier amplitude variations. The applied modulation schemes require high power efficiency with low bit error rate (BER) at mmWave frequencies [3].

On/off keying (OOK) modulation is simple with low implementation costs. It is a two-level ASK modulation, where the transmitter is on when logic "1" is transmitted and it is off during the transmission of logic "0." OOK receivers require an adjustable threshold and automatic gain controller (AGC). There are two types of synchronous (coherent) and envelope (noncoherent) demodulations for OOK [3].

PSK is a large class of digital modulation schemes and its simplest form is binary phase shift keying (BPSK). In BPSK, the carrier phase has two states of 0 and π to transmit logic "1" and "0," while in quadrature phase shift keying (QPSK), the input signal is modulated by 0, $\pi/2$, π and $3\pi/2$ phase shifts. Both BPSK and QPSK modulated signals are demodulated coherently.

In FSK, digital signals are transmitted by modulating the frequency of the carrier. The simplest form of this scheme is when two different frequencies are used to transmit binary information. This kind of modulation can be expanded to an M-ary scheme by employing M different frequencies. Digital signals can be modulated by FSK both coherently with an IQ (in-phase/quadrature) modulator and noncoherently with a voltage-controlled oscillator (VCO).

QAM is a mixture of amplitude and phase shift keyings of a sinusoidal signal. For example, in 4-QAM, four combinations of phase and amplitude of a sinusoidal signal are used to map a 2-bit digital symbol to a specific point on the constellation diagram. In M-QAM, it is possible to send data at higher rates where the number of $\log_2 M$ bits can be represented by a symbol. Although the 64-QAM and 256-QAM are frequently used in modern communication systems, they cannot be applied in mmWave frequencies due to the effect of phase noise and the high power consumption of power amplifiers (PAs) [3].

5.4.2 mmWave Transceiver

Recently, integrated mmWave structures and high-performance components have been announced base on silicon-on-insulator, silicon-germanium, and complementary metal–oxide–semiconductor (CMOS) technologies. The higher level of integration allowed by silicon technologies permits the employment of a mmWave transceiver module with efficient and complex modulation schemes [20]. Figure 5.1 shows a block diagram of a mmWave transceiver. A dual-conversion superheterodyne structure is used with a single frequency synthesizer. In the receiver, the input signal is amplified by a low-noise amplifier (LNA). The amplified signal is then mixed down to the intermediate frequency (IF). At IF, the signal is filtered and amplified by an amplifier. Next, the signal is divided and fed to a couple of double-balanced mixers. The mixers transform the IF signal to I and Q baseband signals. Identical frequency synthesizers are used

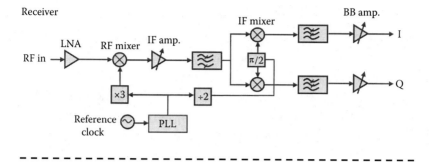

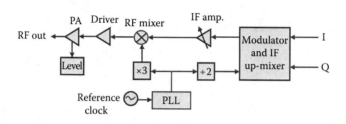

Figure 5.1 Block diagram of a mmWave transceiver.

at both the transmitter and the receiver. The local-oscillator (LO) signal for the RF-to-IF downconverter is produced by a frequency tripler. A divide-by-two operation from the synthesizer output is used to construct the quadrature LO signals for down-conversion from IF to baseband. The resulting signals at the unwanted image frequency are removed by LNA notch filters.

The transmitter is principally a mirror image of the receiver. The baseband I and Q signals are taken in by the transmitter and upconverted to the IF frequency. The IF signal is then filtered, amplified, and upconverted to the RF frequency. Additional amplification is provided before the final PA stage by an image-rejecting predriver (which contains the same notch filters as those in the LNA).

5.4.3 Single-Carrier versus Multicarrier Communications in mmWave Technology

OFDM is a form of multicarrier modulation technique for encoding digital data [21]. This technique is widely used in broadband wireless communication systems to mitigate multipath effects. In the

OFDM technique, a high data rate stream is split into several lower rate streams for simultaneous transmission over individual subcarriers. Each of the OFDM subcarriers can still be modulated by any of the previously mentioned modulation techniques. Since the lower rate parallel data stream can increase the symbol duration, the corresponding time dispersion caused by the multipath phenomenon is decreased. However, some channel measurements conducted in 60 GHz have shown that non-line of sight (NLOS) components suffer from much greater losses than the line of sight (LOS) component. Furthermore, directional antennas and beam steering methods applied in this band can highly reduce the effect of multipath components. As a result, SC transmission is comparable to its multicarrier counterpart from a spectral efficiency viewpoint [19]. Nevertheless, SC transmission is a more cost-effective solution for predominantly LOS conditions with a small delay spread, but OFDM, which has a better ability to mitigate the longer channel delay spread, is more preferable in NLOS conditions.

Multicarrier transmission has other advantages as well. Transceiver implementation is simplified using fast Fourier transform (FFT)/ inverse FFT (IFFT). Furthermore, thanks to the orthogonality of the subcarrier, the frequency-domain equalization is comparatively easier in OFDM than in SC transmission. On the other hand, a time domain equalizer is commonly required in SC transmission to equalize the channel. This makes the receiver more complicated than that for OFDM, especially in multigigabit, high-speed transmission [10].

The peak-to-average power ratio (PAPR) in SC transmission is lower than in OFDM modulation, for the same constellation points. A large PAPR can strongly saturate the output power of the transmitter PA [10]. On the other hand, OFDM transmission must be implemented by robust error control coding schemes. For these reasons, OFDM transmission is preferable for low-rate codes, while the SC system is preferable for high-rate codes [10].

Generally, both OFDM and SC transmissions are candidates for multigigabit wireless communications. However, the selection between the two transmission modes needs a trade-off among numerous factors. Furthermore, if both solutions are accepted, compatibility between them is a challenge [19].

5.5 Millimeter-Wave Antennas for Multigigabit Transmission

Some specific considerations should be addressed for mmWave antenna design. Due to the excess loss at 60 GHz caused by oxygen absorption, a sufficient margin should be considered to overcome this and other losses, such as rain attenuation. Applying antennas with high gain or directivity at the transmitter and/or receiver sides can be helpful in this regard. Additionally, narrow beam antennas are appropriate to minimize the multipath effect. Circular polarization is also suitable to filter out multipath components in the case of LOS. Moreover, the data rate can be improved by using multipolarization [3]. Antenna array beamforming as a technique for directional signal transmission and/or reception can be used in mmWave systems for spatial filtering and interference rejection.

5.5.1 Important Antenna Parameters

5.5.1.1 Gain and Directivity The received signal consists of numerous copies of the transmitted signal with their specific time delays. To overcome the effect of multipath components, it is necessary to focus the radiated power of the antennas in a given direction. Directivity of an antenna is a measure of the antenna radiation power density in the direction of its maximum radiation against the radiation power density of an ideal isotropic antenna with the same total radiation power. Antenna gain shows the ratio of the radiation power of an antenna in a given direction per unit solid angle to the total power accepted from the source by that antenna. The antenna gain and directivity are related by the radiation efficiency of the antenna as

$$G = \eta D, \qquad (5.1)$$

where, G, η and D are gain, efficiency, and directivity, respectively. A typical antenna has an approximate efficiency of 65% [8].

5.5.1.2 Half Power Beamwidth According to Figure 5.2, the 3 dB beamwidth, or half-power beamwidth (HPBW), represents the angle among the half-power locations on the main lobe in an antenna pattern, when referenced to the maximum effective radiation power in the main lobe.

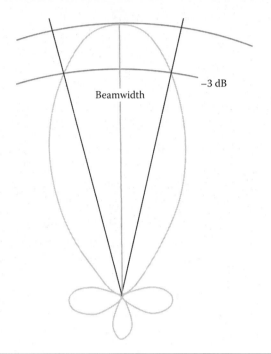

Figure 5.2   Half power beamwidth of an antenna.

5.5.1.3 Gain-to-Q Ratio   The quality factor (Q) of an antenna shows the ratio of the total stored energy in the reactive field to that radiated [22]. For high values of Q, a great part of the reactive energy is stored in the near zone field that results in high loss, narrow antenna bandwidth, and large frequency sensitivity. In mmWave frequencies, antenna gain and bandwidth must be maximized and consequently Q must be minimized to attain a lossless high gain antenna. Thus, in antenna design, optimization of the gain-to-Q ratio is of great importance [8].

5.5.1.4 Polarization   The transverse electromagnetic wave radiated by any antenna to the free space includes an electric and a magnetic field. The electric and magnetic fields are orthogonal to each other and to the direction of propagation. The polarization of an antenna represents the orientation of the radiated electric field with respect to a reference surface (usually Earth). The wave is linearly polarized when it broadcasts on a single plane. Vertical and horizontal polarizations are the two simple polarization forms in the category of linear polarization. Data transmission with independent linearly polarized waves can be used as a frequency reuse technique to double system capacity for a given bandwidth. A wave is circularly polarized if the related electric field composed of two equal amplitude components with a relative phase of ± 90°. In right-hand and left-hand circularly polarized waves, the electric field vector rotates clockwise and anticlockwise, respectively, when looking in the direction of propagation. In the most general case, an elliptically polarized wave is formed by combination of two linear components of the electric field with unequal amplitudes and a constant arbitrary phase relationship.

5.5.2 Antenna Types for mmWave Communications

5.5.2.1 Planar Antennas Microstrip or printed antennas are one of the most common types of planar antennas. These kinds of antennas are used in mmWave applications because of their low cost, low profile and easy integration with other parts of a system. One major advantage of microstrip antenna elements is that they can be simply formed as array structures. They can be applied at 60 GHz and higher frequency broadband communications. They offer some capabilities like beam scanning, high gain equal to 30–36 dBi and moderate to low sidelobe levels. Microstrip antennas show significantly high losses related to both dielectric and conductor parts [23]. They exhibit bandwidths of about a few percent, rising with smaller dielectric constant of the substrate and substrate thickness [24]. A periodic printed array of microstrip antennas can also be used as a planar reflector [8].

The planar slot antennas consist of a radiator designed by cutting a thin slot in a large metallic surface, which is another potential candidate for mmWave applications. The slot has a length of a half-wavelength at the desired frequency and its width is a tiny fraction of the wavelength [8]. The array is constructed by a metallic waveguide with longitudinal or transversal slots. A longitudinal slot located in the wide wall of a waveguide radiates identical to a dipole (with the same dimensions) perpendicular to the slot. The radiation influences from a number of waveguide slots are superimposed in phase to provide a favorable radiation pattern. An easy and low-cost construction of waveguide slotted arrays can be accomplished by a substrate integrated waveguide (SIW). It can be made within a dielectric substrate where the lateral walls are designed by rows of vias [24–26].

5.5.2.2 Lens and Reflector Antennas Lenses are used as reflectors for directing radiation coming from a source, which is typically called the feed. The feed antenna can be any type of antenna, but horns and array of patches are popular [27]. Lens antennas are made from low-loss dielectric materials with higher dielectric constant than air [8]. When an electromagnetic wave encounters an impedance discontinuity caused by a lens, it is partially transmitted and partially reflected. If the wave incidence is not perpendicular to the lens surface, the wave is refracted or bent according to Snell's law. Snell's law states that when electromagnetic wave rays enter a medium with a greater dielectric constant, they are bent toward the normal to the surface, and they are bent away from the normal to the surface when they enter a medium with a lower dielectric constant. The rays are bent again when they emerge from the second surface of the lens. According to Figure 5.3, the converging lens concentrates the rays to its focal point.

Lens and reflector antennas usually show small loss and can be combined with mmWave semiconductor components to generate a moderate power level [24].

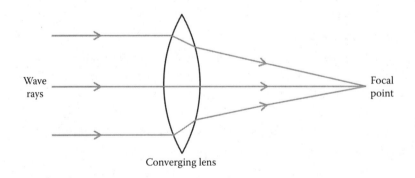

Figure 5.3 Converging lens focusing wave rays at the focal point.

5.5.2.3 Horn Antennas   A horn antenna consists of a rectangular or circular metallic waveguide in the shape of a horn to direct electromagnetic waves in a beam. Horn antennas are widely used as feeders for parabolic or lens antennas at mmWave frequencies. They have moderate directivity and wide bandwidth and can be easily constructed.

5.5.3 Array Beamforming in mmWave Frequencies

Beamforming techniques are used for directional signal transmission or reception in order to control the unfavorable path loss and minimize interference in the mmWave bands. A beamformer adjusts the phase and relative amplitude of the signal at every antenna element to produce the desired pattern. The small wavelengths of the mmWave spectrum simplify the use of a huge number of antennas to produce highly directional beams [28]. Beamforming can be performed either by phase shifters at mmWave frequencies or by signal processing at baseband [8]. Digital beamforming is carried out as digital precoding prior to digital-to-analog conversion (DAC) at the transmitter and subsequent to analog-to-digital conversion (ADC) at the receiver where a specific coefficient is multiplied with the modulated baseband signal [28]. The overall cost of the system is increased because of the need for several radio frequency chains. However, in analog beamforming, either phase shifters or variable gain amplifiers are used to apply complex coefficients to the RF signals. Figure 5.4 shows

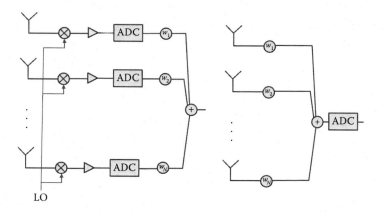

Figure 5.4   Digital versus analog beamforming.

typical structures of both analog and digital beamformers. According to this figure, only one ADC is needed for analog beamforming while multiple ADCs are necessary for digital beamforming [10]. In an OFDM system, digital beamforming is done before the IFFT and after the FFT operations at the transmitter and receiver, respectively.

Generally, digital beamforming offers a greater degree of freedom and superior performance, but is more complex and of higher cost than analog beamforming because of the need to separate FFT/IFFT blocks, DACs, and ADCs for each RF chain. On the other hand, analog beamforming is a simple and effective technique to achieve high beamforming gains; however, it offers less flexibility than digital beamforming. A trade-off between flexibility/performance and simplicity drives the need for hybrid beamforming techniques [28].

5.6 Millimeter-Wave Technology for Upcoming 5G Networks

Mobile communication has been one of the most effective technological inventions. Nowadays, due to the increased popularity of smart phones and other mobile data devices, data traffic is exponentially growing. Special improvements in air interface capacity and allocation of new frequency bands are necessary to handle this huge amount of data traffic [29]. Advanced technologies such as multiple input multiple-output (MIMO), OFDM, multiuser diversity, turbo code, link adaptation, and hybrid automatic repeat request are used by existing fourth-generation (4G) communication systems to realize higher spectral efficiencies [30]. On the other hand, deployment of many smaller cells like femtocells can scale capacity only linearly proportional to the number of cells, and, hence, cannot meet capacity requirements [29].

Recently, the wireless communication community has turned its attention to the mmWave band [31,32]. The huge amount of spectrum available in these frequencies makes them good candidates for use in 5G networks. Larger bandwidth allocations lead to greater data transfer rates [2]. The combination of economical mmWave CMOS technology and steerable high-gain antennas at both mobile and base station (BS) support the feasibility of mmWave

5G networks. Moreover, the accessibility of the 60 GHz unlicensed band has increased attention in multigigabit short-range wireless communications.

5.6.1 Network Architecture

5G networks require a considerable variation in the design of cellular architecture [33]. Existing cellular networks generally use an outdoor BS in the center of a cell communicating with both indoor and outdoor mobile users. However, for indoor users, the signals experience very high penetration loss due to building walls, especially at mmWave frequencies. This considerably reduces the data rate and damages the spectral energy efficiency of the system. According to Figure 5.5, one of the main goals of designing 5G networks is to separate outdoor and indoor environments to avoid the effects of building penetration loss. This can be accomplished by applying a distributed antenna system (DAS) as well as massive MIMO technology. A DAS is deployed with tens or hundreds of spatially separated antenna nodes. Massive MIMO systems can achieve large capacity gains by employing a huge number of antenna elements. The outdoor BS can benefit from both DAS and massive MIMO technologies. Moreover, outdoor users can collaborate to

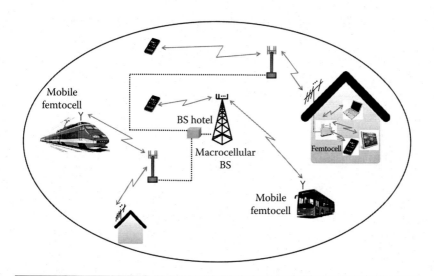

Figure 5.5 Cellular architecture of a 5G network.

form a virtual large antenna array. Furthermore, these arrays will be mounted outside each building and can be connected by cable to the indoor access points. These large antenna arrays also communicate with the outdoor BS or its distributed antennas. Although, this results in more infrastructure cost, it improves the average throughput as well as the spectral and energy efficiencies of the cellular system. On the other hand, as indoor users only require communicating with indoor wireless access points, they can utilize many high-data-rate technologies such as WiFi, femtocell, ultra wideband (UWB), visible light, and mmWave communications that are appropriate for short-range transmission [33]. Furthermore, in order to guarantee good outdoor coverage, outdoor BSs must be deployed with higher density (with the same site-to-site separation distance as microcell or picocell placement in an urban environment) than existing macrocellular BSs [29]. The transmission and/ or reception in outdoor BSs is accomplished by narrow beams to extend the range of the cells and reduce fading, multipath, and interference.

The mobile femtocell concept has been suggested in Haider et al. [34] to accommodate high mobility users in a 5G heterogeneous cellular architecture. These femtocells, which combine the ideas of mobile relay and femtocell, are deployed in vehicles such as buses, trains, and private cars. The femtocell BSs communicate with the users located inside the vehicle; however, the large antenna arrays mounted outside of the vehicle communicate with outdoor BSs. Each mobile femtocell and its corresponding users are all considered as a single unit by the BS, while a user views a mobile femtocell as a regular BS. This can help users of mobile femtocells to receive high-data-rate services with shortened overhead.

As proposed in Dehos et al. [35], mmWave 5G networks can be assisted by the previous cellular technologies to guarantee voice and signaling coverage to mobile users. In this scenario, multiple users are simultaneously connected to the legacy BS, which offers control plane information, while high-speed data connectivity is accomplished by the mmWave BS. This dual connectivity will allow an active user device to perform a fast handoff to the legacy network, for example, when direct LOS with the mmWave BS is obstructed. The legacy BS may activate/deactivate mmWave BSs to save energy and assist the radio resource management to overcome intercell interference.

5.6.2 Air Interface

Beamforming is a key enabling technology of BSs in mmWave 5G networks. The small size of antennas and their small separation distances allow a huge number of antennas to be implemented in the BS. This results in a high beamforming gain in a small area. Both digital and analog beamforming schemes are applicable in mmWave 5G networks. Analog beamforming is usually chosen for low-cost solutions as it requires fewer RF components.

The most important goals of air interface are small latency (about one ms roundtrip) and maximum data rates in excess of 10 Gbps [36]. A frame structure with 100 µs slots is proposed in Cudak et al. [37] to achieve low latency. The high bandwidths available at mmWave frequencies as well as the use of two-stream MIMO are exploited to meet the peak data rate requirements. A particular null cyclic prefix single carrier (NCP-SC) air interface is proposed in Larew et al. [36] for 5G mmWave systems. In NCP-SC, a set of null symbols is added subsequent to a block of data symbols. The null symbols of one data block act as the CP for a subsequent data block. The NCP-SC air interface has some advantages, such as simple frequency domain reception, low PAPR, easy estimation of noise plus interference, adaptation of the CP length without breaking up the timing of the NCP-SC system, and ability of simple processing to overcome extreme Doppler.

The cyclic-prefix-based OFDM (CP-OFDM) has been extensively adopted for current mobile networks, but this traditional multicarrier modulation technique cannot effectively meet the requirements of future 5G networks. One of the main problems of CP-OFDM is that it cannot provide reasonable degrees of system flexibility. However, the filter-bank-based schemes proposed in Schellmann et al. [38] generalize the multicarrier modulation concept. The main concept of the filter-bank-based multicarrier (FBMC) modulation is that the modulated signal on each subcarrier is shaped by a well-designed prototype filter. This provides a new degree of freedom for optimizing the waveform toward various transmission characteristics. Furthermore, the FBMC scheme does not necessarily need a cyclic prefix and hence achieves full spectral efficiency. This scheme acts as a favorable alternative to CP-OFDM with additional flexibility, robustness, and efficiency.

5.6.3 Backhauling

Two business models have been proposed in Dehos et al. [35] for wireless backhaul of mmWave systems. The first one is licensed E-band backhaul, which can be combined with a heterogeneous mmWave 5G network. These solutions that are supported by components and equipment providers, and also operators, intend to decrease the cost of backhaul components. Frequency-division duplexing (FDD) will be presumably realized between the both lower and upper halves of the E-band.

The other business model supported by the members of the Wi-Fi Alliance and Internet service providers (ISPs) is based on upgrading 60 GHz radios. In this case, specific attention should be paid to the oxygen absorption rate for link budget calculations.

High-gain antennas such as horn and/or dielectric lenses solutions are commercially available in both V- and E-bands. However, these products are fairly expensive, bulky, and heavy. A planar array antenna based on substrate-integrated waveguides can be an alternative solution.

Increasing the modulation order to achieve higher spectral efficiency has a significant impact on the linear performance of the transmitter PA. The required linearity can be achieved using CMOS technology for lower-order QAM-modulated signals and bipolar technologies for higher-order signals [35].

5.7 Propagation and Channel Modeling for 5G mmWave Networks

Radio channel characterization and propagation modeling are crucial for the development of mmWave mobile access and backhaul networks [39–41]. These can be helpful in developing new approaches for the air interface, multiple access, innovative architectures for interference mitigation, and additional signal enhancement methods [2].

Extensive measurements must be made in propagation environments to build a reliable statistical model for mmWave multipath environments.

5.7.1 Millimeter-Wave Cellular Propagation Measurements

In Sulyman et al. [42], the 28 and 38 GHz cellular measurements have been made for both LOS and NLOS receiver conditions. The 28 GHz campaign used 10.9° HPBW (with 24.5 dBi gain) and 28.8° HPBW (with 15 dBi gain) antennas at the transmitter and receiver, while the 38 GHz campaign used an antenna with 7.8° HPBW (with 24.5 dBi gain) at the transmitter, and both a 7.8° HPBW (with 25 dBi gain) and a 49.4° HPBW (with 13.3 dBi gain) antennas at the receiver. The power delay profiles (PDPs) of the channel have been recorded by using highly directional antennas [2]. Although, true omnidirectional results were not initially reported, some results have been derived in MacCartney et al. [43] from the directional measurements.

At 28 GHz, the received signal was recorded in Manhattan, for three separate transmitter positions and up to 200 m transmitter–receiver distances. A full azimuth and elevation sweep was accomplished at all receiver locations. The PDPs were measured using an 800 MHz bandwidth channel sounder to determine path loss and multipath effects.

At 38 GHz, measurements were made for 43 transmitter–receiver positions at the University of Texas at Austin. Receiver antenna measurements of 25 dBi and 13.3 dBi were accomplished with a 29–930 and 29–728 m separation, respectively.

5.7.2 Path Loss Modeling

Path loss models are necessary to estimate the link budget and the received signal strength in reliable mmWave 5G cellular networks. Normally, higher frequencies experience higher propagation, penetration, and precipitation, and foliage losses vary depending on the material, rain rates, and the depth of foliage [28]. Path loss, which determines the cell coverage range, is inversely proportional to the square of the carrier frequency. However this must not lead to the most common misunderstanding that these frequencies are not adequate for long-range communications. The lossy mmWave channels can be made more reliable, and perhaps experience lower propagation losses relative to current cellular networks, by making use of highly directional antennas and beamforming techniques at the BS and user devices [28,42].

In cellular planning, path loss predictions are necessary to determine cell coverage [42]. Path loss in urban environments can be determined from a Stanford University Interim (SUI) model when carrier frequency is above 2 GHz [44]:

$$PL(d) = PL(d_0) + 10\gamma \log_{10}\left(\frac{d}{d_0}\right) + X_f + X_b + S, \quad \text{(dB) for } d > d_0 \tag{5.2}$$

where

$$PL(d_0) = 20\log_{10}\left(\frac{4\pi d_0}{\lambda}\right), \tag{5.3}$$

$$\gamma = a - bh_b + \frac{c}{h_b}, \tag{5.4}$$

$$X_f = 6\log_{10}\left(\frac{f}{2000}\right), \tag{5.5}$$

$$X_b = -10.8\log_{10}\left(\frac{h_m}{2}\right), \tag{5.6}$$

λ and f are the carrier wavelength (in m) and frequency (in MHz), respectively
$PL(d_0)$ is the free space path loss in dB at distance
S is the zero-mean lognormal random shadowing variable

Moreover, h_b and h_m are the BS and user antenna heights in meters, respectively. For hilly terrains and thick vegetation conditions, a, b, and c are 4.6, 0.0075 and 12.6, respectively.

Absorption of the millimeter waves by oxygen and water vapor must also be considered for link budget calculations. These losses are higher at resonant frequencies of the gas molecules [45]. The significant absorption peaks occur at 22 and 60 GHz for water vapor and oxygen, respectively. The frequency bands between these absorption peaks have much better propagation conditions.

5.7.3 Attenuation due to Rain, Cloud, and Fog

Millimeter-wave propagation encounters additional loss due to rain. Raindrops are approximately identical in size to radio wavelengths and consequently cause scattering of millimeter waves. This loss is greater during higher rain rate. Rain rate in any location can be determined by referring to a rain region map [45]. The specific attenuation γ_R (dB/km) is calculated from the rain rate R (mm/h) as follows [46]:

$$\gamma_R = kR^\alpha, \tag{5.7}$$

The method of calculating values for the coefficients k and α are given in Recommendation ITU-R P.838-3 [46] at 1–1000 GHz, for both vertical and horizontal polarizations.

For clouds or fog consisting completely of small droplets, normally smaller than 0.01 cm, the Rayleigh approximation can be used at below 200 GHz. The cloud or fog specific attenuation, γ_C (dB/km), can be expressed as [47]

$$\gamma_C = K_l M, \tag{5.8}$$

where K_l and M are specific attenuation coefficient and liquid water density of the cloud or fog.

5.7.4 Penetration Loss

Millimeter waves cannot penetrate brick and concrete very well. In this case, a deep indoor coverage can be achieved using another technique, such as femtocell or WiFi. The human body can also cause poor penetration in mmWave frequencies. However, the penetration loss of millimeter waves is low when they pass through objects like wood, cardboard, and plastic [48].

5.7.5 Foliage Loss

Millimeter waves experience high attenuation in the presence of foliage. The size and distribution of the leaves, branches, and trunks, as well as the height of the trees, are the factors that can influence propagation loss [49]. The empirical foliage loss prediction model is presented in Meng et al. [50]. The foliage loss for the case where the foliage depth is less than 400 m, is given by CCIR [51]

$$L(\text{dB}) = 0.2\, f^{0.3} d^{0.6},\qquad (5.9)$$

where

f is frequency in MHz
d is the foliage thickness in m

This relationship can be used from 200 MHz to 95 GHz.

5.7.6 Scattering/Diffraction

If the LOS path between the transmitter and the receiver is obstructed, the signal arrives at the receiver via reflections from the objects or by diffraction or bending. mmWave frequencies experience lower diffraction but more shadowing and reflection [45] relative to the lower frequencies. Then, the greatest contribution at the receiver is reflected power, which is strongly dependent on the reflectivity of the objects. Furthermore, the shorter wavelengths of the millimeter waves cause the reflecting surfaces to seem rougher. This causes greater diffusion of the signal and less specular reflection.

5.8 Massive MIMO Technology for 5G mmWave Networks

Employing a huge number of antennas at both transmitter and receiver to accommodate more information data is referred to as massive MIMO. Massive MIMO systems have a few hundred antennas that serve tens of users from the same time-frequency resource.

Massive MIMO represents all the advantages of conventional MIMO, but on a much larger scale [52]. This technology allows development of secure and robust broadband networks. This idea can be a solution to achieve considerable enhancements in capacity and spectral efficiency required for 5G networks [48,53]. Massive MIMO can also help achieve more energy-efficient systems and remove the effects of noise, small scale fading, and interference [33]. Additional benefits of massive MIMO are considerable use of low-cost, low-power elements; lower latency; and simplified MAC layer [52]. These benefits of massive MIMO can be attained by straightforward signal processing methods, for example, maximal ratio combining (MRC) and maximal ratio transmission (MRT) [53]. However, channel correlation in mmWave frequencies prevents the use of MRC and MRT, and hence other methods such as zero-forcing (ZF) or minimum mean squared error (MMSE) are applicable.

In a massive MIMO system with Nt antennas at the BS and ideal channel state information (CSI), each user is able to reach identical uplink throughput as the users of a system with a single-antenna BS by consuming $1/N_t$ of their transmit power [54]. In the case of MMSE CSI, each user consumes only $1/\sqrt{N_t}$ of the required power for the same throughput [53]. In massive mmWave MIMO systems, the size of the antenna arrays is significantly smaller than that of current cellular frequencies. The main concern is related to the high path loss and link attenuations at mmWave bands, which can be considerably eliminated by installing an array with a large number of antenna elements.

5.8.1 Considerations of Communication System

A massive MIMO array with a narrow beam can decrease the multipath effects and delay spread of the channel. However, in mmWave systems with a user bandwidth of about hundreds of megahertz up to some gigahertz, the issue of frequency-selective fading investigation is needed [53]. The SC transmission with much better PAPR performance than OFDM can be designed for energy efficient massive MIMO systems. An alternative to SC modulation is constant envelope OFDM (CEOFDM) [55]. CEOFDM transforms the OFDM signal to a signal suitable for a more energy-efficient transmission. A phase demodulation method is required before applying a conventional OFDM demodulator at the receiver side [56].

5.8.2 Considerations for Signal Processing

In massive MIMO systems, the major source of error in CSI estimation is pilot contamination. This occurs if training sequences of neighboring cells are correlated with those transmitted for each cell users. However, in mmWave frequencies, the near-LOS conditions and great path loss can overcome the pilot contamination effect [53]. Moreover, the LOS channel condition and narrow antenna beamwidth can help to use direction-of-arrival (DOA)-based approaches in channel estimation. These approaches eliminate the need for pilot signals and improve the spectral efficiency of the system.

Pico- and femtocells normally do not include high-velocity users. However, at a mmWave spectrum, the low velocity of the users may experience a considerable Doppler shift. Therefore, a Doppler shift similar to that seen in current macrocellular applications can be expected. This means that, if a pilot-based channel estimation method is applied, the CSI would have to be updated at least as frequently as the existing MIMO implementations. Naturally, massive MIMO systems need more frequent updates and thus more computations. The rather slow changes in the direction of users have also motivated applying DOA-based channel estimation approaches. Another important effect is time variations because of the offsets in carrier frequency and LO phase noise. This limiting factor can be compensated by employing sensitive phase and frequency tracking [53].

In mmWave massive MIMO systems, it is hard to explain Rayleigh fading. As a result, user scheduling becomes a serious issue in avoiding high channel correlation in co-channel users. Several specific aspects of mmWave massive MIMO are useful for interference management [53]. Thanks to the high path loss, mmWave frequencies have restricted propagation range and accordingly permit higher frequency reuse. Additionally, the shadowing effects due to LOS or near-LOS propagation can decrease leakage into adjacent cells. The available frequency bands in the mmWave spectrum as well as beamforming techniques with massive MIMO arrays are also helpful for interference management.

5.8.3 Considerations of Transceiver Design

The main challenge of using traditional multiantenna transceivers in a massive MIMO system is the need to replicate several transmit/receive chains [41]. This intensely demands that massive MIMO transceivers be implemented on sharing RF resources. The code-modulated path sharing multiantenna (CPMA) architecture can be used to address this problem [57]. In this architecture, code multiplexing is appropriate to merge the signals from multiple antennas into one RF/IF/baseband/ADC path. This can significantly reduce space and power consumption, while mitigating problems corresponding to LO routing/distribution and cross talk between RF chains [53].

5.9 Conclusions

Worldwide bandwidth scarcity facing wireless communication systems has encouraged the investigation of the mmWave frequency bands for upcoming 5G networks. Although millimeter waves can only propagate a few miles due to atmospheric losses and cannot generally penetrate through solid materials, these losses can be overcome through the use of antenna arrays with a huge number of elements. The specific characteristics of massive MIMO technology along with the use of high directional steerable antennas make mmWave frequencies a good candidate for the next-generation 5G cellular communications. However, some special considerations are needed for massive MIMO system design and signal processing. Besides, for successful deployment of mmWave 5G networks, it is important to keep in mind that various factors can cause propagation loss at these frequencies.

5.10 Future Research Directions

Regardless of opportunities in mmWave cellular systems, there are some crucial challenges to realizing 5G networks [58,59]. As mmWave communication relies on highly directional transmissions, design changes to existing cellular systems are required. One specific problem of highly directional transmission is the design of the synchronization and broadcast signals [58]. Both BSs and mobile users possibly require scanning a wide range of angles before detecting these signals. This scanning may cause delay in BS for handover process. Furthermore, detection of the first random access signals by the user may cause some delay for accurate BS alignment.

Millimeter-wave signals are highly susceptible to shadowing [58] and the presence of obstacles, which cause large fluctuations in path loss. Moreover, channel coherence time is very short in mmWave frequencies, and the channel changes much faster than that of current cellular systems. As a result, developing realistic channel models is crucial for mmWave 5G networks. In mmWave cellular networks, new mechanisms are required to coordinate multiple simultaneous transmissions. Another significant challenge is the power consumption rate by ADCs, which grows considerably by the sampling rate and number of bits for each sample [58].

There are many potential areas for future research on 5G millimeter-wave networks. Since indoor users can be served by mmWave small cells, it would be interesting to evaluate the performance of these systems with indoor infrastructures collocated with buildings [33]. Moreover, as microwave systems are more reliable and less sensitive to blockages, they can coexist with mmWave networks and cover shadowed areas. They can also handle high-mobility users of the mmWave network, which need large amount of overhead due to frequent handovers. In these cases, it is interesting to study how to incorporate microwave cells into the mmWave systems. Flexible user scheduling by high-directive BS antennas to suit 3D system models may also be a topic of interest for future research.

References

1. J. Wells, *Multi-Gigabit Microwave and Millimeter-Wave Wireless Communications*, Artech House, Boston, MA, 2010.
2. T. S. Rappaport, S. Sun, R. Mayzus, H. Zhao, Y. Azar, K. Wang, G. N. Wong, J. K. Schulz, M. Samimi, and F. Gutierrez, Millimeter wave mobile communications for 5G cellular: It will work!, *IEEE Access*, 1, 335–349, 2013.
3. K. C. Huang and Z. Wang, *Millimeter Wave Communication Systems*, John Wiley & Sons, Hoboken, NJ, 2011.
4. C. W. Tolbert, A. W. Straiton, and C. O. Britt, Phantom radar targets at millimeter radio wavelengths, *IRE Transactions on Antennas and Propagation*, 6(4), 380–384, 1958.

5. J. W. Meyer, Radar astronomy at millimeter and submillimeter wavelengths, *Proceedings of the IEEE*, 54(4), 484–492, 1966.

6. F. B. Dyer and E. K. Reedy, Millimeter RADAR at Georgia Tech, in *Proceedings of the S-MTT International Microwave Symposium Digest*, Atlanta, GA, pp. 152, June 12–14, 1974.

7. P. Adhikari, *Understanding Millimeter Wave Wireless Communication*, Loea Corporation, San Diego, CA, 2008.

8. K. C. Huang and D. J. Edwards, *Millimetre Wave Antennas for Gigabit Wireless Communications*, John Wiley & Sons, Hoboken, NJ, 2008.

9. F. Lin, W. Hu, and A. Li, Millimeter-wave technology for medical applications, in *Proceedings of the IEEE MTT-S International Microwave Workshop Series on Millimeter Wave Wireless Technology and Applications*, Nanjing, China, September 18–20, 2012.

10. S.-K. Yong, P. Xia, and A. Valdes-Garcia, *60 GHz Technology for Gbps WLAN and WPAN: From Theory to Practice*, John Wiley & Sons, Chichester, U.K., 2011.

11. J. Wells, *Licensing and License Fee Considerations for E-band 71-76 GHz and 81-86 GHz Wireless Systems*, E-Band Communications Corporation, San Diego, CA, 2009.

12. IEEE 802.15.3c, IEEE Standard for information technology— Telecommunications and information exchange between systems— Local and metropolitan area networks—Specific requirements. Part 15.3: Wireless Medium Access Control (MAC) and Physical Layer (PHY) Specifications for High Rate Wireless Personal Area Networks (WPANs) Amendment 2: Millimeter-wave-based Alternative Physical Layer Extension, October 2009.

13. T. Baykas, C.-S. Sum, Z. Lan, J. Wang, M. Azizur Rahman, and H. Harada, IEEE 802.15.3c: The first IEEE wireless standard for data rates over 1 Gb/s, *IEEE Communications Magazine*, 49(7), 114–121, July 2011.

14. Standard ECMA-387, High rate 60GHz PHY, MAC and HDMI PAL, 2nd edn., December 2010.

15. ISO/IEC 13156, Information technology—Telecommunications and information exchange between systems—High rate 60 GHz PHY, MAC and HDMI PAL, 2st edn., September 2011.

16. IEEE 802.11ad, IEEE Standard for information technology— Telecommunications and information exchange between systems— Local and metropolitan area networks—Specific requirements-Part 11: Wireless LAN Medium Access Control (MAC) and Physical Layer (PHY) Specifications Amendment 3: Enhancements for Very High Throughput in the 60 GHz Band, December 2012.

17. Wireless Gigabit Alliance, Defining the future of multi-gigabit wireless communications, July, 2010. Available at: http://wilocity.com/resources/WiGigWhitepaper_FINAL5.pdf.

18. WirelessHD Specification Version 1.1 Overview, May, 2010. Available at: http://www.wirelesshd.org/pdfs/WirelessHD-Specification-Overview-v1.1May2010.pdf.

19. N. Guo, R. C. Qiu, S. S. Mo, and K. Takahashi, 60-GHz millimeter-wave radio: Principle, technology, and new results, *EURASIP Journal on Wireless Communications and Networking*, 2007(1), January 2007.

20. D. Liu, B. Gaucher, U. Pfeiffer, and J. Grzyb, *Advanced Millimeter-Wave Technologies Antennas, Packaging and Circuits*, John Wiley & Sons, Chichester, U.K., 2009.

21. Y. Wu and W. Y. Zou, Orthogonal frequency division multiplexing: A multi-carrier modulation scheme, *IEEE Transactions on Consumer Electronics*, 41(3), 392–399, August 1995.

22. M. Gustafsson and S. Nordeb, Bandwidth, Q factor, and resonance models of antennas, Technical report, Lund Institute of Technology, Lund, Sweden, 2005.

23. J. Schoebel and P. Herrero, Planar antenna technology for mm-wave automotive radar, sensing, and communications, in *Radar Technology*, ed. G. Kouemou, InTech, Rijeka, Croatia, January 2010, pp. 297–318.

24. W. Menzel and A. Moebius, Antenna concepts for millimeter-wave automotive radar sensors, *Proceedings of the IEEE*, 100(7), 2372–2379, July 2012.

25. J. Hirokawa and M. Ando, 76 GHz post-wall waveguide fed parallel plate slot arrays for car-radar applications, in *Proceedings of the IEEE International Symposium Antennas Propagation*, Salt Lake City, UT, Vol. 1, pp. 98–101, July 16–20, 2000.

26. J. F. Xu, W. Hong, P. Chen, and K. Wu, Design and implementation of low sidelobe substrate integrated waveguide longitudinal slot array antennas, *IET Microwaves, Antennas & Propagation*, 3(5), 790–797, August 2009.

27. T. Komljenovic, Lens antennas—Analysis and synthesis at mm-waves. Available at: http://www.fer.hr/download/repository/kvalifikascijski rad komljenvic.pdf.

28. W. Roh, J.-Y. Seol, J. H. Park, B. Lee, J. Lee, Y. Kim, J. Cho, and K. Cheun, Millimeter-wave beamforming as an enabling technology for 5G cellular communications: Theoretical feasibility and prototype results, *IEEE Communications Magazine*, 52(2), 106–113, February 2014.

29. Z. Pi and F. Khan, An introduction to millimeter-wave mobile broadband systems, *IEEE Communications Magazine*, 49(6), 101–107, June 2011.

30. F. Khan, *LTE for 4G Mobile Broadband: Air Interface Technologies and Performance*, Cambridge University Press, Cambridge, U.K., 2009.

31. A. Bleicher, Millimeter waves may be the future of 5G phones, *IEEE Spectrum*, February 2014.

32. M. Elkashlan, T. Q. Duong, and H.-H. Chen, Millimeter-wave communications for 5G: Fundamentals: Part I, *IEEE Communications Magazine*, 52(9), 52–54, September 2014.

33. C.-X. Wang, F. Haider, X. Gao, X.-H. You, Y. Yang, D. Yuan, H. M. Aggoune, H. Haas, S. Fletcher, and E. Hepsaydir, Cellular architecture and key technologies for 5G wireless communication networks, *IEEE Communications Magazine*, 52(2), 122–130, February 2014.

34. F. Haider, C.-X. Wang, H. Haas, D. Yuan, H. Wang, X. Gao, X.-H. You, and E. Hepsaydir, Spectral efficiency analysis of mobile femto-cell based cellular systems, in *Proceedings of the IEEE 13th International Conference on Communication Technology (ICCT)*, Jinan, China, pp. 347–351, September 25–28, 2011.

35. C. Dehos, J. Luis González, A. De Domenico, D. Kténas, and L. Dussopt, Millimeter-wave access and backhauling: The solution to the exponential data traffic increase in 5G mobile communications systems?, *IEEE Communications Magazine*, 52(9), 88–95, September 2014.

36. S. G. Larew, T. A. Thomas, M. Cudak, and A. Ghosh, Air interface design and ray tracing study for 5G millimeter wave communications, in *Proceedings of the IEEE Globecom Workshops*, Atlanta, GA, pp. 117–122, December 9–13, 2013.

37. M. Cudak, A. Ghosh, T. Kovarik, R. Ratasuk, T. A. Thomas, F. W. Vook, and P. Moorut, Moving towards mm wave-based beyond-4G (B-4G) technology, in *Proceedings of the IEEE 77th Vehicular Technology Conference (VTC Spring)*, Dresden, Germany, June 2–5, 2013.

38. M. Schellmann, Z. Zhao, H. Lin, P. Siohan, N. Rajatheva, V. Luecken, and A. Ishaque, FBMC-based air interface for 5G mobile: Challenges and proposed solutions, in *Proceedings of the Ninth International Conference on Cognitive Radio Oriented Wireless Networks and Communications (CROWNCOM)*, Oulu, Finland, pp. 102–107, June 2–4, 2014.

39. G. R. MacCartney, Jr., J. Zhang, S. Nie, and T. S. Rappaport, Path loss models for 5G millimeter wave propagation channels in urban microcells, in *Proceedings of the IEEE Global Communications Conference, Exhibition & Industry Forum*, Atlanta, GA, pp. 3948–3953, December 9–13, 2013.

40. T. S. Rappaport, F. Gutierrez, E. Ben-Dor, J. N. Murdock, Y. Qiao, and J. I. Tamir, Broadband millimeter-wave propagation measurements and models using adaptive-beam antennas for outdoor urban cellular communications, *IEEE Transactions on Antennas and Propagation*, 61(4), 1850–1859, April 2013.

41. J. N. Murdock, E. Ben-Dor, Y. Qiao, J. I. Tamir, and T. S. Rappaport, A 38 GHz cellular outage study for an urban campus environment, in *Proceedings of the IEEE Wireless Communications and Networking Conference (WCNC)*, Paris, France, pp. 3085–3090, April 1–4, 2012.

42. A. I. Sulyman, A. T. Nassar, M. K. Samimi, G. R. MacCartney, Jr., T. S. Rappaport, and A. Alsanie, Radio propagation path loss models for 5G cellular networks in the 28 GHz and 38 GHz millimeter-wave bands, *IEEE Communications Magazine*, 52(9), 78–86, September 2014.

43. G. R. MacCartney, M. K. Samimi, and T. S. Rappaport, Omnidirectional path loss models in New York City at 28 GHz and 73 GHz, in *Proceedings of the IEEE Personal, Indoor, and Mobile Radio Communications (PIMRC)*, Washington, DC, pp. 227–231, September 2–5, 2014.

44. Y. Zhang, *WiMAX Network Planning and Optimization*, Taylor & Francis Group, New York, 2009.

45. M. Marcus and B. Pattan, Millimeter wave propagation: Spectrum management implications, *IEEE Microwave Magazine*, 6(2), 54–62, June 2005.

46. Recommendation ITU-R P.838-3, Specific attenuation model for rain for use in prediction methods, International Telecommunication Union, Geneva, Switzerland, September 2005.

47. Recommendation ITU-R P.840-4, Attenuation due to clouds and fog, International Telecommunication Union, Geneva, Switzerland, November 2009.

48. Z. Pi and F. Khan, A millimeter-wave massive MIMO system for next generation mobile broadband, in *Proceedings of the 46th Asilomar Conference on Signals, Systems and Computers (ASILOMAR)*, Pacific Grove, CA, pp. 693–698, November 4–7, 2012.

49. D. R. Godara, S. K. Modi, and R. K. Rawat, Study of millimeter wave scattering from ground & vegetation at 35 GHz, *International Journal of Soft Computing and Engineering*, 1(6), 202–204, January 2012.

50. Y. S. Meng, Y. H. Lee, and B. C. Ng, Empirical near ground path loss modeling in a forest at VHF and UHF bands, *IEEE Transactions on Antennas and Propagation*, 57(5), 1461–1468, May 2009.

51. CCIR, Influences of terrain irregularities and vegetation on troposphere propagation, CCIR report, Geneva, Switzerland, pp. 235–236, 1986.

52. E. G. Larsson, O. Edfors, F. Tufvesson, and T. L. Marzetta, Massive MIMO for next generation wireless systems, *IEEE Communications Magazine*, 52(2), 186–195, February 2014.

53. A. L. Swindlehurst, E. Ayanoglu, P. Heydari, and F. Capolino, Millimeter-wave massive MIMO: The next wireless revolution?, *IEEE Communications Magazine*, 52(9), 56–62, September 2014.

54. H. Ngo, E. Larsson, and T. Marzetta, Energy and spectral efficiency of very large multiuser MIMO systems, *IEEE Transactions on Communications*, 61, 1436–49, April 2013.

55. T. S. Rappaport, R. W. Heath, Jr., R. C. Daniels, and J. N. Murdock, *Millimeter Wave Wireless Communications*, Prentice Hall, Upper Saddle River, NJ, 2014.

56. S. C. Thompson, A. U. Ahmed, J. G. Proakis, J. R. Zeidler, and M. J. Geile, Constant envelope OFDM, *IEEE Transactions on Communications*, 56, 1300–1312, August 2008.

57. A. Jahanian, F. Tzeng, and P. Heydari, Code-modulated path-sharing multi-antenna receivers: Theory and analysis, *IEEE Transactions on Wireless Communications*, 8(5), 2193–2201, May 2009.

58. S. Rangan, T. S. Rappaport, and E. Erkip, Millimeter-wave cellular wireless networks: Potentials and challenges, *Proceedings of the IEEE*, 3, 366–385, March 2014.

59. J. Karjalaineny, M. Nekoveey, H. Benny, W. Kimy, J. Parkz, and H. Sungsooz, Challenges and opportunities of mm-wave communication in 5G networks, in *Proceedings of the Ninth International Conference on Cognitive Radio Oriented Wireless Networks and Communications (CROWNCOM)*, Oulu, Finland, pp. 372–376, June 2–4, 2014.

6

MULTICELLULAR HETEROGENEOUS NETWORKS

A 5G Perspective

M. BALA KRISHNA

Contents

6.1 Introduction

Multicellular heterogeneous networks serve the demands of high-speed Internet services, streaming audio/video applications, social networking applications, device-to-device communication systems, and machine-to-machine communication systems. Heterogeneous networks (HetNets) consist of high-power nodes (such as macrocells, microcells, remote relay head nodes [RRHNs]) and low-power nodes (such as picocells and femtocells). A macro base station (MaBS) covers the network area but cannot reach the demands of user nodes in dense traffic zones. Hence, pico base stations (PiBSs) and femto base stations (FeBSs) that cover a few meters resolve the issue of connectivity and serve the user nodes in dense traffic areas and hotspot regions of the network. Dense deployment of picocells and femtocells in hotspot zones cause signal interference that further leads to the drop in performance gain. Techniques such as beamforming, carrier aggregation, coordinated multipoint transmission (CoMP) and sub-frame scheduling increase the spectral efficiency and achieve high data transfer rates in HetNets. Orthogonal frequency division multiplexing (OFDM) and orthogonal frequency division multiple access (OFDMA) implemented with CoMP and enhanced intercell interference coordination (eICIC) reduce the effects of interference in small-cell HetNets. Cloud-based radio access network (C-RAN) [1] is a fronthaul network comprising multiple remote relay head nodes connected to a baseband unit located near the user nodes. C-RAN implements a centralized approach for resource allocation and CoMP-based transmissions to minimize the ICI effect.

6.1.1 Sleep/Active State Management

Sleep and active modes of picocells and femtocells are adjusted with respect to incoming mobile user nodes and transmission of cell discovery signals of small cells. Nanoscale accuracy in small-cell synchronization is achieved through a global navigation satellite system and a global positioning system that locate the transceiver antennas in the neighborhood. Based on transmission power levels, the cells are grouped into layers of varying frame lengths to transmit or receive the data packets.

6.2 OFDM and OFDMA Techniques in HetNets

HetNets implement OFDM/OFDMA schemes for efficient resource allocation and channel scheduling in high-speed networks. OFDM and OFDMA techniques enable simultaneous usage of the frequency spectrum for macro-cell and small-cell user nodes in real-time broadband, Wi-Fi, and WiMAX networks. The OFDMA technique applied for wireless broadband applications provides better results but reduces the performance of dense and small-cell HetNets. Signal-to-interference-plus-noise ratio (SINR), cell-specific reference signals (CRS), and downlink control information govern the reliability and power variations of data transmission over the physical downlink control channel and control channels [2]. The number of control channel elements allocated per transmitted frames is based on SINR levels of the corresponding channel. The user deploys the femtocells with the same or different frequencies of MaBS, and the service provider deploys the picocells at hotspot zones for yielding cost-effective services. The upper limits of transmission power [3] for the femtocell are measured with respect to the nearest MaBS. Due to high interferences between picocells or femtocells and adjacent high-power nodes (micro base station [MiBS] and MaBS), the OFDMA approach leads to packet losses and reduction in throughput. With an adaptive small-cell approach [4], the pico cells and femto cells can be deployed in two modes as follows:

1. In half-functional mode, separate baseband units are used to process RF generation.
2. In full-functional mode, the RF generation and baseband units are embedded together.

The set of pico cells and femto cells along with co-located baseband components is connected with the serving gateway node that functions as a switch tunable in varying network traffic conditions. Serving gateway node supports operation and maintenance services to generate a number of subbands for each active small-cell node (pico/femto cells) in the network. Further, these subbands are converted to a discrete number of OFDM subcarriers with a predefined spacing. The allocation of subbands for baseband components depends on the number of active user nodes present in the network. Congestion at the baseband is reduced by allocating distributed frequency bands to small-cell nodes. The adjacent nodes are selected based on the laws of maximal cliques to achieve optimum fairness in the network.

6.3 Dense HetNets

Dense HetNets comprise a large number of small-cell networks, operate at high frequency, and synchronize [5] with small cells to enhance the services in the network. Due to the dense population of pico and femto nodes, high ICI is experienced by user nodes that are associated with one or more base stations within the neighborhood of MaBS. The parameters such as beamforming capacities, carrier aggregation, cell outage, and optimal usage of channel state information (CSI) is quite significant to reduce ICI in HetNets. Reducing the transmission power of pico cells and femto cells, and adjusting the beamforming weights below the threshold level optimizes the performance of small-cell networks. ICIC and CoMP transmission reduce the effects of ICI. Based on the levels of ICI and network traffic conditions, the MaBS regulates the coverage levels so that the macro user nodes can be associated with the neighboring PiBS or FeBS. Network operators offload the cell outage (interruptions in services due to coverage losses) without degrading the quality of service for licensed and unlicensed spectrum users associated with base stations.

6.3.1 MIMO-OFDM for Dense HetNets

Multiple-input multiple-output (MIMO) techniques are used to enhance the performance of communication channels by defining the optimum number of subchannels, number of subframes, and the allocation of physical resource blocks. The 3D channel coefficient matrix is defined in terms of the number of transceiver antennas per the number of carrier components and the number of elevated antenna elements per each of the transmitting antenna. Antenna elevation factor defines the height of the PiBS and FeBS to be mounted. The CSI parameter varies for each cell and determines the corresponding of ICI levels (high-frequency reuse factor indicating high intercell interference and vice versa). Based on these values [5], the OFDM method is then applied to determine the number of subframes for the control and transmission channels. The number of transitional frames to be used in between the transmission and reception of the data packets is varied as per the total subframes.

6.4 Components of Multicellular Heterogeneous Networks

6.4.1 Macrocells

Macrocells comprise high-power radio access nodes, covering a maximum area of the network. A macrocell consists of a base station with high-power large coverage antenna (40–46 dB) [6] and multiple gateway nodes and server nodes. Macro cell nodes avail the networking resources and act as the central coordinate system in HetNets. MaBS are mainly deployed at essential locations of city or suburban areas with eminent infrastructure. A macro base station antenna is mounted on the top of a building that uses the spectrum bandwidth efficiently and covers the MiBS and users in the neighborhood area. In high traffic zones, the coverage capability of macrocell becomes less prominent due to long-awaited incoming/outgoing user services. Similarly, a MaBS cannot provide optimum services to end users at remote places. Hence, remote relay nodes and micro-, pico-, and femtocells are used to address the issues of coverage and connectivity in remote and dense traffic regions of the network. Figure 6.1 illustrates the components of multicellular HetNets comprising MaBSs, MiBSs, PiBSs, and FeBSs. Deployment of a large MaBS has many constraints such as space, permissions from licensing authorities, and a coordinated design. However, small cells such as pico- and femtocells are less expensive and are adaptive to network variations.

6.4.2 Microcells

Microcells comprise high-power nodes with low transmission antennas as compared with MaBS. MiBSs are deployed by network operators to improve the coverage of HetNets. MiBSs work in coordination with the nearest MaBS to support users in densely populated areas such as stadiums, universities, and business centers [7].

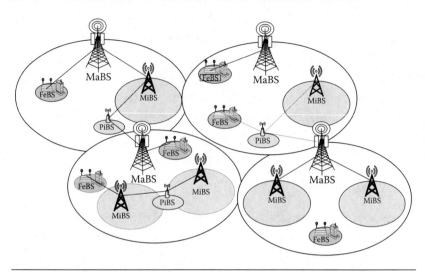

Figure 6.1   Multicellular heterogeneous network comprising macro base station, micro base station, pico base station, and femto base station.

Microcell towers are installed in remote and rural areas [8] to deliver the services to remote users.

6.4.3 Picocells

Picocells comprise low-power nodes, deployed by network operators in hotspot zones such as indoor stadiums and shopping malls. Picocells cover a few meters, and the signal strength within coverage area has less interference with adjacent macro users in the network. The effect of ICI from adjacent macro users reduces the quality of signals for pico users and decreases network performance.

6.4.4 Femtocells

Femtocells comprise low-power nodes and cover less network area as compared with picocells and microcells. FeBSs are deployed in two modes—(1) in the normal mode, potential users extend the services of home network and small office, and (2) in the enhanced mode, network operators support 10–20 users that can further extend the network services. Femto access points provide services for closed access group and registered members of co-channel deployment. In a dense network scenario, the femto signals are strongly affected due to the interferences from adjacent femto-, pico-, and macrocells.

6.4.5 Remote Relay Head Node

RRHNs are deployed at remote places to enhance the coverage of MaBSs. By deriving their energy and network resources from adjacent base stations, RRHNs extend services to rural and remote areas where the usage of spectral resources is minimal. MiBSs are deployed in high-traffic zones to extend the limitations of RRHNs.

6.5 Software-Defined Cellular Networks

Software-defined networking (SDN) is an innovative paradigm for simplifying the task of controlling, scheduling, and managing the control and data planes of the network. The data plane regulates the packet forwarding devices [9] and optimizes the resources through network virtualization. Intelligent and self-organizing techniques enhance the resource-sharing features of small-cell networks and efficiently uses the networking devices. Mobile virtual networks use the services of SDNs to optimize the spectrum bandwidth and facilitate services to mobile end users.

6.5.1 Architecture of Software-Defined Cellular Network

SDN architecture in multicellular HetNets comprises data plane, control plane, and business service plane. The salient features of each plane are explained as follows:

6.5.1.1 Data Plane This plane uses network management and coordination functions to forward the data toward destination. Based on the size of network traffic, a relevant flow-based packet forwarding mechanism is adopted in SDN. The data plane forwards the data packets through dedicated route paths that are regulated by the control plane.

6.5.1.2 Control Plane This plane schedules the network resources, traffic flow and coordinates with gateway nodes, virtual network access points, data-forwarding nodes, and relay nodes of the network. The control plane is backward commutable with openflow architecture and decides the role of data and virtual planes of SDN. Advance reprogrammable and intelligent algorithms are applied to the resource pool to achieve optimum device coordination in the network. Global information and service policies are updated periodically using distributed and centralized techniques in SDNs.

6.5.1.3 Business Service Plane/Interactive Frontal Plane This plane coordinates with service providers, vendors, and users to establish a efficient business service model. The users are connected based on an application and networking interface that can be configured across multiple platforms simultaneously. Based on user requirements and application specifications [10], SDNs implement a contextual approach and define the service policies in the business plane. SDNs support global and local operating systems that coordinate with network services. Thereby, the hotspot user density at cellular level is effectively handled by the adjacent base stations.

6.5.2 Cognitive Small Cells in HetNets

Cognitive radios (CRs) provide promising solutions to spectral scarcities of ever-increasing network traffic conditions in HetNets. In cognitive networks, the secondary users adapt to quick probing of the communication channel and avail the information of available spectral holes in the channel. Macro- and small-cell radios are embedded with cognitive features to improve spectral sensing, spectral analysis, and spectral allocation. This further avoids congestion in dense traffic zones and reduces the effect of ICI. With an efficient allocation method of spectral holes to users in high-traffic areas, the cognitive small cells improve the energy efficiency and achieve optimum throughput in HetNets. The cognitive central base station is capable of sensing and identifying the underutilized and overutilized regions of licensed spectrum and provides an opportunistic access to secondary user nodes. This spectrum information is conveyed to small cells to determine the size of mobile users in high-traffic and ICI conditions. The cognitive femto and pico access points then adjust the sleep and active modes of user nodes irrespective of their location position. The effect of aggregated interference [11] determines the performance in dense HetNets. Small cells enhanced with CRs [12] can also be used to estimate the available spectrum in adjacent tiers and calculate the inter-tier interference. This interference information increases the degree of coordination between macro and small cells in HetNets. Deploying cognitive relay nodes in dense HetNets resolves the issue spectrum holes and contributes to enhanced coordination and reduced ICI in the network.

6.5.3 System Architecture Evolution for 5G

C-RAN technology supports high-speed data traffic and improves the performance of multicellular HetNets. 5G networks are densely deployed small-cell networks with embedded C-RAN and support millions of devices with enabled licensed and unlicensed spectrum bands. 5G networks aim at coordinated fronthaul and backhaul networks with centralized and distributed approach to establish the connectivity in the network. System architecture evolution for 5G networks is reinforced with C-RAN [1] and multi-RAT architectures in multi-small-cell HetNets. Five backbone technologies used in 5G networks [13,14] are (1) C-RAN with CoMP and eICIC, (2) massive MIMO systems, (3) millimeter-wave (mmWave) technology, (4) context-aware user-centric application, and (5) smart devices supporting device-to-device (D2D) and machine to machine (M2M) services. Figure 6.2 illustrates the C-RAN-based system architecture for 5G networks. Backbone technologies used in 5G networks are explained as follows:

1. *C-RAN with CoMP and eICIC*: C-RAN is a low-cost, device-centric, and fronthaul wireless network [14] that comprises the centralized base station and coordinates with distributed antennas of multiple small cells. The coverage area is extended with respect to available resources and the operational cost of the network. Usage of the single band by small cell reduces the complexity of CoMP and eICIC techniques. Network operators and service providers evaluate the network traffic conditions and forward the information to the central base station. This technology allows C-RAN to apply device-centric approach for dynamic allocation of available resources to multicellular nodes.

   CoMP [15] in intra-tier and inter-tier modes potentially improves the beamforming abilities in 5G. Intra-tier CoMP with end user nodes are connected to RRHNs to reduce the

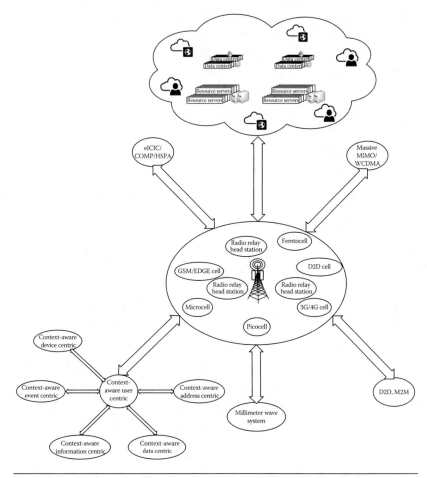

Figure 6.2   Cloud-based radio access network system architecture for 5G networks.

complexity of resource allocation in the network. C-RAN reduces the size of adjacency list (that is limited to nearest neighbors) based on CSI, queue state information, and overhead signal parameters.

2. *Massive MIMO systems*: Multiuser massive MIMO systems create large arrays of coverage envelopes with optimized transmit power and high multiplexing antenna gain. C-RAN systems use distributed antenna systems under a single large base station and envisage the usage of wide range of frequencies extended from macro-meter to mmWave dimensions.

Massive MIMO techniques enable the MaBS and MiBS to work in correspondence with RRHNs to extend the coverage capacities of 5G networks and reduce the size of the antenna array.

3. *Millimeter-wave technology*: mmWave technology supports high-frequency (3–300 GHz) data transmission over a short distance range that is used in the design of 5G communication networks. The frequency range is varied with the magnitude of applications in multicellular networks.

4. *Context-aware user-centric applications*: 5G networks intend to deal with millions of devices with varying context-aware services for user-centric applications. Context awareness technology [16] deals with attributes related to (1) device specifications, (2) event specifications, and (3) application specifications. Context-aware services and interface at user level refers to resource sharing, optimal connectivity and coordination across the devices in the network. Context-aware services at the device level refer to sharing of resources and information between multilayer networks. Context-aware services at the application level refer to sharing of data such as audio, video, and images through social networking applications. Context-aware services at the network level refer to resources availability, quality of communication links, network traffic at interface modules, and the throughput rate. Context-aware services at the physical level refer to infrastructural requirement, geographical location, and environmental conditions.

5. *Smart devices supporting D2D and M2M services*: D2D communication systems are clusters of multicellular devices enabled with device connectivity using a social networking platform. D2D communication achieves enhanced spectral efficiency and decreases the network latency and complexity in network traffic. M2M communication systems are a group of server nodes that store (and forward) the data and establish services with heterogeneous devices in the network.

6.6 Mobile Cloud Computing in Multicellular HetNets

Mobile cloud computing in HetNets enhances the computational capabilities of small cells by allocating the local and global networking resources based on network traffic. Mobile cloud computing supports traffic-offload and enables high-speed packet transmission. Coding and computational distributions in cloud network and active device [17] connectivity optimize the performance of HetNets. Software clones and ternary offload decision systems reduce the problems related to mobile users associated with multiple small cells. The Network Implementation Testbed using Open Source platforms [18] estimates the schedules and reservations of available resources for efficient data management in cloud environment. Energy computation, network performance, and offloading the remote factors with respect to the local coordinates reduce the network complexity.

6.7 Multitier Architecture of Cloud RAN for Efficient Data Management in HetNets

The multitier architecture of Cloud RAN for efficient data management (MTADM) in HetNets comprises of five tiers. Multitier architecture of C-RAN explores the data obtained from small-cell database, macrocell database, and internetwork server database. Knowledge discovery in data rules; extraction, transformation, and loading (ETL); and data mining algorithms are applied to the data before uploading it to C-RAN. The data are further explored using intelligent-based rules and soft computing techniques. Finally, the processed data are sent back to C-RAN tier that distributes the information to macro database, small-cell database, and Internet server database.

Tier-wise architecture of MTADM as shown in Figure 6.3 is elucidated as follows:

- *Network database layer (Tier I)*: This tier comprises the small-cell HetNet database, mobile database, and local network database. The network entities received are sent to the macro database and Internetwork server database.
- *Data preprocessing and management layer (Tier II)*: This tier comprises database and data mining components such as

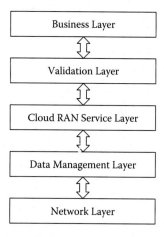

Figure 6.3   Multitier cloud radio access network.

ETL, data aggregation, and clustering. This tier filters the data, reduces data anomalies, and eliminates redundant data from the server.

- *Cloud RAN service layer (Tier III)*: This tier comprises data obtained from the network database, internetwork database, business database, vendor database, virtual network operator's database, and application service provider's database. Cloud RAN database support secure transactions that can be accessed through private, public, and hybrid modes.

- *Computational and validation layer (Tier IV)*: This tier comprises computational algorithms that apply intelligent methods to study the behavior of network model and verify the consistency level. The network model is trained and validated to meet the predictions and recommendations of customers. The revised updated network models are sent to Cloud RAN and macro database to optimize the accessibility with customers.

- *Impact evaluation and business layer (Tier V)*: This tier retrieves the performance assessment from customers and evaluates the impact of revised data obtained from Cloud RAN. The final business model is sent to vendors and networking service providers. The business models approved by the designers, developers, service providers, and vendors are released to facilitate services to customers in multicellular HetNets.

6.7.1 Big Data in HetNets

Big data in HetNets deals with filling the gap in existing analytical models, building innovative business models [19] that are applicable to real-time HetNets. Big data is characterized by its size (MB/GB/TB), type (historical, reviews, current data, etc.), generation, processing speed (byte/minutes, bytes/hours, etc.), degree of trust (password enabled data access), and vitality of information (emerging market trends, impacts in the future).

6.8 Internet of Things in LTE/HetNets

Internet of Things (IoT) is a paradigm that brings various heterogeneous computing and connecting entities into a single large platform and facilitate services to real-time applications such as industrial, healthcare, civil, military, and environmental monitoring. IoT is featured by smart, autonomous, flexible, and interactive social objects that provide trusted services to connect a large number of users across the world. IoT deals with huge data and issues related to context awareness, resource management, multioperating system supportability, interoperability, and trust across the users. The social objects [20] in IoT are classified as follows:

1. *Social-conscious object*: Smart objects coordinate and communicate with the external environment such as human social networks and real-time applications. Emerging trends in smart objects (D2D, M2M, and so on) connect the technical aspects of applications (e.g., healthcare monitoring differs from environmental monitoring). 5G networks are designed with smart objects that are flexible and adaptable in multiple platforms. Object isolation and standalone device data do not provide exact information. Hence, application program interfaces and socially-conscious objects interact with the surrounding environment and derive the contextual information.

2. *Smart-acting object*: IoT connects smart devices that are embedded in monitoring and surveillance applications,

and these smart devices vary in size, topology, and operating conditions. Smart-acting objects use the pseudo-social approach in functional environment and provide services to users. Smart objects connect devices at all levels (machine to users, machines to machines, users to users, and so on) and generate spontaneous actions pertinent to working conditions of the network. Smart objects apply the user-centric approach to coordinate between the users and machines in the network.

3. *Social network building object*: IoT supports a three-layer architecture that comprises the following components: (1) fixed standalone sensors, (2) fixed standalone sensors with mobile sinks, and (3) multiple fixed standalone sensors and mobile sensor nodes with static and dynamic sink nodes. Social objects capable of building context-aware networks improve the quality of service. Registered social network members are recognized by a unique identity (such as active digital identity) [20] to ensure secured and authentic communication.

6.9 Inband/Outband Vehicular Communication in Small-Cell HetNets

HetNets extend physical and MAC services that provide inband (same operational frequencies for macro- and small cells) and outband (diverse operational frequencies for macro- and small cells) spectrum services [21]. The inband underlay spectrum uses limited resources and minimizes the interference levels in sparse vehicular traffic conditions, whereas the overlay inband communications use the adequate number of resources in dense vehicular traffic conditions. For outband vehicular services using the licensed spectrum, the base station controls and coordinates the devices addressed by the small cell. Excess vehicular traffic is redirected to an adjacent Wi-Fi service point to extend the services of small-cell HetNets. Cognitive vehicular users utilizing the unlicensed spectrum bands are controlled and restricted by services of Wi-Fi, NFC, and so on. Table 6.1 enumerates the features of inband and outband device communications.

Table 6.1 Features of Inband and Outband Device Communications

TYPE OF DEVICE COMMUNICATION	SERVICE	ADVANTAGES	DISADVANTAGES
Inband device communication	Overlay	1. Spatial diversities increase spectral efficiency.	1. Leads to underutilization of spectral resources.
		2. Supports only inband frequencies.	2. Allows either cellular or D2D communication.
		3. BS controls and increases quality of vehicular services.	3. Complexity power and interference management.
	Underlay	1. Performance is enhanced since the BS controls and coordinates with the beacon and relay nodes.	1. Interference issues are not addressed adequately.
		2. Multicasting and incremental relay techniques are used to increase the throughput rate.	2. Complexity increases in large-scale HetNets.
Outband device communication	Controlled	1. Dedicating resources are not required; there is minimum interference in vehicular traffic network.	1. Interference due to unlicensed spectrum is not controlled by the BS.
		2. Resource allocation for cellular and vehicular communication is easier.	2. The devices need to support both cellular and Wi-Fi services.
	Autonomous	1. Simultaneous cellular and Wi-Fi services are enabled in vehicular traffic network.	1. Power and interference management is a significant issue; otherwise it leads to wastage of resources.
		2. Reduces the overhead of cellular networks by diverting the additional vehicular network based on adjacent Wi-Fi service points.	2. Encoding and decoding of packets need to be precise and efficient.

Source: Asadi, A. et al., *IEEE Commun. Sur. Tutor.,* 16(4), 1801, 2014.

6.10 Conclusions

Multicellular HetNets consist of varying high-power nodes (e.g., macrocells, microcells, and remote relay head nodes) and low-power nodes (e.g., picocells and femtocells). Beamforming, carrier aggregation, CoMP, and subframe scheduling increase spectral efficiency and achieve high data rates in HetNets. CoMP and eICIC together reduce the effects of interference in small-cell HetNets.

Cognitive radios provide promising solutions to spectral scarcities of ever-increasing network traffic conditions. In multi–small cell HetNets, the system architecture evolution of 5G networks is reinforced with C-RAN and multi-RAT architectures. Multitier C-RAN and mobile cloud computing in HetNets enhance the business and data services by allocating local and global networking resources. Internet of Things for LTE HetNets supports social-conscious object, smart-acting object, and social network building object to extend the functionality and services of multicellular HetNets.

6.11 Future Research Directions

Further research includes the design and deployment strategies for increasing the number of multicellular users. The implementation of software-defined networks and cognitive radio technologies at successive macro- and microbase stations improves the usage of available primary spectrum for secondary users. Resource partitioning methods for eICIC and CoMP and resource constraints for wireless fronthaul and backhaul need to be considered in multicellular HetNets. Multilayer splitting schemes to resolve the interference, and game theory approach for spectrum sensing and spectrum sharing, need to be considered in small-cell networks. Vehicular communications with varying mobility patterns need to be designed for multicellular HetNets. Multitier Cloud RAT/RAN and IoT features to extend the functionality of HetNets need to be considered.

List of Abbreviations

3D	Three-dimensional
5G	Fifth-Generation Networks
BS	Base Station
CDMA	Code Division Multiple Access
CoMP	Coordinated Multipoint
CR	Cognitive Radio
C-RAN	Cloud-Radio Access Network
CRS	Cell-specific Reference Signals
CSI	Channel State Information
D2D	Device-to-Device

DCI	Downlink Control Information
eICIC	enhanced Intercell Interference Coordination
eNBs	Evolved NodeBs
ETL	Extraction, Transformation, and Loading
HeNB	Home eNodeB
HetNet	Heterogeneous Network
ICI	Intercell Interference
ICIC	Intercell Interference Coordination
IoT	Internet of Things
LTE	Long-Term Evolution
LTE-A	LTE Advanced
M2M	Machine-to-Machine
MAC	Medium Access Control
MIMO	Multiple-Input Multiple-Output
MTADM	Multitier Architecture of Cloud RAN for Efficient Data Management
NFC	Near Field Communication
OFDM	Orthogonal Frequency Division Multiplexing
OFDMA	Orthogonal Frequency Division Multiple Access
RAT	Radio Access Technology
RF	Radio Frequency
RRHNs	Remote Radio Head Nodes
SDMN	Software-Defined Mobile Networks
SDN	Software-Defined Network
SDR	Software-Defined Radio
SINR	Signal-to-Interference-Plus-Noise Ratio
Wi-Fi	Wireless Fidelity
WiMAX	Worldwide Interoperability for Microwave Access
WLAN	Wireless Local Area Network

References

1. Peng, M., Li, Y., Zhao, Z., and Wang, C., System architecture and key technologies for 5G heterogeneous cloud radio access networks, *IEEE Networks*, March–April 2015, 29(2), 6–14.
2. Chen, M., Huang, A., and Xie, L., Power synergy to enhance DCI reliability for OFDM-based mobile system optimization, *IEEE Global Communications Conference (GLOBECOM)—Wireless Networking Symposium*, Austin, TX, December 8–12, 2014, pp. 5011–5017.

3. Yun, J.-H. and Shin, K. G., Adaptive interference management of OFDMA femtocells for co-channel deployment, *IEEE Journal on Selected Areas in Communications*, June 2011, 29(6), 1225–1241.
4. Ni, W. and Collings, I. B., A new adaptive small-cell architecture, *IEEE Journal on Selected Areas in Communications*, May 2013, 31(5), 829–839.
5. Zou, K. J., Yang, K. W., Wang, M., Ren, B., Hu, J., Zhang, J., Hua, M., and You, X., Network synchronization for dense small cell networks, *IEEE Journal of Wireless Communications*, April 2015, 22(2), 108–117.
6. Tian, P., Tian, H., Gao, L., Wang, J., She, X., and Chen, L., Deployment analysis and optimization of macro-pico heterogeneous networks in LTE-A system, *IEEE 15th International Symposium on Wireless Personal Multimedia Communications (WPMC)*, Taipei, Taiwan, September 24–27, 2012, pp. 246–250.
7. Dufkova, K., Popovic, M., Khalili, R., Boudec, J.-Y. L., Bjelica, M., and Kencl, L., Energy consumption comparison between macro-micro and public femto deployment in a plausible LTE network, *ACM Second International Conference on Energy-Efficient Computing and Networking (e-Energy)*, Columbia University, New York, May 31 to June 1, 2011, pp. 67–76.
8. Zhang, L., Gang, F., and Shuang, Q., A comparison study of coupled and decoupled uplink heterogeneous cellular networks, *Computing Research Repository (CoRR)*, cs.NIs, arXiv:1502.01887 [cs.NI], February 6, 2015, pp. 1–7.
9. Yang, M., Li, Y., Jin, D., Zeng, L., Wu, X., and Vasilakos, A. V., Software-defined and virtualized future mobile and wireless networks: A survey, *Springer Journal of Mobile Networks and Applications*, February 2015, 20(1), 4–18.
10. Kreutz, D., Ramos, F. M. V., Verissimo, P., Rothenberg, C. E., Azodolmolky, S., and Uhlig, S., Software-defined networking: A comprehensive survey, *Proceedings of the IEEE*, January 2015, 103(1), 14–76.
11. Wildemeersch, M., Quek, T. Q. S., Slump, C. H., and Rabbachin, A., Cognitive small cell networks: Energy efficiency and trade-offs, *IEEE Transactions on Communications*, September 2013, 61(9), 4016–4029.
12. Wang, W., Yu, G., and Huang, A., Cognitive radio enhanced interference coordination for femtocell networks, *IEEE Communications Magazine, Series: Heterogeneous and Small Cell Networks*, June 2013, 51(6), 37–43.
13. Bangerter, B., Talwar, S., Arefi, R., and Stewart, K., Networks and devices for the 5G era, *IEEE Communications Magazine, Topic—5G Wireless Communication Systems: Prospects and Challenges*, February 2014, 52(2), 90–96.
14. Boccardi, F., Jr., Heath, R. W., Lozano, A., Marzetta, T. L., and Popovski, P., Five disruptive technology directions for 5G, *IEEE Communications Magazine*, February 2014, 52(2), 74–80.
15. Peng, M., Li, Y., Jiang, J., Li, J., and Wang, C., Heterogeneous cloud radio access networks: A new perspective for enhancing spectral and energy efficiencies, *IEEE Wireless Communications*, December 2014, 21(6), 126–135.

16. Perera, C., Zaslavsky, A., Christen, P., and Georgakopoulos, D., Context aware computing for the internet of things: A survey, *IEEE Communications Surveys & Tutorials*, First Quarter 2014, 16(1), 414–454.

17. Mazza, D., Tarchi, D., and Corazza, G. E., A partial offloading technique for wireless mobile cloud computing in smart cities, *IEEE European Conference on Networks and Communications (EuCNC)*, Bologna, Italy, June 23–26, 2014, pp. 1–5.

18. Pechlivanidou, K., Katsalis, K., Igoumenos, I., Katsaros, D., Korakis, T., and Tassiulas, L., NITOS testbed: A cloud based wireless experimentation facility, *IEEE 26th International Teletraffic Congress (ITC)*, Karlskrona, Sweden, September 9–11, 2014, pp. 1–6.

19. Assuncao, M. D., Calheiros, R. N., Bianchi, S., Netto, M. A. S., and Buyya, R., Big data computing and clouds: Trends and future directions, *Elsevier Journal of Parallel and Distributed Computing*, August 25, 2014, 79–80, 3–15.

20. Atzori, L., Iera, A., and Morabito, G., From "smart objects" to "social objects": The next evolutionary step of the internet of things, *IEEE Communications Magazine, Series: Ad Hoc and Sensor Networks*, January 2014, 52(1), 97–105.

21. Asadi, A., Wang, Q., and Mancuso, V., A survey on device-to-device communication in cellular networks, *IEEE Communications Surveys & Tutorials*, Fourth Quarter 2014, 16(4), 1801–1819.

7

ENERGY-EFFICIENT MOBILE WIRELESS NETWORK OPERATIONS FOR 4G AND BEYOND USING HETNETS

ISAAC WOUNGANG,
GLAUCIO H.S. CARVALHO,
ALAGAN ANPALAGAN, SANJAY KUMAR
DHURANDHER, MD MIZANUR RAHMAN,
AND KAZI R. ISLAM

Contents

7.1 Introduction

The last decade witnessed an astronomical growth in the demand for wireless access. The earlier predominant service was mobile telephony, but with the rapid development of the information and communications technology (ICT) sector, the focus of wireless access has progressively been shifted to mobile Internet due to driving force applications such as Facebook, Google Suite, and YouTube, to name a few. Some recent statistics [1] have revealed that (1) in the first half of 2014 the number of smartphones in use has surpassed the number of PCs accessing web services; and (2) the ICT data traffic is expected to drastically increase by about 1000-fold by 2020, with an annual growth rate of 66% [2], and this increasing demand will quickly drain the scarce radio resources. As new applications emerge on a daily basis, the number of mobile users will continue to augment, leading to potential challenges for mobile network operators (MNOs) in terms of achieving the delivery of their services with appropriate quality of service (QoS) while at the same time maintaining an acceptable level of energy consumption resulting from their operations.

According to the teletraffic theory, as long as the offered traffic increases, the number of resources should also increase in order to maintain the same QoS level. For mobile networks, 4G long-term evolution (LTE)-based networks, and future 5G standards-based mobile networks, this requirement is equivalent to that of acquiring more energy-efficient wireless infrastructures from an MNO perspective. As stated in [3], base stations (BSs) are the major consumers of energy by MNOs and the use of more active BSs implies that more energy is expected to be consumed when operating the system, which may lead to increased greenhouse gas emission in the atmosphere. On the other hand, envisaging a network expansion will also have a huge impact on MNOs' operational expenditures since doing so may lead to the use of more human resources by MNOs for network operations, administration, and maintenance activities [4]. For these reasons, there is a clear demand for more sustainable green network design and deployments from an MNO's perspective.

With focus on the aforementioned target, an interesting cost-effective solution would be to promote some kind of cooperation between MNOs so that they can share their respective core wireless access infrastructures since a mobile user equipped with a multiradio interface device such as an Apple iPhone or a mobile PDA device can be connected to any of the available MNOs' wireless access networks. This vision of cooperation is supported by the heterogeneous wireless networks (HetNet) paradigm [5], which has arisen from the request that MNOs must be able to operate their core wireless access networks in a manner that is truly seamless and transparent to their respective clients. These wireless access networks may be composed of various different radio access technologies (RATs) and formats of cells, to name a few. As such, for HetNets to become truly operational from a deployment perspective, LTE HetNets have been introduced [6] in order to enable MNOs to meet the requirements of a host of scenarios that may occur within these networks. These requirements are yet to be investigated; and they can be formulated in terms of providing the proper level of coverage and performance at the cell edges, or in terms of putting together the small bands in order to provide the bandwidth needed for 4G LTE, or in terms of providing the same level of service quality independently of the wireless access network technologies in presence and the areas within the cells and so on.

According to [7], there is currently no proposed comprehensive framework standard that describes how MNOs should be prepared for HetNet deployments that also help automate the upgrading of 4G technologies and standards to future 5G technologies and standards when these will become available. Due to their design features, HetNets can be exploited by MNOs to achieve some significant gains in terms of energy savings during their operations. Assuming that there is some form of cooperation among MNOs forming the HetNet, their wireless access resources will also be commonly shared among their subscribers. Thus, radio resource management (RRM) algorithms for HetNets [8] should be designed with the goal of managing the allocation of resources among all incoming service requests while minimizing the power consumption of each of the wireless access infrastructures supported by MNOs.

In this chapter, the focus is on RRM algorithms that can provide some design guidelines to orient wireless engineers in achieving greater energy efficiency on MNOs' operations. To the best of our knowledge, HetNets have been extensively investigated in the recent years [8,9], but there is no prior work that specifically discusses the integration of HetNets with emerging spectrum technologies and/or RRM techniques, and then analyzes their potential in terms of energy consumption saving.

The remainder of the chapter is organized as follows. In Section 7.2, the concept of HetNets is overviewed, and the design components and considerations for 4G LTE and future 5G-based mobile systems are discussed. In Section 7.3, some RRM schemes for HetNets are discussed. In Section 7.4, some energy-efficient schemes for HetNets are presented. In Section 7.5, the proposed energy-efficient RRM design for HetNets is presented. In Section 7.6, some discussions on how HetNets can be integrated with emerging spectrum technologies or RRM schemes to achieve green designs are presented. In Section 7.7, preliminary numerical results validating the effectiveness of the proposed RRM design are presented. Section 7.8 concludes the chapter. Finally, Section 7.9 highlights some future research directions.

7.2 4G and 5G HetNets System Design Components and Considerations

This section introduces the HetNet paradigm [5] as well as the design components and considerations for 4G LTE and future 5G-based mobile systems from a resource management perspective.

7.2.1 HetNets

HetNets [5] are typically composed of a conglomeration of multiple wireless access networks and technologies (RATs) that are expected to operate together in a complementary and collaborative manner, while leveraging their features for the benefit of the users. Examples of RATs include Wi-Fi, Bluetooth, Zigbee, WiMAX, and IEEE 802 WLANs, to name a few. An interesting feature of HetNets is that some of these wireless access networks are overlaid by others, forming a multilayer structure as shown in Figure 7.1. In Figure 7.1, K RATs coexist in the same environment, where the jth RAT, $j = 1, ..., K$ is composed of a wireless link with B_j radio resources. These radio resources are shared by incoming service connections, and it should be noted that a unit of resource has a physical meaning, which is dependent on the implementation of the radio interface. On the other hand, independently of the type of access technology used (orthogonal frequency-division multiplexing, time division multiple access, code division multiple access, frequency division multiple access, etc.), the system capacity can be interpreted in terms of effective bandwidth requirement [10].

When an incoming service connection requires access to the HetNet, an RRM software agent referred to as optimal joint call admission control (JCAC) has to decide on its acceptance or rejection, and, if accepted, to which RAT it should be directed for processing. Since every user equipment (UE) subscribes with one MNO through a home network, the selected RAT must belong to that home network or an alternative access network belonging to another MNO, assuming that a service-level agreement [11] prevails between these networks, which enables the UE to still access the same services offered by its home network when roaming in the alternative network area.

An example of HetNet structure where the wireless access networks of two MNOs are expected to share their wireless infrastructures in order to satisfy the growth in offered traffic is shown

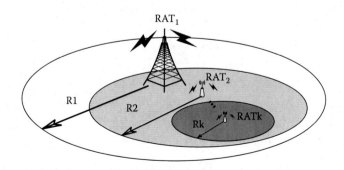

Figure 7.1 HetNet multilayer structure.

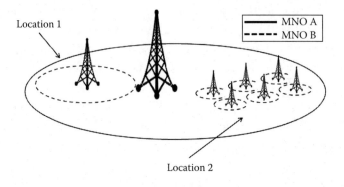

Figure 7.2 Example of HetNet with two MNOs.

in Figure 7.2. In Figure 7.2, the wireless access network of MNO A can be an IEEE 802.16 WiMAX-based solution with a large coverage area, whereas that of MNO B can be a 3G cellular mobile network with macrocells (shown in location 1) and a dense deployment of smaller cells, micro-, pico-, or femtocells (shown in location 2). In both locations, MNOs can benefit each other by sharing their wireless infrastructures. For instance, in location 1, the users of both MNOs can enjoy the large coverage area and the higher date rates that are provided by 3G solutions and WiMAX. This type of multilayer architecture can be effectively explored to match multiple design characteristics, including

- System capacity boost—By combining the capacity of each individual wireless access network within the intended coverage region, the whole system may support many more users.
- Increased system coverage—For instance, by combining WiMAX and cellular mobile networks, a large geographical area can be covered.
- Enhanced user satisfaction—Given the differences in technologies and data rates, each wireless network can be utilized to satisfy a specific user's desired target.
- Offered access cost options to end users—MNOs will have the flexibility in designing the market strategies that are appealing to their end users.

Of course, to take full advantage of the aforementioned multilayer architecture, a mobile user must be equipped with the appropriate multi-interface device capable of sensing and connecting the wireless access network that matches the best his/her personal expectations and application requirements.

7.2.2 4G and 5G HetNets System Design Components

4G-based mobile systems (so-called 4G HetNets) have been designed as multilayered cellular topologies that are made of self-organizing, user-operated indoor and low-power small cells such as femtocells, picocells, microcells, metrocells, and WLANs. These cells can be designed using a variety of architectures [12] that are overlaid by a macro cell with interfering wireless links and high powers (typically 20–40 W per carrier).

This sophisticated combination of heterogeneous small and large cells that share the same radio spectrum in the same geographical area are expected to be managed by the same MNO, leading to several challenges in terms of traffic management over the multilayered cell boundaries, in particular, the challenge of radio resource sharing and co-channel interference management, for which adaptive and dynamic channel allocation approaches such as opportunistic spectrum access and dynamic spectrum access (DSA) should be designed to cope with.

One of the solution strategies that have been proposed in the literature has been to exploit the spatiotemporal diversities among transmissions in the orthogonal frequency division multiple access-based medium access in 4G HetNets. Following this trend, a framework (referred to as hybrid radio resource sharing scheme) was proposed in [13]. This scheme uses a dynamic switching algorithm to enable a link adaptation module and a cognitive radio (CR)-based resource sharing module to work in coordination to determine whether the orthogonal channelization or the nonorthogonal channelization process should be invoked, based on some constraints such as availability of radio resources, realization of all traffic demands, and mobility handoff, to name a few.

According to [14], 5G-based HetNets and networks are expected to support a converged, unified, and flexible control plane for a heterogeneous access network environment based on the required characteristics of functions in a 5G system. Such control plane considers different users and data plane concepts. It also includes some network control functions for managing the next-generation devices and services while maintaining the support for 4G-based systems and their currently known features. Examples of data plane concepts that are based on some implementation principles are software-defined networks (SDNs) and network functions virtualization (NFV) methods [15], which can both support the development of 5G technologies.

SDNs are meant to enable the network control plane (operated, for instance, by MNOs) to be managed separately from the forwarding plane. For instance, using SDNs, the LTE Evolved Packet Core in heterogeneous 5G mobile networks can be implemented by splitting its key functions between an SDN-based transport infrastructure and a virtual cloud environment, providing that a method be introduced to optimize the virtual resource allocation and usage. The benefit of this mechanism is that it can facilitate the load balancing over the different wireless technologies comprising the system.

On the other hand, using an NFV methodology will allow MNOs to virtualize various software components based on some network functions instead of having to manage the dedicated hardware infrastructures of the system. According to [15], this method can help facilitating the inter-domain QoS and management of resources based on traffic patterns and services requirements. For instance, using NFV, multiple instances of the network functions, each running its own VM, can be co-located on the same hardware, hence facilitating the dynamic allocation of resources and functions to the different components of the 5G-based system.

7.3 RRM Schemes for HetNets

To take advantage of the HetNet characteristics as discussed in Section 7.2.1, an RRM algorithm for HetNets must be designed to meet the following goals: (1) determine whether an incoming service request is to be accepted or denied and (2) select the appropriate wireless access network (RAT) that can accommodate the service request while achieving a given performance target (or combination of targets) such as network access cost, security, data rate, mobility, and energy efficiency, to name a few. Typically, the RRM algorithm achieves these tasks by following a two-step process referred to as network selection and RRM processes.

7.3.1 *Network Selection Process*

The network selection process is invoked by the RRM algorithm during two distinctive phases. The first one (called initial network selection phase) refers to when a user wants to gain access to the network resources by initiating an incoming service request. In wireless networks, the users migrate from one point to another, making handoffs. This corresponds to the second phase where the network selection is invoked and the best wireless network option to host the mobile user has to be selected. Although both phases stand for network selection, they are quite different in terms of priorities and objectives. For instance, handoff calls usually have priority over new calls as long as there are previous calls in progress in the system; thus dropping them is more annoying for users than denying their first access service requests. Since smaller cells consume less power and have high capacity due to a higher signal-to-noise ratio, they are the best candidates to be selected as nodes of the initial wireless access network. In this case, it is not mandatory to address the mobility issue. On the other hand, for a user crossing the coverage area, mobility is the most important goal to be satisfied in order to guarantee a seamless connectivity.

7.3.2 *RRM Process*

The RRM process is meant to efficiently distribute the radio resources among different users of the network while achieving a balance in terms of each user's requirements and MNO's profitability. More precisely, the RRM problem consists of assigning a server for the incoming service requests as long as the service provisioning of the other ongoing users is not violated and there is enough capacity left for accommodating these requests. Approaches used to solve this problem involve (1) designing the traffic management mechanisms (e.g., call admission control [CAC] schemes) that are able to maintain the QoS provisioning and requirements, and (2) designing scheduling mechanisms that can help deciding whether a data packet accepted by the CAC scheme should be transmitted or not in the radio resources.

In the HetNets context, traditional RRM frameworks for homogeneous networks are not suitable for use since the information pertaining to the whole vision of the system is not necessarily taken into account in their original designs [8,16]. Also, the radio resources of the wireless access networks composing the HetNet may be extremely scarce, and the QoS requirements of their targeted applications may be diverse and often conflicting. For these reasons, in a typical multi-layer HetNet architecture (Figure 7.1), a JCAC mechanism such as the one illustrated in Figure 7.3 should be invoked to efficiently distribute the incoming service requests among the available RATs based on criteria such as blocking probability of the incoming service request, user satisfaction in terms of QoS demands, power consumption at each RAT, delay in service request transaction, communication overhead, and traffic load distribution, to name a few [5], assuming that the ith RAT ($1 \leq i \leq K$) contributes B_i radio channels to the common pool of resources to be managed. These radio channels can be assumed to be orthogonal so that cross-RAT interference is avoided.

As shown in Figure 7.3, the JCAC mechanism attempts to accept as many mobile users as possible in order to optimize system utilization, even though a portion of the incoming service requests will be blocked due to insufficient available radio resources in the associated RATs. For the JCAC mechanism to work effectively, a level of coupling [5] must exist that enforces the cooperation among the available RATs. This requirement is realized by implementing a decision-making algorithm within the JCAC scheme, which takes into account the aforementioned criteria.

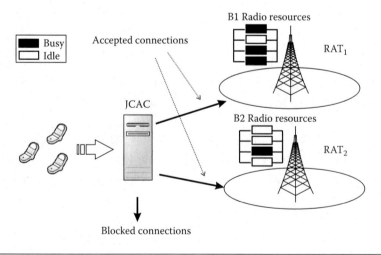

Figure 7.3 JCAC in a multi-RAT environment.

7.3.2.1 JCAC-Based RRM Schemes Several JCAC-based RRM schemes for HetNets have been investigated in the literature. Representative ones are discussed as follows. In [8], the key challenges that constitute an efficient RRM design for HetNets are discussed, and a taxonomy of RRM solutions in terms of decision-making criteria is proposed. Similarly, in [5,9], some JCAC-based approaches for HetNets are proposed, each of which is meant to perform a joint cooperative management of all resources involved in the participant RATs. In [17], Chen et al. proposed a DCM scheme for HetNets made of macro and smaller cells. This algorithm makes use of a queuing theory framework to analyze the trade-offs that should be considered when dealing with green RRM designs for HetNets, namely, deployment efficiency versus energy efficiency, spectrum efficiency versus bandwidth efficiency, power consumption versus delay, to name a few. Some discussions on how these trade-offs can impact the performance and operations of future wireless systems from an MNO's perspective are also highlighted.

In [18], Lorincz et al. proposed a DCM-based green RRM scheme in HetNets that can guarantee a certain QoS when optimizing the power consumption. Their proposed scheme is shown to optimize the management of the on/off state and transmission power of access stations based on traffic requirements and application scenarios, achieving up to 50% of energy consumption saving. In [19], Cheng et al. proposed the use of CR to address the interference mitigation problem caused by a random deployment of femtocells inside macrocells. A game theory method is used to solve this problem, and some discussions on how the use of CR can alleviate the interference mitigation problem in two-tier HetNets are also provided.

In [20], Gur et al. proposed a green RRM design for HetNets where femto BSs are equipped with a CR and work as secondary users (SUs) in a DSA strategy. Their proposed scheme combines an infrastructure-based overlay CR network with a femtocell-based CR scheme to provide dynamic spectrum access and sensing capabilities that are better than what would have been obtained if a single constituent of the architecture was considered. In [21], Bennis and Niyato proposed a reinforcement learning-based self-optimization, organization, and self-healing (SON) algorithm that runs in each femto BS, and showed that when only the local information is used, femtocells are capable of achieving self-organization. Their proposed algorithm is tested, showing a considerable interference reduction among femto-cells and the macrocell.

In [22], Morosi et al. proposed two joint RRM (JRRM) schemes, namely, the so-called forecasting-based sleep mode algorithm and the so-called real-time traffic measurements for sleep mode algorithm. For each of these algorithms, the energy saving is achieved by turning on/off base transceiver stations' power amplifiers only. In [23], Bousia et al. proposed an RRM scheme that involves the use of base trans-ceiver stations' power amplifiers only switching on/off procedure in a dual-carrier UMTS network. Their scheme is shown to considerably reduce the network power consumption even when a less aggressive power-saving strategy is employed.

In [24], Carvalho et al. proposed a green JRRM scheme for HetNets that uses the DCM-based algorithm as decision maker to determine when a microcell should be switched on/off to save some energy. Several scenario applications are investigated, showing prom-ising results from an MNO's operations perspective in terms of effi-ciency in energy consumption. Following the same trend, this chapter continues the investigation of the initial RAT selection method for HetNets introduced in [25]. In this method, the optimal JCAC is formulated within the framework of semi-Markov decision process (SMDP), using a HetNet model composed of two RATs as will be described in Section 7.5.

7.3.2.2 RRM Design Perspective for 5G HetNets   From the 5G-based HetNets perspective, HetNets composed of macrocells and high density of deployed small cells are currently envisaged as the most suitable solution to satisfy the expected average area spectral effi-ciency, cognitive radio resource management (CRRM), and energy dissipation requirements, for 4G/future 5G standards–based mobile networks [26]. In this regard, new network architectures and RRM schemes should be investigated. Following this trend, Omar et al. [26] have proposed an architecture for 5G HetNets, made of two components: a data collection system and a decision support sys-tem, meant to control and manage the spectrum allocation strat-egy. In [27], the challenge for designing a CRRM scheme on top of the resource allocation scheme in order to control Layer 1 and

Layer 2 radio operations while adapting to diverse communication paradigms in LTE-A/LTE-B is addressed. The most recent developments of the CRRM into a software-defined design are also discussed, leading to the design of an SDN processor architecture for the LTE family of systems. Some suggestions on how this architecture can be adapted for the next evolution of cellular network designs are also highlighted.

7.3.2.3 Power Consumption Models for HetNets An example JCAC-based JRRM scheme is given in Figure 7.7, where each RAT is associated with a controller. Typically, the appropriate specification of a JRRM scheme for a HetNet depends on the knowledge of the traffic model, the locations of the mobile users, and the time periods when these users have issued their service requests to the system or have completed them, to name a few. Based on these parameters, a HetNet model for energy/power consumption can be designed. Examples of such designs are provided in [28–30], each of which is a function of various system parameters such as number of users at busy hour, power consumed at each site, and bit rate, to name a few. It should be noticed that the appropriate selection of these parameters is hard to determine [16]. For this reason, linear power models such as the one introduced in [31] have been suggested, with an example implementation given in [32].

In the aforementioned JCAC decision-making algorithm, depending on the type of offered traffic model considered, the UE can choose to operate either in connected mode or in idle mode. Therefore, using some suggested energy models, an MNO can establish a power-saving strategy at the network level. For instance, a threshold mechanism can be set on the traffic destined to specific selected RATs in such a way as to control the macrocell radio resource occupancy and the operations of RATs for energy-saving purpose [33].

7.4 Energy-Efficient Schemes for HetNets

Green design in the ICT sector is an emerging topic that has gained a lot of attention in the recent years. Several initiatives for improving the energy efficiency of communication systems have been proposed [5,34,35]. An example of such initiative in the context of HetNets [35] has been to minimize the energy consumption through reducing the network scanning operations that take place when the network selection procedure starts. Another initiative [36] has been to determine the most energy-efficient wireless access network that can be used to run applications such as FTP, web, and video conference, to name a few. This decision is made by considering the fact that real and non-real-time applications have different characteristics not only in terms of QoS requirements, but also in terms of energy consumption. For real-time applications, the energy consumption matches the power consumption, whereas for non-real-time applications, that observation is not always true as long as the energy consumption is proportional to the power consumption divided by the data rate. In that case, the network with the higher power consumption may be selected as long as it has the highest possible data rate.

Given the fact that the majority of power consumption by MNOs is attributed to the operations of BSs in their wireless access infrastructures, it is believed that the greater achievements in terms of green MNO operations can be realized by properly managing these network elements [37]. Furthermore, the size of the cells has a deep impact on the power consumed by a BS in the sense that the larger the cell coverage is, the more power the BS will consume in order to transmit its signals. Thus, the design of energy-efficient RRM algorithms by wireless engineers must be guided by the fact that the more active the BSs are, the more power they will consume, which may be dependent on the cell size. With this principle in mind, wireless engineers can establish a network selection process within RRM algorithms, with the goal that the most energy-efficient wireless access network will be selected for an incoming service request. Of course, this decision-making process must also take into account other systems objectives and aspects. For instance, GSM networks (known as 2G technology) do not support the delivery of higher date rates; thus, they cannot be used for carrying out the Internet services. In summary, addressing the energy-efficiency concern at the management level of the radio resources (from the BSs side) is a driving force for the design of eco-friendly MNO operations. To achieve this goal, wireless engineers must consider the pros and cons of various RATs that formed the HetNet architecture, thereby contributing to its operations.

Energy-efficient schemes for HetNets so far proposed in the literature can be grouped as follows:

- Schemes that involve powerful hardware designs meant to significantly reduce the power consumption of BSs, for instance, cooling hardware solutions, where the BS components in fast sleep modes are implemented to adjust the traffic load levels for reducing the energy consumption [38].
- Energy-aware radio link–based schemes that enable a BS to decide on when and how data should be transmitted to the users of the HetNet within its cell assuming the knowledge of the channel and load conditions of the surrounding cells.
- Prediction models to estimate the total power consumption in the wireless access networks composing the HetNet using bottom-up approaches, such as the ones proposed in [39,40].
- Schemes that are designed with the goal to optimize the cell size and deployment in terms of energy consumption based on some conditions [41].
- Schemes that enable the possibility of offloading the data traffic of the indoor users to femtocells [42], resulting in a kind of joint macro-femto cell deployment scenario, where more outdoor users are served with equivalent system resources.
- Schemes that enable the wireless access networks to adapt their characteristics to the incurred dynamic load variations or to reduce the active nodes in the network. For instance, such goal can be achieved by studying the spatiotemporal variation of mobile subscribers along with their traffic demands [43], by shutting down BSs when the traffic load is low using suitable vertical handover schemes [44], by using optimization solutions [45]; schemes that implement a bandwidth adaptation approach, which rely on the idea that less frequency resources tend to reduce the node's transmission power [46].

A comprehensive up-to-date classification of energy-aware schemes that aim to lower the energy consumption of wireless access networks is provided in [47].

7.5 Proposed Energy-Efficient RRM Design for HetNets

Several RAT selection methods for HetNets have been proposed in the literature. Representative ones are described as follows. In [5], Lopez-Benitez and Gozalvez proposed the CRRM scheme, which helps distribute the heterogeneous traffic among the available RATs while accounting for the radio resources available at each RAT. Their algorithms are shown to achieve the appropriate user/service QoS levels based on predefined decision criteria used to determine the most suitable user-to-RAT assignment. Similarly, in [48], the problem of convergence of wireless access networks was investigated and an RRM architecture for HetNets was proposed, which relies on the IEEE 802.21 framework. In [37], Suleiman et al. proposed a JRRM scheme for HetNet that relies on several attributes of the constituent wireless access technologies, such as load balancing, variability of network resources based on traffic conditions, and asymmetry of access networks overlap, which are used as design constraints. In [49], Kajioka et al. proposed an adaptive resource allocation scheme for HetNets, in which each node determines the wireless network resources to be assigned to every application that it supports. In [50], Giupponi et al. proposed a JRRM scheme that can achieve an efficient usage of a joint pool of resources belonging to different RATs. In [51], Falowo and Anthony Chan proposed a dynamic RAT selection scheme for assigning a multimode terminal with a single or group of calls to the most suitable RAT in HetNets. In [52], Haldar et al. proposed a cross-layer RRM scheme for network and channel selection in heterogeneous cognitive wireless networks, which classifies the user's application based on the so-called analytic hierarchy process algorithm and selects the most suitable channel according to the user's requirements. In [53], Porjazoski and Popovski proposed an RAT selection method for choosing the new incoming and handover calls based on service type, network load, and mobility of the user. In [54], Zhu et al. investigated the JCAC problem in HetNets and proposed an immune optimization algorithm to solve it. In their approach, the user's preference and traffic load distribution are optimized based on the dynamic pricing concept. In [55], Si et al. proposed an optimal RAT selection method for HetNets using the application layer QoS and network access price as criteria. A selection policy scheme is also introduced that assigns some indices to the candidate wireless access networks. These indices are then used for selecting the best possible wireless access network that can suitably handle the user's requests.

In this chapter, an RRM approach for HetNets is proposed, in which the JCAC-based decision process relies on the design of a cost function. The details description of the proposed method follows.

7.5.1 HetNet Model

The considered HetNet architecture is composed of two co-located RATs as shown in Figure 7.4, where the jth RAT (j = 1, 2) has N_j radio resources. We assume that there are two types of service connections (although more types of service connections could be considered), where the ith service connection follows a Poisson process with parameter λ_i and requires b_i radio resources. The channel holding time is assumed to follow an exponential distribution with mean rate μ_i and traffic intensity $\rho_i = \mu_i/\lambda_i$.

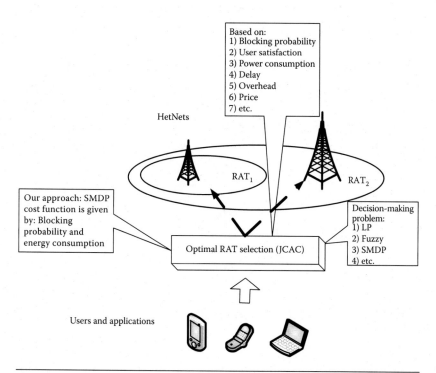

Figure 7.4 Proposed approach.

When an incoming service connection (from a user) requires access to the network, it is expected that the RAT selection method will decide on its acceptance or rejection. If acceptance is granted, the network selection method will decide on which RAT the incoming service connection must be directed to for processing. To address this challenge, we have modified the network selection method (Figure 7.4) introduced in [25]. This method formulates the RAT selection problem in HetNet using an SMDP-based model, for which the derived JCAC optimal policy determines the optimal RAT, that is, the RAT that should be used to handle the incoming service connection.

Using the same SMDP model, the contribution of this chapter consists in deriving the JCAC optimal policy based on the introduction of a new network cost function that weights three criteria: the blocking cost—that is, the cost incurred whenever an incoming service connection is blocked by the system; an access cost—that is, the cost of accessing a certain type of service within a given RAT; and the energy consumption cost—that is, the amount of energy needed to operate a BS or access point. These weighted costs provide the opportunity to MNOs to decide on the relative importance of each of the system's objectives. For instance, if an MNO's objective is to reduce the overall network energy consumption cost as much as possible, then the weight assigned to the energy consumption cost will be much higher than that assigned to the access cost and blocking cost, with the expectation that the end result will be an energy-efficient RRM scheme.

7.5.2 SMDP Model

The state space of the SMDP is defined as a five-tuple [25]:

$$S = (n_{11}, n_{21}, n_{12}, n_{22}, e) \tag{7.1}$$

where $e = [0\ 1\ 2]^T$, Z^T denotes the transpose of matrix Z, and for each RAT, the following constraints are satisfied:

$$0 \leq n_{ij} \leq \left\lceil \frac{N_j}{b_i} \right\rceil, \quad e = [0\ 1\ 2]^T \tag{7.2}$$

where

n_{ij} is the number of calls of type i connection in RATj
N_j is the capacity of RATj
b_i is the bandwidth required by type i connection
$e = 0$ is the departure of a connection
$e = 1$ is the arrival of a connection of the type 1
$e = 2$ is the arrival of a connection of type 2

In every state $x \in S$, the controller can choose one of the following possible actions upon the arrival of a new incoming request:

$$A(x) = \begin{cases} B, & \text{if } e = 0,1,2 \\ AR1, & \text{if } e = 1,2, \text{ and } b_i + b_1 n_{11} + b_2 n_{21} \leq N_1, i = 1,2 \\ AR2, & \text{if } e = 1,2, \text{ and } b_i + b_1 n_{12} + b_2 n_{22} \leq N_2, i = 1,2 \end{cases} \quad (7.3)$$

When the system is in state $x \in S$ and the action $a \in A(x)$ is chosen, the expected time until the next decision epoch is determined by [25]

$$\tau(x,a) = \frac{1}{\lambda_1 + \lambda_2 + n_{11}\mu_1 + n_{21}\mu_2 + n_{12}\mu_1 + n_{22}\mu_2} \quad (7.4)$$

where

λ_i is the arrival rate for type i call
n_{ij} is the number of calls of type i connection in RATj
μ_i is the channel holding time for service class i, $i = 1,2$

If $p(x, y, a)$ denotes the probability that at the next decision epoch, the system will be in state $y \in S$ if action $a \in A(x)$ is chosen in state x, the transition probabilities in case of arrival of type 1 call are obtained as [25]

$$tp(x,y,a) = \begin{cases} \lambda_1 \tau(x,a), x = (n_{11}, n_{21}, n_{12}, n_{22}, 1) \\ \quad \Rightarrow y = (n_{11} + 1, n_{21}, n_{12}, n_{22}, e), & \text{if } a = AR1 \in A(x) \\ \lambda_1 \tau(x,a), x = (n_{11}, n_{21}, n_{12}, n_{22}, 1) \\ \quad \Rightarrow y = (n_{11}, n_{21}, n_{12} + 1, n_{22}, e), & \text{if } a = AR2 \in A(x) \\ \lambda_1 \tau(x,a), x = (n_{11}, n_{21}, n_{12}, n_{22}, 1) \\ \quad \Rightarrow y = x, & \text{if } a = B \in A(x) \end{cases}$$

$$(7.5)$$

The transition probabilities in case of arrival of type 2 call (respectively in case of departure of both types of calls) are obtained in a similar way.

7.5.3 JCAC Optimal Policy

If the system is in state $x \in S$ and the action $a \in A(x)$ is chosen, the admission by the JCAC scheme is determined based on the following cost function:

$$C(x,a) = \omega_1 g_{bc}(x,a) + \omega_2 g_{ac}(x,a) + \omega_3 g_{ec}(x,a) \tag{7.6}$$

where

ω_k, $k = 1, 2, 3$ are some weights, with $\omega_1 + \omega_2 + \omega_3 = 1$
$g_{bc}(x, a)$ is the blocking cost function
$g_{ac}(x, a)$ is the network access cost function
$g_{ec}(x, a)$ is the energy consumption cost

It should be noticed that the values assigned to the above weights can be adjusted according to the MNO's need.

In Equation 7.6, $g_{bc}(x, a)$ is obtained as

$$g_{bc}(x,a) = \begin{cases} BC_i, & e = 1,2 \text{ and } a = B \\ 0, & \text{otherwise} \end{cases} \tag{7.7}$$

where BC_i is the blocking cost of the ith service class; $g_{ac}(x, a)$ is obtained as

$$g_{ac}(x,a) = \begin{cases} \dfrac{A_{ij} - \min_{1 \le i, j \le 2}(A_{ij})}{\max_{1 \le i, j \le 2}(A_{ij}) - \min_{1 \le i, j \le 2}(A_{ij})}, & \text{if } e = 1,2 \text{ and } a = AR_j \\ 0, & \text{otherwise} \end{cases} \tag{7.8}$$

where A_{ij} is the access cost for the service of type i for RATj; and $g_{ec}(x, a)$ is obtained as

$$g_{ec}(x,a) = \begin{cases} \dfrac{E_j - \min_{1 \le j \le 2}(E_j)}{\max_{1 \le j \le 2}(E_j) - \min_{1 \le j \le 2}(E_j)}, & \text{if } e = 1,2 \text{ and } a = AR_j \\ 0, & \text{otherwise} \end{cases} \tag{7.9}$$

where E_j is the energy consumed by the jth RAT.

The JCAC policy is an n-tuple vector specifying for each state of the MDP the action to be selected in that state. Here, we consider a stationary and deterministic policy, that is, the policy does not change in time and in a given state. Note that for a Markov decision model with finite state space and finite action sets, there exists an optimal policy that is stationary and deterministic. Such a policy is an application from S to A, which associates at each state x an action in $A(x)$, that is, $\forall\, x \in S$, $R_x \in A(x)$. It should be noted that for the derivation of the performance parameters for a given policy, the SMDP model with transition probabilities $p(x, y, R_x)$ is a traditional continuous time Markov chain.

To derive the optimal JCAC policy, the continuous time SMDP model is first converted into a discrete time MDP model so that for each stationary policy, the average cost per time unit in the discrete-time Markov model is the same as that in the semi-Markov model. This approach is referred to as data-transformation method [56], which works as follows. Let $x, y \in S$ and $a \in A(x)$ be an action, choose a number τ such that

$$0 < \tau \leq \min_{x,a} \tau(x,a) \qquad (7.10)$$

then perform the following transformations:

$$\overline{S} = S$$

$$\overline{A}(x) = A(x), \quad x \in \overline{S}$$

$$\overline{C}(x,a) = \frac{C(x,a)}{\tau(x,a)}, \quad x \in \overline{S} \text{ and } a \in \overline{A}(x)$$

$$\overline{p}(x,y,a) = \begin{cases} \dfrac{\tau}{\tau(x,a)} p(x,y,a), & x \neq y, x \in \overline{S} \text{ and } a \in \overline{A}(x) \\[2ex] \dfrac{\tau}{\tau(x,a)} p(x,y,a) + \left[1 - \dfrac{\tau}{\tau(x,a)}\right], & x = y, x \in \overline{S} \text{ and } a \in \overline{A}(x) \end{cases}$$

$$(7.11)$$

where $\overline{i}$ denotes the converted component. After turning the continuous time SMDP model into a discrete time MDP model, the value iteration algorithm [56] is invoked to obtain the optimal JCAC policy $R(n)$, whose average cost function is given by $g_i(R(n))$ such that

$$0 \leq \frac{g_i(R(n)) - g^*}{g^*} \leq \epsilon \tag{7.12}$$

where g^* denotes the minimal average cost per time unit [25]. The pseudo-code of the value iteration algorithm [56] is described in Algorithm 7.1.

Algorithm 7.1: Value iteration algorithm [56]

Step 0: Initialization. Choose $V_0(i)$ such that $0 \leq V_0(i) \leq \min_a \{c(i, a)/\tau(i, a)\}$ for all i. Choose a number τ where $0 < \tau < \min_{i,a} \tau(i, a)$. Let $n = 1$.

Step 1: Value iteration step. Compute the function $V_n(i)$, $i \in I$ using

$$V_n(i) = \min_{a \in A(i)} \left[\frac{c(i,a)}{\tau(i,a)} + \frac{\tau}{\tau(i,a)} \sum_{j \in I} p(i,j,a) V_{n-1}(j) + \left(1 - \frac{\tau}{\tau(i,a)} \right) V_{n-1}(i) \right] \tag{7.13}$$

Let $R(n)$ be a stationary policy whose actions minimize the right-hand side of Equation 7.13.

Step 2: Compute the bounds m_n on the minimal cost using

$$m_n = \min_{j \in I} \{V_n(i) - V_{n-1}(i)\}, \quad M_n = \max_{j \in I} \{V_n(i) - V_{n-1}(i)\} \tag{7.14}$$

Step 3: Stopping condition. The algorithm is stopped when policy $R(n)$ is obtained such that $0 \leq (M_n - m_n) \leq \epsilon m_n$ where ϵ is a predefined accuracy threshold. Otherwise, go to step 4.

Step 4: Continue $n = n + 1$ and go back to step 1.

7.6 Energy-Efficiency Improvements Using HetNets

This section discusses some opportunities that may result in energy savings from MNO operations when HetNets are combined with some emerging spectrum technologies and/or RRM approaches.

7.6.1 *Cognitive Radio HetNets*

Cognitive radio HetNet (CR-HetNet) stands for an integration of HetNet with CR. It is a technology that promises to lighten the issue of spectrum scarcity in wireless networks. The idea behind CR [57] is to enable unlicensed users (called SUs) to opportunistically access the wireless spectrum allocated to licensed or primary users (PUs) as long as the interference perceived by the PUs stays below a tolerable threshold. In other words, PUs do not see SUs because they have a preemptive priority that allow them to get access to the radio resources as soon as these are requested, regardless of whether the radio resources are in use or not by SUs. In this case, an SU can sense the idle radio channels and perform the spectrum handoff to them in order to resume its service. This can be achieved by using a spectrum management algorithm such as DSA. In terms of network functionality, some DSA strategies can be implemented in the form of CAC algorithms, that is, algorithms that deal with the acceptance/rejection of incoming requests. One of the fundamental issues that may arise when accepting more traffic in an already crowded network is that it may considerably degrade the QoS as perceived by the subscribers of the network. A possible way to overcome this burden is to design RRM algorithms in HetNets that can bolster DSA strategies, thereby turning the system into a CR-HetNet. The advantage of such approach is obvious since users will have the opportunity to share the wireless infrastructures of MNOs while these MNOs will keep enjoying an increase in their network utilization. Figure 7.5 illustrates the types of possible connections and spectrum management strategies that can be adopted for CR-HetNets.

In event 1 of Figure 7.5, the mobile user tries to connect to its home network and due to the temporary lack of radio resources, its request is denied. Afterward, in event 2, the user opportunistically accesses the radio resources of wireless network 2 as an SU. Next, in event 3, upon an arrival of a PU requesting its radio channels, the SU has to vacate the requested radio resource and perform the spectrum handoff by moving to another radio channel in the same spectrum (referred to as event 4) in order to continue its service. However, when this network becomes overloaded and its home network is still unavailable, it makes the handoff for wireless network 3 (referred to as event 5).

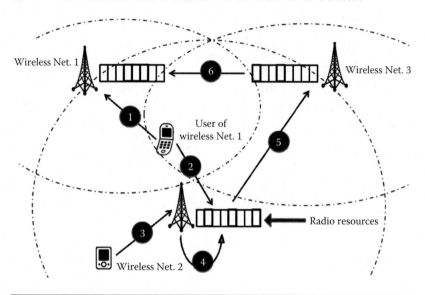

Figure 7.5   Example of CR HetNet, where the events are numbered.

Finally, it switches to its home network to resume its service as soon as a radio resource becomes idle (referred to as event 6).

7.6.2 HetNets and Dynamic Coverage Management Algorithms

In a wireless network, BSs are the elements that consume most of the network energy [24]. Thus, an effective way of optimizing the power usage at BSs so as to achieve a green design is to properly manage the operations of BSs. This can be done by exploring the multilayer architecture of the HetNet. Using this architecture, it is possible to dynamically manage the geographical area being covered, switching off some layers (or cells inside them) according to traffic load fluctuations [17] as long as the other layers can handle the total offered traffic load. The network functionality that is responsible for this type of coverage control can be realized by the DCM algorithm, which runs inside the RRM algorithm. Typically, the overlay architecture of a HetNet can be exploited to minimize the power consumption via the use of DCM algorithms, the main idea behind the use of these algorithms being the spatiotemporal variations of the offered traffic load. Indeed, due to load fluctuations, some cells on the system may experimentally underutilize their wireless resources during some periods of the day while others may not. As a consequence, the DCM algorithm can switch on/off a BS regardless of the cell size when it experiences a light traffic load [17], as long as another cell (or layer) can carry its traffic. This mechanism would help save some energy.

In general, the design of a green HetNet using DCM algorithms relies on the considered energy model [58], which in turn is a function of various system parameters such as power consumed at each site, traffic bit rate, and number of users at busy or peak hours, to name a few. A combination of these system parameters that result in an energy model that can help in achieving a green HetNet design is difficult to determine. However, designing a suitable JRRM optimization process may help in this regard. As an example, an architecture illustrating the application of DCM algorithms within the RRM algorithm [24] is depicted in Figure 7.6.

The DCM algorithm used in Figure 7.6 has two main modules, namely, the load control module and the DCM script. The load control module is meant to periodically monitor the radio resource occupancies of the wireless access networks that composed the HetNets and periodically transmits them to the DCM module. When the DCM module gets the information about the system load, it decides about which layer (or cell) in the system should be switched off/on, based on

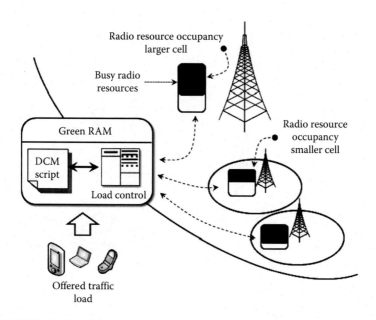

Figure 7.6 A system architecture for the DCM algorithm.

certain criteria, which are themselves based on MNOs' level of cooperation and network architectures. For instance, in a scenario such as the one shown in location 1 of Figure 7.2, MNOs might decide to turn on/off the cell with smaller coverage size and keep the macrocell always active in order to guarantee the user's mobility. In the scenario presented in location 2 of Figure 7.2, the opposite situation may occur; that is, MNOs may decide to turn on/off the macrocell and use only femto BSs to provide the wireless access because of their lower power consumption and higher capacity, which is appealing from a green perspective. It is worth mentioning that in Figure 7.6, cells with different sizes are utilized, but this is not a mandatory requirement [18].

An example of a JRRM scheme for HetNets is shown in Figure 7.7, where each RAT is associated with a controller, and these controllers are equipped with different technologies such as radio network controllers for UTRAN systems, access service network gateways for WiMAX, BS controllers for GERAN systems, and multiple enhanced node Bs (eNBs) for LTE systems, to name a few. From a practical perspective, the JRRM server will reside in any of the HetNets controllers or will be handled by a dedicated third-party server. This server will monitor the network continuously and collect information on the traffic load at each RAT. Whenever an event occurs in the network, the

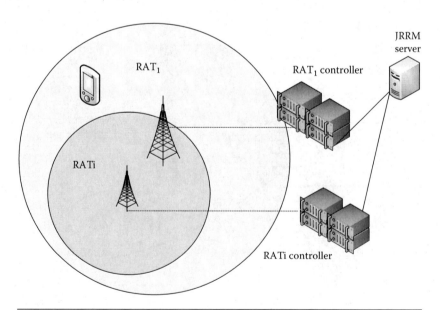

Figure 7.7   Example of JRRM scheme.

appropriate RAT controller will be responsible for reporting this event to the JRRM server. This way, the server will be able to take appropriate resource allocation decisions based on such information and then will execute them using a decision maker such as the one proposed in [16].

7.6.3 HetNets and Self-Optimization, Organization, and Self-Healing Algorithms

With the expectation that MNOs will adhere to the use of HetNets for their network operations, network engineers will soon face the practical problem of how to manage the huge amount of data that are exchanged among MNOs to keep the system operational. Apart from HetNets, network engineers will still have to manage their own wireless access networks, which include core networks and service delivery platforms, to name a few. To efficiently address this increase in network complexity, the 3GPP LTE standardization [59] has introduced the concept of SON algorithms. These algorithms have primarily been designed with the goal to enhance the operational efficiency of next-generation 4G and future 5G standards–based mobile networks since they are expected to autonomously perform underlying managing tasks.

The application of SON algorithms in HetNets can contribute to the improvement of the operations of MNOs. It can also help achieve an efficient green network design. For instance, a SON algorithm combined with a DCM algorithm can help (1) retrieve the information on the network load through traffic statistics, (2) assist the load control module, (3) empty the cells with light traffic load by handing their ongoing calls off to other cells, and so on. SON algorithms can also be applied in conjunction with DSA strategies to optimize the decision about the network selection in overlapping coverage. Since the network conditions can change quickly, some unused radio resources for SUs in any of the available cells can be reserved automatically in order to guarantee the service continuity during the handoff process. SON algorithms can also be considered as potential models for controlling, distributing, and managing efficiently the frequency spectrum and available resources [60], so as to maximize the benefit of the available limited spectrum, providing that a suitable RRM model and network architecture be designed.

7.7 Numerical Results

Due to the state–space explosion problem as shown in Equation 7.1, the discrete time Markov model can be solved only for small values of N_j (we have used $N_1 = 24$ and $N_2 = 12$). The proposed RRM scheme is implemented as an event-driven system written in Borland C++ version 5. Each system configuration is run for a sufficiently small precision value ($1.0e - 12$). In the considered HetNet architecture, the two co-located networks, that is, RAT_1 and RAT_2, are considered as representatives of GSM and 4G/UMTS technologies. We have also considered two types of service classes (class 1 and class 2) for each RAT. Table 7.1 shows the fixed parameters that have been used in our analysis.

The following performance metrics are considered:

- Mean carried traffic M_e^a, defined as

$$M_e^a = \sum_{x \in S; e=1,2; a=AR1,AR2 \in A(x)} \left(\sum_{j=1}^{2} \lambda_j + \sum_{j=1}^{2} \sum_{i=1}^{2} n_{ij}\mu_i \right) \pi_x \quad (7.15)$$

where

π_x; $x \in S$ is the continuous time Markov chain steady-state probability distribution in the optimal policy

n_{ij} is the number of calls of type i connection for RATj

- Connection blocking probability Pb_i of a new ith service class, defined as

$$Pb_i = 1 - \frac{M_i^a}{\lambda_i} \quad (7.16)$$

It should be noted that for MNOs' benefits, Pb_i must be as low as possible.

Table 7.1 Fixed Simulation Parameters

PARAMETER	VALUE	PARAMETER	VALUE	PARAMETER	VALUE
N_1	24 channels	μ_1	1/120s (voice)	E_1	3802 W
N_2	12 channels	μ_2	1/120s (voice)	E_2	300 W
A_{11}	20	A_{12}	10	BC_1	1.0
A_{21}	10	A_{22}	5	BC_2	0.8

- Bandwidth utilization U_j of the jth RAT, defined as

$$U_j = \frac{1}{N_j} \sum_{x \in S; a \in A(x); \forall i; \forall j; n_{ij} > 0} b_i n_{ij} \pi_x \qquad (7.17)$$

where

b_i is the bandwidth of *type i* call connection
N_j is the capacity of RATj

The following scenarios have been considered:

- *Scenario I*: The traffic intensity is defined as the number of calls received by the network elements in a unit area at a given time interval. In this scenario, the traffic intensity is varied and the impact of this variation on blocking probabilities and RAT utilization are studied.
- *Scenario II*: The bandwidth of each class of traffic is varied and the impact of this variation on blocking probabilities and RAT utilization are studied.
- *Scenario III*: The weight of the energy consumption (ω_3) is varied and the impact of this variation on blocking probabilities and RAT utilization are studied. The goal is to illustrate the relative importance of the energy consumption cost on the system's performance.

For these scenarios, the parameters that have been used are given in Table 7.2.

Table 7.2 Parameters Used in the Considered Scenarios

PARAMETER FOR SCENARIO I	VALUE	PARAMETER FOR SCENARIO I	VALUE
Bandwidth for service class 1 (b_1)	2 channels	Weight 1 of blocking cost (ω_1)	0.6
Bandwidth for service class 2 (b_2)	1 channel	Weight 2 of access price cost (ω_2)	0.2
		Weight 3 of energy consumption cost (ω_3)	0.2
PARAMETER SCENARIO II	VALUE	PARAMETER FOR SCENARIO II	VALUE
Intensities for service class 1 (ρ_1)	7 channels	Weight 1 of blocking cost (ω_1)	0.6
Intensities for service class 2 (ρ_2)	3 channels	Weight 2 of access price cost (ω_2)	0.2
		Weight 3 of energy consumption cost (ω_3)	0.2
PARAMETER SCENARIO III	VALUE	PARAMETER FOR SCENARIO III	VALUE
Bandwidth for service class 1 (b_1)	2 channels	Intensities for service class 1 (ρ_1)	7
Bandwidth for service class 2 (b_2)	1 channel	Intensities for service class 1 (ρ_2)	3
		Weight 2 of access price (ω_2)	0.2

7.7.1 Results for Scenario I

These results are captured in Figures 7.8 and 7.9. In Figure 7.8a and b, it is observed that for a fixed system capacity, when the traffic intensities increase, the blocking probabilities also increase. Also, more traffic of class 1 are blocked compared to traffic of class 2. This behavior

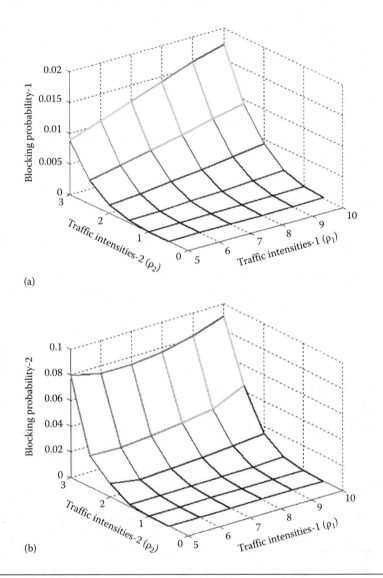

(a)

(b)

Figure 7.8 Scenario I: blocking probability versus traffic intensities: (a) case of class 1 calls and (b) case of class 2 calls.

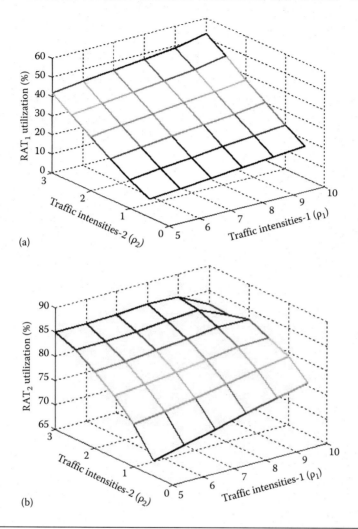

Figure 7.9 Scenario I: RAT utilization versus traffic intensities: (a) case of RAT$_1$ (denoted RAT$_1$ in the figure) and (b) case of RAT$_2$ (denoted RAT$_2$ in the figure).

is attributed to the fact that class 1 calls necessitate more bandwidth than class 2 calls. In Figure 7.9a and b, it is observed that when traffic intensities are smaller, the JCAC policy accepts more calls in the less energy consumed RAT (i.e., RAT$_2$), resulting in some energy saving. Because of this, the utilization of RAT$_2$ is much more better than that of RAT$_1$. On the other hand, when traffic intensities are high, more calls are accepted in RAT$_1$ (compared to RAT$_2$), leading to a higher utilization of RAT$_1$.

7.7.2 Results for Scenario II

These results are captured in Figures 7.10 and 7.11. In Figure 7.10a and b, it is observed that when the bandwidth of class 1 calls (resp. class 2 calls) increases, less vacant channels are available to accommodate new incoming calls, thus the blocking probabilities of each traffic class become high, which was expected. In Figure 7.11a and b, it is observed that RAT_2 utilization is much better than RAT_1 utilization,

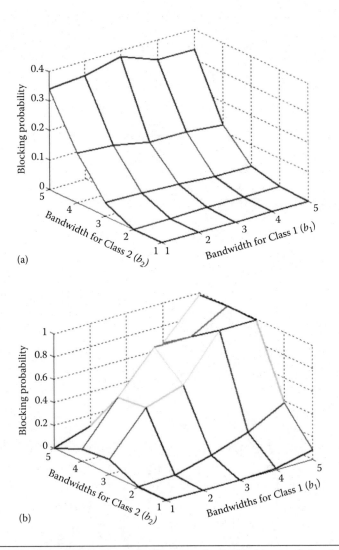

Figure 7.10   Scenario II: blocking probability versus bandwidth: (a) case of class 1 call and (b) case of class 2 call.

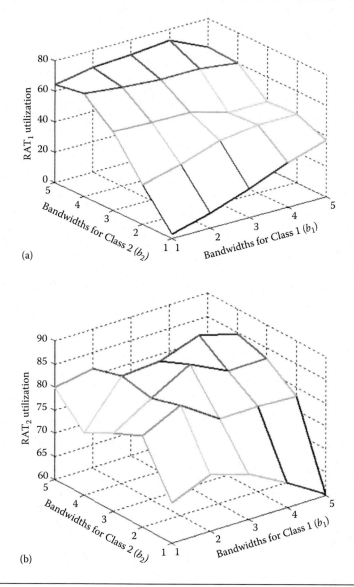

Figure 7.11   Scenario II: RAT utilization versus bandwidth: (a) case of RAT_1 (denoted RAT_1 in the figure) and (b) case of RAT_2 (denoted RAT_2 in the figure).

especially when the bandwidth of each call is low. This is attributed to the fact that RAT_2 consumes less energy compared to RAT_1; thus, it is more often selected by the JCAC policy. However, when the required bandwidth for each call increases, the optimal policy accepts more and more calls in RAT_1 when RAT_2 can no longer accommodate new incoming calls (i.e., when it is fully utilized).

7.7.3 Results for Scenario III

These results are captured in Figures 7.12 and 7.13. In Figure 7.12a, it is observed that when the weight of the energy cost (ω_3) is set to a value in the range [0.1, 0.25], the optimal policy accepts incoming calls from both class 1 and class 2. However, when $\omega_3 \geq 0.25$, the optimal

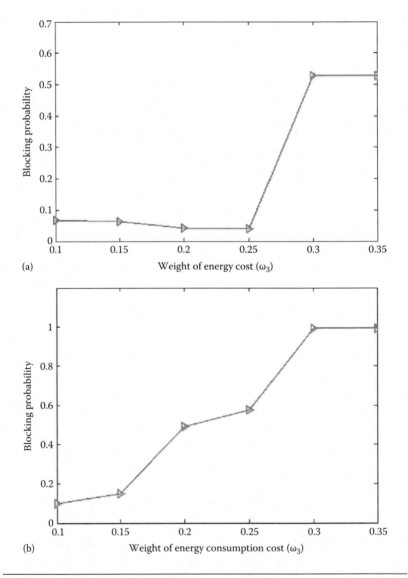

Figure 7.12   Scenario III: blocking probability versus weight of energy consumption cost in the total cost: (a) case of class 1 calls and (b) case of class 2 calls.

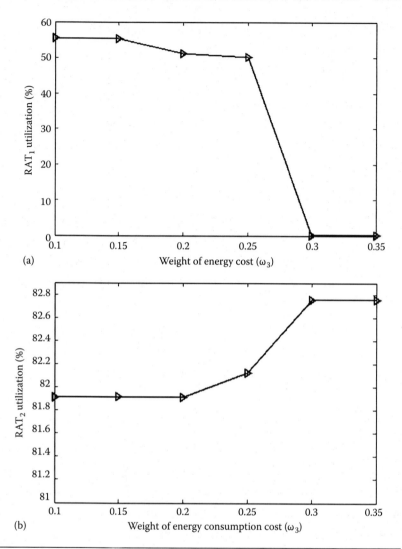

Figure 7.13 Scenario III: RAT utilization versus weight of energy consumption cost in the total cost: (a) case of RAT₁ and (b) case of RAT₂.

policy starts to reject both types of calls. The same behavior prevails in Figure 7.12b. It is also observed that more class 1 calls are blocked by the optimal policy compared to class 2 calls. For the results shown in Figure 7.13, the weight of the access price (ω_2) is fixed to 20% of the total cost and that of the blocking cost (ω_1) varies from 45% to 70%. In Figures 7.13a and b, it is observed that when the weight of the energy consumption cost (ω_3) is varied from 10% to 35%, the JCAC policy utilizes more channels of RAT_2 compared to that of RAT_1. This might be due to the fact that RAT_1 necessitates more energy consumption for its operations than RAT_2 does. However, when ω_3 is set to (30%), the RAT_1 utilization reaches almost 0% (Figure 7.13a), whereas RAT_2 channel utilization increases slightly (Figure 7.13b). This behavior is attributed to the fact that more calls that were supposed to be served by RAT_1 have been handled by RAT_2.

7.7.4 Analysis of the Optimal Policy

In this section, the behavior of the optimal policy is analyzed using the system configuration in Table 7.3. The focus here is on determining how the policy allocates the different calls over existing RATs under different loads in order to achieve optimality. The following convention is adopted:

- "+" means class 1 call is accepted in RAT_1.
- "*" means class 1 call is accepted in RAT_2.
- "x, y" means class 1 call usually accepted in RAT_2, but RAT_1 starts taking class 1 call when the number of class 1 calls in RAT_1 is equal to x or higher and the number of class 2 calls in RAT_1 is equal to y or lower, that is, (x, y) or $(x + 1, y - 1)$.
- "B" means the blocking of a call.

For the cases when the RAT_1 channel load is less than or equal to 10, the results are shown in Tables 7.4 and 7.5. It is clear that when the number of class 2 calls in RAT_2 is 0, regardless of the RAT_1 resource occupancy, the optimal JCAC decides to accept class 1 calls in RAT_2 until RAT_2 resources are fully occupied. Also, when the number of class 2 calls in RAT_2 is greater than zero and the radio resource occupancy of RAT_1 is low, the optimal JCAC decides to accept class 1 call in RAT_1. From Table 7.6, it can is observed that when the radio resource occupancy of RAT_1 is moderate (e.g., when RAT_1 load is 14), the optimal JCAC decides to accept class 1 calls in RAT_2, but also to accept class 1 calls in RAT_1 when the total number of calls (i.e., class 1 and class 2 calls) in RAT_1 becomes low. From Tables 7.7 and 7.8, it is observed that when the radio resource occupancy of RAT_1 is high (e.g., when RAT_1 load is greater than or equal to 18), the optimal JCAC decides to accept class 1 calls in RAT_2 until RAT_2 resources are fully occupied, but it also

Table 7.3 System Configuration

PARAMETER	VALUE	PARAMETER	VALUE	PARAMETER	VALUE	PARAMETER	VALUE	PARAMETER	VALUE
N_1	20 channels	μ_1	1/120s (voice)	b_1	2 channels	ρ_1	5	ω_1	0.8
N_2	10 channels	μ_2	1/120s (voice)	b_2	1 channel	ρ_2	3	ω_3	0.2
								ω_2	0

Table 7.4 When RAT$_1$ Channel Load Is Less Than 10

		\multicolumn{11}{c}{NO. OF CLASS 2 CALLS IN RAT$_2$}									
		0	1	2	3	4	5	6	7	8	9	10
NO. OF CLASS 1 CALLS	0	*	+	+	+	+	+	+	+	+	+	+
IN RAT$_2$	1	*	+	+	+	+	+	+	+	+		
	2	*	+	+	+	+	+	+				
	3	*	+	+	+	+						
	4	*	+	+								
	5	+										

Table 7.5 When RAT$_1$ Channel Load Is 10

		\multicolumn{11}{c}{NO. OF CLASS 2 CALLS IN RAT$_2$}										
		0	1	2	3	4	5	6	7	8	9	10
NO. OF CLASS 1 CALLS	0	*	2, 6	+	+	+	+	+	+	+	+	+
IN RAT$_2$	1	*	+	+	+	+	+	+	+	+		
	2	*	+	+	+	+	+	+				
	3	*	+	+	+	+						
	4	*	+	+								
	5	+										

Table 7.6 When RAT$_1$ Channel Load Is 14

		\multicolumn{11}{c}{NO. OF CLASS 2 CALLS IN RAT$_2$}										
		0	1	2	3	4	5	6	7	8	9	10
NO. OF CLASS 1 CALLS	0	*	7, 0	7, 0	5, 4	5, 4	3, 8	3, 8	+	+	+	+
IN RAT$_2$	1	*	7, 0	7, 0	5, 4	5, 4	2, 10	2, 10	+	+		
	2	*	6, 2	6, 2	4, 6	4, 6	+	+				
	3	*	5, 4	6, 2	+	+						
	4	*	+	+								
	5	+										

Table 7.7 When RAT$_1$ Channel Load Is 18

		\multicolumn{11}{c}{NO. OF CLASS 2 CALLS IN RAT$_2$}										
		0	1	2	3	4	5	6	7	8	9	10
NO. OF CLASS 1 CALLS	0	*	*	9, 0	9, 0	8, 2	7, 4	7, 4	6, 6	5, 8	+	+
IN RAT$_2$	1	*	9, 0	9, 0	8, 2	8, 2	7, 4	7, 4	+	+		
	2	*	9, 0	9, 0	8, 2	8, 2	+	+				
	3	*	9, 0	9, 0	+	+						
	4	*	+	+								
	5	+										

Table 7.8 When RAT_1 Channel Load Is Greater Than 18

		NO. OF CLASS 2 CALLS IN RAT_2										
		0	1	2	3	4	5	6	7	8	9	10
NO. OF CLASS 1 CALLS	0	*	*	*	*	*	*	*	*	*	B	B
IN RAT_2	1	*	*	*	*	*	*	*	B	B		
	2	*	*	*	*	*	B	B				
	3	*	*	*	B	B						
	4	*	B	B								
	5	B										

gradually accepts class 1 calls in RAT_1 when the total number of calls (i.e., class 1 and class 2 calls) in RAT_1 become low. It is also observed that when there is no more occupancy to accept class 1 calls in RAT_1 (i.e., when RAT_1 load is greater than or equal to 18), the optimal JCAC decides to accept class 1 calls in RAT_2 until the RAT_2 resources are fully occupied. Also, the JCAC blocks the class 1 calls when there is no more occupancy to accept class 1 calls in RAT_1 or RAT_2.

7.8 Conclusions

In this chapter, we have discussed some ideas on how HetNets can be combined with emerging wireless spectrum and/or network management approaches to help reduce the energy consumption resulting from MNO operations. Following that trend, we have provided an example of a green JRRM-based HetNets design, where the HetNet is composed of two co-located RATs. The RAT selection problem in such HetNet is formulated as an SMDP model, whose derived JCAC optimal policy is used to determine the less energy consuming RAT for handling the user's incoming request. Given the simplicity and robustness of our JCAC-based decision-making strategy, we believe that our proposed RAT selection method can be used by MNOs as an eco-friendly design solution to successfully address the escalation problem of RAT selection deployment in urban areas.

7.9 Future Research Directions

The proposed SMDP model can be explored further to support more sophisticated HetNets architectures, that is, architectures with more than two RATs supporting several service classes. Of course, such a system will involve a huge number of states and actions to be dealt with. We also believe that the inter-RAT handover problem in such an architecture is yet to be investigated. For instance, this problem can be modeled as a stochastic process in which the state variables are treated as handover traffic classes (both vertical and horizontal) and the considered actions are nothing more than the priority among these classes.

In the proposed RRM scheme, the discovery and selection of the wireless access network to accommodate the incoming service request are somewhat dependent on the user's subscription since this contains some precious information such as the types of networks as well as the types of allowable services they can support. This set of information was not considered in our design as additional entries to the implemented JCAC-based decision-making model. Doing so will strengthen the practicality aspect of the proposed system, which is considered as future work.

Acknowledgments

This work is partially supported by a grant from the National Science and Engineering Research Council of Canada (NSERC), Ref # RGPIN/293233-2011, held by the first author, and a grant from the Conselho Nacional de Desenvolvimento Científico e Tecnológico (CNPq), Brazil, held by the second author.

References

1. Gartner Inc., Top predictions for IT organizations and users, 2010 and beyond: A new balance, 2014, https://www.gartner.com/doc/1268513/gartners-top-predictions-it-organizations (accessed November 11, 2014).
2. Cisco Systems Inc., Cisco visual networking index: Global mobile data traffic forecast update, 2013–2018, http://www.cisco.com (accessed November 11, 2014).
3. T. Chen, Y. Yang, H. Zhang, H. Kim, and K. Horneman, Network energy saving technologies for green wireless access networks, *IEEE Wireless Communications*, 18(5), 30–38, October 2011.
4. M. M. S. Marwangi et al., Challenges and practical implementation of self-organizing networks in LTE/LTE-advanced systems, in: *IEEE International Conference on IT & Multimedia*, Kuala Lumpur, Malaysia, November 14–16, 2011, pp. 1–5.
5. M. Lopez-Benitez and J. Gozalvez, Common radio resource management algorithms for multimedia heterogeneous wireless networks, *IEEE Transactions on Mobile Computing*, 10(9), 1201–1213, September 2011.

6. 3GPP, The mobile broadband standard, 2014, http://www.3gorg/technologies/keywords-acronyms/98-lte (accessed November 10, 2014).

7. Mind Commerce Publishing, HetNet deployment and operational strategies for 4G and beyond, March 19, 2014, http://www.prweb.com/pdfdownload/11671141.pdf (accessed November 10, 2014).

8. K. Piamrat, A. Ksentini, J.-M. Bonnin, and C. Viho, Radio resource management in emerging heterogeneous wireless networks, *Computer Communications*, 34(9), 1066–1076, June 2011.

9. O. E. Falowo and H. A. Chan, Joint call admission control algorithms: Requirements, approaches, and design considerations, *Computer Communications*, 31(6), 1200–1217, April 2008.

10. N. Nasser and H. Hassanein, Dynamic threshold-based call admission framework for prioritized multimedia traffic in wireless cellular networks, in: *Proceedings of the IEEE Global Telecommunications Conference (GLOBECOM 2004)*, Dallas, TX, November 29 to December 3, 2004, pp. 644–649.

11. J. Sachs and M. Olsson, Access network discovery and selection in the evolved 3GPP multi-access system architecture, *European Transactions on Telecommunications*, 21(6), 544–557, 2010.

12. D. N. Knisely and F. Favichia, Standardization of femtocells in 3GPP2, *IEEE Communications Magazine*, 47(9), 76–82, September 2009.

13. I. A. Qaimkhani, Improving frequency reuse and cochannel interference coordination in 4G HetNets, MSc thesis report, Electrical and Computer Engineering Department, University of Waterloo, Waterloo, Ontario, Canada, 2013.

14. The 5G-Infra Public Private Partnership, 5G-infra-PPP pre-structuring model v2.0, 2014, http://5g-peu/wp-content/uploads/2014/03/5G-Infra-PPP_Pre-structuring-Model_v2.0.pdf (accessed November 10, 2014).

15. A. Manzalini et al., Software defined networks for future networks and services, in: White Paper based on *the IEEE Workshop SDN4FNS*, Trento, Italy, November 11–13, January 29, 2014.

16. G. H. S. Carvalho, I. Woungang, A. Anpalagan, R. W. L. Coutinho, and J. C. W. A. Costa, A semi-Markov decision process-based joint call admission control for inter-RAT cell re-selection in next generation wireless networks, *Computer Networks*, 57, 3545–3562, 2013.

17. Y. Chen, S. Zhang, S. Xu, and G. Y. Li, Fundamental trade-offs on green wireless networks, *IEEE Communications Magazine*, 49(6), 30–37, June 2011.

18. J. Lorincz, A. Capone, and D. Begu, Optimized network management for energy savings of wireless access networks, *Computer Networks*, 55(3), 514–540, February 2011.

19. S.-M. Cheng, S.-Y. Lien, F.-S. Chu, and K.-C. Chen, On exploiting cognitive radio to mitigate interference in macro/femto heterogeneous networks, *IEEE Wireless Communications Magazine*, 18(3), 40–47, June 2011.

20. G. Gur, S. Bayhan, and F. Alagoz, Cognitive femtocell networks: An overlay architecture for localized dynamic spectrum access, *IEEE Wireless Communications Magazine*, 17(4), 62–70, August 2010.

21. M. Bennis and D. Niyato, A Q-learning based approach to interference avoidance in self-organized femtocell networks, in: *IEEE International Workshop on Femtocell Networks*, Miami, FL. 2010.
22. S. Morosi, E. D. Re, and P. Piunti, Traffic based energy saving strategies for green cellular networks, in: *Proceedings of the 18th European Wireless Conference*, Poznan, Poland, April 18–20, 2012.
23. A. Bousia, E. Kartsakli, L. Alonso, and C. Verikoukis, Energy efficient base station maximization switch off scheme for LTE-advanced, in: *Proceedings of the IEEE 17th International Workshop on Computer Aided Modeling and Design of Communication Links and Networks (CAMAD)*, Barcelona, Spain, September 17–19, 2012, pp. 256–260.
24. G. H. S. Carvalho, I. Woungang, and A. Anpalagan, Towards energy efficiency in next generation green mobile networks: A queueing theory approach, Chapter 27, in: M. Obaidat et al. (eds.), *Handbook on Green Communication and Systems*, Elsevier, November 6, 2012, pp. 687–719.
25. Md M. Rahman, An optimal initial radio access technology selection method for heterogeneous wireless networks, Master thesis, Department of Computer Science, Ryerson University, Toronto, Ontario, Canada, December 2012.
26. T. R. Omar, A. E. Kamal, and J. M. Chang, Downlink spectrum allocation in 5G HetNets, in: *Proceedings of the IEEE International Wireless Communications and Mobile Computing Conference (IWCMC 2014)*, Nicosia, Cyprus, August 4–8, 2014.
27. S.-Y. Lien, K.-C. Chen, Y.-C. Liang, and Y. Lin, Cognitive radio resource management for future cellular networks, *IEEE Wireless Communications*, 21(1), 70–79, February 2014.
28. M. Deruyck, W. Joseph, and L. Martens, Power consumption model for macrocell and microcell base stations, *Transactions on Emerging Telecommunications Technologies*, 25(3), 320–333, March 2014.
29. G. Araniti, J. Cosmas, A. Iera, A. Loiacono, A. Molinaro, and A. Orsino, Power consumption model using green policies in heterogeneous networks, in: *Proceedings of the IEEE International Symposium on Broadband Multimedia Systems and Broadcasting (BMSB)*, Beijing, China, June 25–27, 2014, pp. 1–5.
30. P. Dini, M. Miozzo, N. Bui, and N. Baldo, A model to analyze the energy savings of base station sleep mode in LTE HetNets, in: *Proceedings of the IEEE International Conference on Green Computing and Communications (GreenCom 2013)*, Beijing, China, August 20–23, 2013, pp. 1375–1380.
31. L. M. Correia, D. Zeller, O. Blume, D. Ferling, A. Kangas, Y. Jading, I. Gódor, G. Auer, and L. V. D. Perre, Challenges and enabling technologies for energy aware mobile radio networks, *IEEE Communications Magazine*, 48(11), 66–72, November 2010.

32. M. Ismail and W. Zhuang, Network cooperation for energy saving in green radio communications, *IEEE Wireless Communications*, 18(5), 76–81, October 2011.

33. G. H. S. Carvalho, I. Woungang, A. Anpalagan, and S. K. Dhurandher, Energy-efficient radio resource management scheme for heterogeneous wireless networks: A queueing theory perspective, *Journal of Convergence* (FTRA, Thomson Router ISI), 3(4), 15–22, December 2012.

34. S. Murugesan and P. A. Laplante, IT for a Greener Planet Guest editors' introduction, *IT Professional*, 13(1), 16–18, January to February 2011.

35. H. Liu et al., Energy efficient network selection and seamless handovers in mixed networks, in: *IEEE International Symposium on WWMMN*, Kos, Greece, June 15–19, 2009, pp. 1–9.

36. I. Chamodrak and D. Martakos, A utility-based fuzzy TOPSIS method for energy efficient network selection in heterogeneous wireless networks, *Applied Soft Computing*, 11(4), 3734–3743, June 2011.

37. K. Suleiman, H. Chan, and M. Dlodlo, Issues in designing joint radio resource management for heterogeneous wireless networks, in: *Proceedings of the IEEE Seventh International Conference on Wireless Communications, Networking and Mobile Computing (WiCOM)*, Kunming, China, August 8–10, 2011, p. 15.

38. K. Cho, J. Kim, and S. Stapleton, A highly efficient Doherty feed forward linear power amplifier for W-CDMA base-station applications, *IEEE Transactions on Microwave Theory and Techniques*, 53(1), 292–300, January 2005.

39. G. Auer et al., How much energy is needed to run a wireless network? *IEEE Wireless Communications Magazine*, 18(5), 4049, October 2011.

40. M. Deruyck, E. Tanghe, W. Joseph, and L. Martens, Modelling and optimization of power consumption in wireless access networks, *Elsevier Computer Communications*, 34(17), 2036–2046, 2011.

41. M. Ericson, Total network base station energy cost vs. deployment, in: *Proceedings of the IEEE Vehicular Technology Conference (VTC Spring)*, Budapest, Hungary, May 2011.

42. F. Cao and Z. Fan, The tradeoff between energy efficiency and system performance of femtocell deployment, in: *Proceedings of International Symposium on Wireless Communication Systems (ISWCS)*, York, U.K., September 2010.

43. D. Zeller, M. Olsson, O. Blume, A. Fehske, D. Ferling, W. Tomaselli, and I. Gdor, Sustainable wireless broadband access to the future Internet—The EARTH Project, in: A. Gales and A. Gavras (eds.), *The Future Internet*, Vol. 7858, Springer, 2013, pp. 249–271.

44. J.-M. Kelif and M. Coupechoux, Cell breathing, sectorization and densification in cellular networks, in: *International Symposium on Modeling and Optimization in Mobile, Ad Hoc, and Wireless Networks (WiOPT)*, Seoul, South Korea, June 2009.

45. E. Oh, K. Son, and B. Krishnamachari, Dynamic base station switching on/off strategies for green cellular networks, *IEEE Transactions on Wireless Communications*, 12(5), 2126–2136, 2013.

46. P. Frenger, P. Moberg, J. Malmodin, Y. Jading, and I. Godor, Reducing energy consumption in LTE with cell DTX, in: *Proceedings of the IEEE Vehicular Technology Conference (VTC Spring)*, Yokohama, Japan, May 15–18, 2011, pp. 1–5.

47. S. Tombaz, On the design of energy efficient wireless access networks, Doctoral thesis, Information and Communication Technology, KTH Royal Institute of Technology, Stockholm, Sweden, TRITAICTCOS1403, SE-100 44, May 2014.

48. V. Atanasovski, V. Rakovic, and L. Gavrilovska, Efficient resource management in future heterogeneous wireless networks: The RIWCOS approach, in: *Proceedings of the IEEE Military Communications Conference (MILCOM 2010)*, San Jose, CA, October 31 to November 3, 2010, pp. 2286–2291.

49. S. Kajioka, N. Wakamiya, and M. Murata, Autonomous and adaptive resource allocation among multiple nodes and multiple applications in heterogeneous wireless networks, *Journal of Computer and System Sciences*, 78(6), 1673–1685, November 2012.

50. L. Giupponi, R. Agusti, J. Perez-Romero, and O. Sallen, A novel joint radio resource management approach with reinforcement learning mechanisms, in: *Twenty-Fourth IEEE International Conference on Performance, Computing, and Communications*, Phoenix, AZ, April 7–9, 2005, pp. 621–626.

51. O. Falowo and H. Anthony Chan, Dynamic rat selection for multiple calls in heterogeneous wireless networks using group decision-making technique, *Computer Networks*, 56(4), 1390–1401, March 16, 2012.

52. K. Haldar, C. Ghosh, and D. Agrawal, Dynamic spectrum access and network selection in heterogeneous cognitive wireless networks, *Pervasive and Mobile Computing*, 9(4), 484–497, August 2013.

53. M. Porjazoski and B. Popovski, Radio access technology selection algorithm for heterogeneous wireless networks based on service type, user mobility and network load, in: *Tenth International Conference on Telecommunication in Modern Satellite Cable and Broadcasting Services (TELSIKS)*, Nis, Serbia, Vol. 2, October 5–8, 2011, pp. 475–478.

54. S. Zhu, F. Liu, Y. Qi, Z. Chai, and J. Wu, Immune optimization algorithm for solving joint call admission control problem in next-generation wireless network, *Engineering Applications of Artificial Intelligence*, 25, 1395–1402, 2012.

55. P. Si, H. Ji, and F. Yu, Optimal network selection in heterogeneous wireless multimedia networks, *Wireless Networks*, 16(5), 1277–1288, 2010.

56. H. C. Tijms, *A First Course in Stochastic Models*, Wiley & Sons, April 2003, 492pp.

57. E. Hossain, D. Niyato, and Z. Han, *Dynamic Spectrum Access and Management in Cognitive Radio Networks*, London, UK: Cambridge University Press, 2009.

58. Z. Hasan, H. Boostanimehr, and V. K. Bhargava, Green cellular networks: A survey, some research issues and challenges, *IEEE Communications Surveys and Tutorials*, 13(4), 524–540, Fourth Quarter 2011.
59. K. N. Premnath and S. Rajavelu, Challenges in self organizing networks for wireless telecommunications, in: *IEEE International Conference on Recent Trends in Information Technology*, Chennai, India, June 3–5, 2011.
60. U. Barth and E. Kuehn, Self-organization in 4G mobile networks: Motivation and vision, in: *Seventh IEEE International Symposium on Wireless Communication Systems (ISWCS)*, York, U.K., September 19–22, 2010, pp. 731–735.

8

OPPORTUNISTIC MULTICONNECT WITH P2P WIFI AND CELLULAR PROVIDERS

MARAT ZHANIKEEV

Contents

The technologies discussed in this chapter have a direct relation to 4G and 5G wireless communications where Long-Term Evolution Advanced (LTE-A) plays a key role [5] while specific technologies like femtocells [7], multiple-input multiple-output (MIMO), self-organized networks (SONs), and coordinated multipoint (CoMP) [6] add important details to the overall picture. The MultiConnect and GroupConnect technologies discussed in this chapter can be viewed as a practical tool that implements a combination of features from MIMO, SON, and CoMP technologies.

4G and beyond networks pursue two main goals: *higher transmission rates* and *efficient offload*. 5G networks are expected to support data rates up to 1 Gbps under the LTE-A technology [8]. Technological details of current 4G and its future evolution can be found in [5]. Specific technologies like MIMO or the generic concept of multistream aggregation are expected to help achieve this goal [6]. The offload part of the technology is more complex and involves several distinct approaches. On one hand, the offload helps solve the problem of interference under higher target rates. On the other hand, implementation of the offload at a large scale may require the use of local connectivity where the specific technologies considered in current research are the traditional WiFi, CoMP, and SON [6]. LTE-A also has the provider-based offload solution that involves a new layer between base stations (BSs) and users in the form of a high number of small BSs—the generic concept of femtocells or the more specific evolved NodeB (eNB)—with a much more active control function between cellular network provider and the population of large and small BSs [6]. Another chapter in this book focuses on the control function through the practical example of video streaming over wireless channels.

This chapter discusses a specific offload technology where the offload is achieved by using a hybrid of cellular and local WiFi connectivity. The core technology can already be implemented in practice using WiFi Direct together with any existing cellular connectivity.

Note that, by contrast, CoMP and SON technologies are not found in commodity devices at the time of this writing. However, the basic concepts are similar.

This chapter pays special attention to various multitechnologies. While traditional wireless research focuses on *multihop* and *multipath* features of end-to-end (e2e) routing [14], this chapter discusses *MultiConnect* and *GroupConnect* concepts [1]. The specific practical case considered in this chapter is a two-connect between a cellular (3G/4G) and the WiFi Direct connectivities on a given device. WiFi Direct is relatively much faster and therefore can be used for internal connectivity among members of a group—hence the GroupConnect technology based on the base concept of MultiConnect. This two-connect has multiple practical uses, namely, *sharing cellular connectivity by group members* or the opposite usecase of *pooling cellular resources of the entire group* [1].

Note that similar trends are found in recent Mobile Access NETwork (MANET) research that also discusses logical grouping of wireless devices [22,23]. Assuming that people (devices' owners) are expected to willingly participate in a virtual group, social aspects of group communication become important [24]. Broadly speaking, this branch of MANET research is moving toward full convergence with cellular technologies that is expected to happen at 5G [6].

The structure of this chapter is as follows: Section 8.1 discusses various multitechnologies and introduces the MultiConnect and GroupConnect concepts used throughout this chapter. These key concepts are discussed from the viewpoint of network resource virtualization in Section 8.2. Section 8.3 presents the practical setting of a university campus where students use GroupConnect to increase individual throughput. This practical setting is analyzed using two practical models and trace-based analysis using Crawdad traces [55] in Section 8.4. The chapter is concluded in Section 8.5.

8.1 MultiHop, MultiPath, and MultiConnect

This chapter takes a detour away from traditional MANETs in the understanding contained in numerous surveys [14]. Traditional MANETs are forced to tackle many problems, among which the problems of *multihop connections* and *end-to-end efficiency* are arguably the most prominent ones. This chapter is different in that it does talk about efficiency but is not concerned with either multihop or e2e aspects. This argument is unfolded in detail further on.

8.1.1 Evolution of Wireless Technologies

In place of a full introduction—which is unnecessary in the first place and can be found in several other chapters in this book—this section follows the evolution of various wireless technologies over the years, leading up to the alternate formulations in the focal point of this chapter.

The evolution of wireless technologies can (arguably) be jammed into the following three main epochs:

Epoch 1: MultiHop MANETs and QoS. MANETs always needed to be multihop, following the core idea that there would be an all-wireless world where people would only be able to reach each other multihop connections. Note that this very dream might not be pursued as strongly nowadays as it used to be. This chapter shows that research is converging on a new dream—a semi-wireless world where relatively few-hop connections are routed via nonwireless or at least more reliable wireless switches. When we talk of multihop connections, we always take into account the problem of quality of service (QoS) or, more generally, efficiency of multihop e2e connections. Research shows that e2e QoS grows increasingly difficult to support with longer paths [14].

Epoch 2: Energy efficiency. The e2e QoS problem can possibly be solved if one forces all the nodes to be always on and to service all the paths that get requested, without the option to refuse. Research on MANETs quickly understood that such an ideal world would not work in practice. Therefore, this epoch in MANET evolution is all about *energy efficiency.* The optimization problem here gets more complicated. Where it was all about finding the shortest route between wireless nodes A and B (shortest naturally meaning best QoS), the route now also has to represent a good trade-off between energy efficiency and good e2e QoS. Optimizations here are more complex, multiparametric, and difficult to solve. The fact that due to mobility of wireless nodes e2e paths have to deal with poor reliability only complicates things further.

Epoch 3: Social components. It can be argued that the growing optimizational complexity in the two previous epochs was the primary cause of the need to find a solution in the social realm. It is also possible that social aspects were introduced simply because MANETs are more often than not created by people and are therefore intrinsically social. This epoch has only recently started and is still ongoing. In incorporates methods that have to do with grouping (flocking, etc.) of nodes [22], which would ideally simplify e2e communications simply because groups/flocks can be used as discrete units of addressing and routing. Many methods work on purely social aspects such as *incentive to participate and opportunism.*

The cellular world has been undergoing a different evolution that mostly focuses on increasing throughput and efficiency of cellular access networks. The ongoing standards focus on the LTE-A technology as the main candidate for evolution between 4G and 5G [6]. According to current standards, 5G will implement a two-tier wireless access network (current one is one-tier) where the top tier will be formed by BSs while the bottom tier will consist of a large number of very small BSs. Research often refers to the small BSs using general terms like *microcells, picocells,* or *femtocells,* while technical documentation prefers to build on the terminology that exists in current 3G and 4G networks and calls them eNB [6]. The main purpose of eNBs is to create a dynamic access network that can be controlled at a relatively small scale. The much smaller grain also makes it possible to avoid congestion at BSs by offloading traffic to eNBs.

Compared with MANETs, cellular technology does not have the problem of multihop and much less concern for energy efficiency. However, social components discussed earlier (Epoch 3 of MANET evolution) will exist in 5G as well. The two particular features that are somewhat related to the MultiConnect and GroupConnect functionality are *device-to-device (d2d)* and *machine-to-machine (m2m)* communications defined in LTE-A [6]. The d2d happens when two devices communicate either directly or use eNBs (femtocells, WiFi, etc.) to create direct routes between each other. The m2m is a broader concept related to Internet-of-Things where any device (machine) can communicate over the 5G network. Obviously, both *d2d* and *m2m* modes are vivid examples of traffic offload in action. Moreover, using d2d as the basic unit of communication one can easily scale up to group communication, thus approaching the GroupConnect usecases discussed in this chapter.

Abruptly changing the narrative, there is an alternative viewpoint with wireless technologies. They can be viewed as the *local versus remote* argument [13]. This argument is based on current practice where 3G/LTE connections are widespread in practice but are very slow. MANETs might help but they are not as widespread and might be fast in the first one to two hops, but multihop connections suffer from low reliability [15]. Additionally, real content—and this chapter talks about cloud services—is rarely carried over MANET-powered infrastructure, thus further limiting the reach.

The local part of local versus remote makes sense even in the framework of MANET evolution. In fact, the social approach mentioned earlier creates groups of mobile nodes that share the same geographical location [22]. There are many other practical uses of the *local* aspect. Cloud services are recently given attention in form of a *mobile cloud* [11]. Practice shows that local throughput is at least 10× better than 3G/LTE or even two-hop MANETs [11].

The grouping/social aspect leads to the term *GroupConnect* coined by this author in an earlier work [1]. GroupConnect can be viewed as a kind of virtualization where the entire network, computing, and so on, set of resources of the entire group of nodes is pooled and abstracted for a number of practical usecases. GroupConnect is perfect for throughput-hungry cloud services [2].

This chapter mostly focuses on the local versus remote viewpoint at practical wireless technologies. The volume of material does not allow for an extensive overview of traditional wireless technologies, MANETs in particular. However, all the necessary references on MANET counterparts and related technologies and methods are provided abundantly throughout this chapter.

Finally, it should be noted that this chapter works with a set of *multi-x* technologies, where *x* can be -hop, -path, -connect, and others. Specifically, the two core concepts are MultiConnect and

GroupConnect. Note that both these terms cannot be found in traditional MANET resources, which support the claim on the *brand new viewpoint*.

8.1.2 MultiHop: Traditional MANETs

As was said before, MANETs are largely concerned with multihop connections [14]. There is a huge bulk of literature on the various multihop routing methods [15], and specifically protocols [16]. These two references provide excellent surveys on the topic.

Recently, more attention is being spent on e2e QoS and efficiency of multihop paths [18]. Even this narrow problem is multifaceted. For example, some research is concerned with e2e performance of traditional communications protocols like TCP [17], whose performance in wireless domain differs from that in the traditional settings. The presence of a connection is not a guarantee of good e2e QoS, so some research is concerned with bandwidth-optimized e2e routing [19], meaning that paths are optimized for the largest possible traffic they can carry e2e. As mentioned earlier, energy efficiency is a major aspect of e2e efficiency [20,21].

The recently developed social viewpoint at e2e routing also has multiple aspects. The basic argument is that mobile nodes—people in most cases but could also be sensors, etc.—are grouped and routing algorithms treat groups as sources and destinations when establishing e2e routes [23]. This brings several advantages. For example, two or more people used to have to support separate e2e paths for communication in pairs, but now groups only need one connection to pass information between each other while individuals are resolved and addressed within the group at each end. Flocking—another term used for grouping—also helps with improved e2e connectivity, when the latter is viewed as connectivity between the groups rather than individuals [22]. When a connection between groups/flocks is disrupted, it is likely that another pair of group/flock members can be found in each group to establish a new connection.

Social aspects are not only about grouping or flocking of mobile nodes. There is some literature of the social aspects proper. For example, the entire concept of *opportunistic* communications is a social concept. Mobile nodes *grab opportunities* they find based on their current geographical location and ultimate goals [24]. Some research in this area is focused on incentivizing participants to contribute their resources to a given group/flock/community of mobile nodes [23,24]. Note that some of these features are found in the local versus remote argument studied in this chapter.

Vehicular Access NETworks (VANETs) require separate attention [25]. They are basically MANETs; in fact, many of the same methods are used in both areas. However, the biggest difference in VANETs is in the mobility part. This is to be expected simply because cars move differently from people. This is why research on VANETs spends considerable attention to studying the difference in mobility and what this difference means for application of MANET methods in VANETs [25].

Before we go into the details of the local versus remote argument, it is important to list all the specific problems covered by MANETs and where solutions to these problems generate even more problems down the road.

> *Problem 1: End-to-End Wireless Connectivity.* This problem is all about multihop connections [14]. The main problem here is reliability of e2e connections [14]. It is notoriously bad for traditional applications [17], which warrants creating new wireless-specific protocols and applications. There is high overhead from maintenance traffic that is necessary to support e2e paths in a constantly changing environment [27]. The energy problem is not only about the energy spent by receiving or sending end, but also incorporates all relay nodes [20,21]. It is well known that the energy problem today cannot really be solved simply by increasing the battery size [26], which is why most methods assume that energy is an extremely limited resource.
>
> *Problem 2: Energy Efficiency in MANETs.* It should be noted that the energy problem is still the hot research topic today with no clear winners or killer-methods. Which is why it is its own major problem. Again, given that increasing battery size is not a solution [26], the research looks into other solutions. Two main solutions are offered. One is the *energy-aware routing*, with excellent survey on the topic in [20]. The other method is the *on-demand routing*, where routes are not supported continuously but are created for a short lifespan on the spot [21].

Problem 3: e2e Resilience in MANETs. The earlier two problems are actually in conflict. On the one hand, acceptable e2e routes can be created and maintained; on the other hand, QoS and energy constraints greatly reduce both the selection options and the period of time during which a given selection remains valid. The solutions include backup routes, fast recovery after disruptions, and so on [16]. There are also methods that optimize e2e path creation for multiple parameters including QoS, energy efficiency, and resilience [18]. Note that the *social approach* is also part of a solution that has resilience in mind [22,23]. Opportunistic routing in socially aware scenarios [24] is also good for resilience because disruptions are resolved faster via various other parallel opportunities made available to each mobile node, depending on environment. Practical applications for such solutions are numerous, for example, disaster relief [13,27].

In view of the aforementioned problems and solutions, delay tolerant network (DTN) appears to be an independent topic [40]. However, the demarcation line between DTN and other wireless research is fading away. For example, while traditionally DTN is considered to be a non-real-time technology, real-time methods are already discussed in literature [41]. Note that disaster relief is also a valid application in DTNs [40]. This chapter revisits DTN several times, slightly short of placing the technology itself in the main scope.

The aforementioned problem-solution listing is made with one purpose in mind: many aspects of recently developed methods (e.g., social aspects) bring MANETs closer to the local versus remote viewpoint discussed in this chapter. This connection is revisited several times throughout this chapter, for example, when discussing applications in context of *mobile clouds* [28].

8.1.3 MultiConnect and Other Throughput Boosters

One problem that remains unsolved—or even tackled—by MANETs is the problem of practical e2e throughput [19]. MANETs are notoriously bad at supporting high e2e throughput and perform steadily worse with increasing hop-length of e2e paths. What can be done

Table 8.1 Four Possible Technologies

	SINGLE CONNECTION	MULTIPATH
Singular connectivity	Traditional applications	Traditional multipath
Multiple connectivity	*No known cases (wasted potential)*	Group communication *3G/LTE + WiFi Direct* This proposal

Note: Two of them currently exist. The Group Communication technology is facilitated via wireless MultiConnect and is the key point of this chapter.

about it? It happens that *throughput boosting* has already existed as a valid research subject for several years. See Table 8.1 for a simple taxonomy of such research.

The taxonomy is primitive in that it is based on only two parameters: the number of parallel paths versus parallel connectivities. Note that this is a very uncommon classification and would probably not appear in traditional literature on MANETs of wireless technologies, in general. However, it is important to understand the fundamental premises of this chapter. Also note that the terms *connection* and *connectivity* are different.

> *Traditional technologies* are single connection and single connectivity. These are all traditional technologies and require no effort to make them work in practice. It is possible that a device has more than one connectivities—3G with WiFi with Bluetooth, etc.—but in the traditional setting the device *defaults* to one of them and does not allow for parallel use. This chapter spends some more time on the default behavior further on. There is no throughput boosting in traditional technologies.
>
> *MultiPath technologies* are all about multiple paths over one available (or default) connectivity. This is a hot research topic, with a good survey of various multitechnologies available in [30], and practical software implementations [3]. MultiPath TCP (MPTCP) is arguably the most popular example [30] and is the version of TCP that allows for segmentation and aggregation of TCP sessions over multiple packet streams carried over multiple distinct routing paths. MPTCP is well studied in theory and field tests [31] and even implemented in a Linux kernel [46]. There are various optimizations, methods, and algorithms pertaining to the concept [32]. Also, although

not labeled as multipath specifically, there is a set of methods that solve the same basic problem. For example, m2m sync or distributed sync is a problem that involves the multipath concept in some form [33]. Multipath fits well into e2e QoS in MANETs, where it can make use of path redundancy [18], where the redundancy itself is replaced by parallel use. Multipath can help boost throughput but only within the physical limits of a given technology used for connectivity. Specific field tests are discussed further in this chapter.

Multiple connectivity for a single connection is nonexistent because such a technology would be forced to default to one of the available connectivities and thus climb up one cell in Table 8.1. Otherwise, in the best case it can be considered as a case of the last class later.

MultiConnect with MultiPath is the ultimate goal and application environment discussed in this chapter. Table 8.1 shows one of many practical examples—a 3G/LTE + WiFi Direct [45]. This very combination is discussed in a practical application later in this chapter. Note that *GroupConnect* is the new related term coined in an early work by this author in [1]. Here, MultiConnect + MultiPath combination describes the environment (settings, etc.), based on which an application environment like GroupConnect can exist. All these concepts are discussed in detail throughout this chapter.

It should be obvious that the GroupConnect kind of technology has the biggest capability of boosting application throughput.

Now, let us put the three models in the context of 4G and 5G technologies. Obviously, current 3G/4G technologies are classified as *Traditional.* With 5G technologies, due to the incompleteness of current standards, the overall picture is spotty. For example, the concept of *aggregation* [9] normally refers to multichannel aggregation of content over the same connectivity, where MIMO is arguably the most common and currently existing technology. 5G looks forward to a broader notion of aggregation (*carrier aggregation* [9]) where individual substreams can come from different base stations or micro stations.

The *d2d* and *m2m* technologies are currently discussed under 5G but not in the context of one-to-many or m2m (group) communications [5].

Instead, current context for *d2d* and *m2m* is closer to the single-connection single-connectivity class in Table 8.1. However, technologically speaking, once the basic *d2d* functionality becomes available, it can be used in the context of the *MultiConnect with MultiPath* model, where it will seamlessly merge into the concept becoming one of the practical tools that help achieve the functionality expressed by the model.

8.1.4 Bluetooth, WiFi Direct, and Others: Capacity Trends

There is an ongoing capacity race in the wireless world. As 3G is migrating to LTE, WiFi and Bluetooth also migrate to their respective new, more effective versions. However, this evolution followed by migration does not necessary result in increased throughput. See [11] for a good survey on capacity limitations of local and long-haul technologies, including both old and new ones. A previous study by this author also measured practical throughput on several wireless service providers (WSPs) in Japan [1]. The latter study complements with very little overlap the data presented in [11].

Figure 8.1 shows the overall trends extracted from raw data in [1]. The following measurement setting was used. Three real WSPs in Japan were selected, of them one LTE and two 3G providers. One 3G provider offered a data plan where traffic was limited at 300 kbps. A simple download application ran in browser, measuring download time for each request. Downloaded file was of a fixed size, sufficient to require about 30 s of download time at the best possible performance. In terms of network performance, the application measured *available bandwidth* at the time of the measurement.

As is shown in Figure 8.1, data are separated into *workdays* (week days), *Saturdays*, *Sundays* (holiday), and *Holidays* (3+ days). Specifically, Golden Week in Japan (beginning of May) was in the middle of the field study and is the 3+ day holiday showing up in data. Results are shown as a distribution over time of day, with 30 m step.

What can be concluded from Figure 8.1? First, we can see that all WSPs offer very bad throughput regardless of type of day, hour, or LTE/3G technologies. Interestingly, the best provider (cross bullets) is not LTE, but a 3G provider that only recently entered the market at the time of the field tests. This shows that performance does not

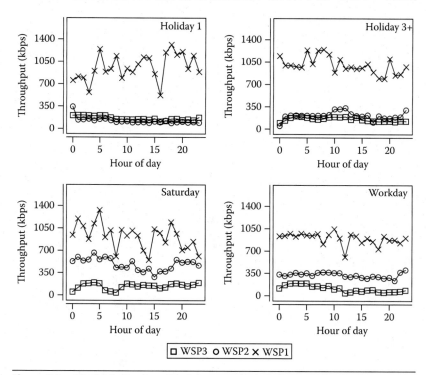

Figure 8.1 Actual data from 3 months of measurements in April–June 2013 using three separate LTE/3G providers in Japan.

cluster based on the type of technology, but the level of penetration, or, in other words, the number of people using the services of a given provider at the time.

Another way to put this peculiar artifact is to state that *crowd beats technology*, meaning that with large populations of users it does not really matter how new or advantageous (capacity-wise) a given technology is. From the figure, we can see that crowds especially overwhelm providers on holidays. We also see little effect of rate limitation—the data plan with 300 kbps rate limit performs nearly identically to an LTE provider. We can see some trends in relation to time of day, which is natural because most users sleep at night and the measurement was conducted in only one country that fits entirely into one time belt.

The take-home lesson from Figure 8.1 is that average expected throughput, regardless of the underlying technology, is below 300 kbps. It is very different from the claims made by LTE providers today. Note that these findings are not original and can be found in field tests in [11,12] with roughly the same basic outcome. This paints a rather pessimistic overall picture in terms of performance, especially given that penetration of wireless technology is expected to grow further. Moreover, this pessimism goes against the main target of the local versus remote argument whose entire purpose is to boost e2e throughput for individual users.

Let us return to the local versus remote argument. Figure 8.2 shows results from local tests where local means that the wireless technologies are available only for local communication, normally 30 m and under. More details on the setup and detailed raw data can be found in an earlier study by this author in [1]. Two best candidates— Bluetooth 4.0 and WiFi Direct (v 1.0)—were selected. Bluetooth 4.0 is supposed to be energy-efficient while offering improved through-put. WiFi Direct [45] is a relatively little-known technology that this chapter places in the center of the narrative.

Throughput was measured by sending 1, 5, or 10 files each 200 MB from device A to device B at the distances of 1 and 10 m. Multiple tests were conducted and the average value plotted in Figure 8.2.

The following conclusions can be offered. Average throughput of Bluetooth is 1 Mbps, which is unacceptably low. It depends little on distance and number of files, probably because it is already slow and environment facts play little role. Average throughput on WiFi Direct

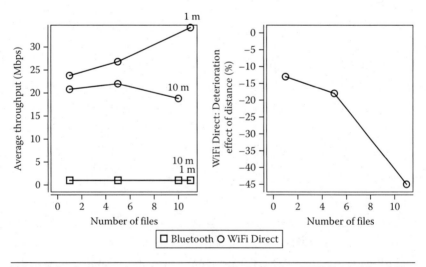

Figure 8.2   Comparison in performance between Bluetooth 4.0 and WiFi Direct at 1 and 10 m distances and various number of transferred files.

is 25 Mbps. It depends on distance and number of files. Large bulk seems to experience better throughput (probably because of higher stability) at short distance. Right side of the figure shows how distance (going from 1 to 10 m) and number of files affect throughput. In the worst case (very large downloads), one can expect a 40% drop in throughput down to about 15 Mbps.

Take-home from this data is as follows: WiFi Direct is a very good technology because its local throughput is at least 10× better than any remote technology. In fact, it is better when compared with multihop connections in MANETs.

8.1.5 P2P WiFi and Specifically WiFi Direct

Another name for WiFi Direct (shorter form of WiFi Direct) is P2P WiFi, according to the specs in [45]. Technologically, WiFi Direct is a direct communication between two wireless nodes without using an access point (AP) in between—the traditional way of communicating between two nodes. Internally, each node implements WiFi Hotspot and P2P WiFi specs in one node [45], meaning that each node is both a server and a client in a two-way communication session. However, the technology is an abstraction for its applications and technical details about hotspots and clients are hidden from application layer where apps simply create connections between two nodes directly.

There are two obvious benefits from using WiFi Direct:

1. The high throughput it can sustain—see the aforementioned measurement data
2. Ability to use WiFi Direct in parallel with any other connectivity, that is, the MultiConnect feature

Figure 8.3 visualizes the MultiConnect aspect of WiFi Direct in the following visualization. Two classes of pairs of technologies are possible—*winner* and *join* classes. *Winner* means that from a given pair of technology, only one of them can be used as a default technology. For example, with both LAN and Wireless present, LAN is the winner. *Join* means that both connectivities can be used together. WiFi Direct in this classification takes a special place of a technology that can *join* with all other available technologies. Remembering the

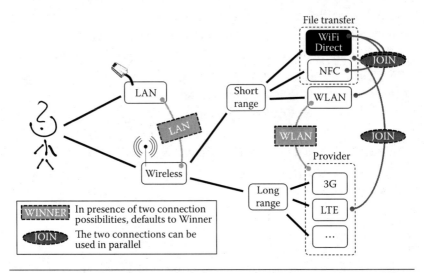

Figure 8.3 The place WiFi Direct takes among other technologies and specifically its ability to operate in parallel with others.

connection–connectivity classification earlier, this makes WiFi Direct a means toward a *MultiPath* + *MultiConnect* technology. This is another key premise in this chapter.

On the way toward the 5G suite of technologies, the figure can go through several major changes. For example, software-defined radio [6] can add even more flexibility, making it possible to have parallel transmissions (JOIN) even within the 3G (LTE, 4G, and 5G) module where it can be accomplished, for example, by controlling antenna arrays in MIMO in software. It is also likely that the short-range branch of wireless technologies will expand (future versions of near field communication [NFC] and WiFi Direct) and even become compatible with cellular short-range communications that would make it possible for this module to talk directly to 5G base cells or microcells. While these changes will result in even higher flexibility of the classification in Figure 8.3, the overall notion of ability (or inability) of a device to use two or more connectivities in parallel will remain the same.

Judging from the current state of 5G standards, there is clear understanding that 5G networks will be a mixture of technologies. For example, the concept of environments with different radio access technologies is currently discussed [9]. However, current standards do not clearly specify the necessity to implement the ability to use multiple technologies in parallel.

Let us return to the local versus remote argument. In the context of Figure 8.3, it could be changed to *local with remote* because the visualization shows that WiFi Direct is a local technology that can be used in parallel (with) many remote technologies. These concepts are all new, but some aspects have already been considered in literature in the context of mobile clouds [11].

The remaining measurement is now to measure performance of MultiConnect, that is, WiFi Direct used together with a remote connectivity.

Figure 8.4 shows performance from parallel connection of WiFi Direct with either 3G/LTE (marked as G) or conventional WiFi (marked as W). All the components are the same as before, except for the conventional WiFi, which is simply a WiFi connection to a local AP and then to the Internet. Naturally, WiFi Direct is running in parallel with either G or W downloads. The plots are slightly different in that they plot remote throughput versus local throughput.

The following lessons can be made from Figure 8.4. There is seemingly an unordered scattering of bullets across the plots. However, there is an order to the distribution of the data points. Specifically, WiFi throughput is generally higher than that of 3G/LTE connections. However, there is roughly a downsloping trend that shows that WiFi Direct throughput is inversely proportional to throughput on remote connections. The worst case is at 7–8 Mbps at 180 kbps of remote throughput. It is also evident that physical distance is less of a problem than cross-interfering between WiFi Direct and the parallel technology. Yet, note that 8 Mbps is still 15×–20× better than throughput on a remote connection.

Figure 8.5 performs cross-correlation function (CCF) analysis of the entire bulk of raw data from Figure 8.4. CCF is executed as follows. Lists of WiFi Direct throughput and one of the metrics on the horizontal scale are formed in their natural order, and CCF is calculated on the trends of values in the two lists. This directly shows whether WiFi Direct throughput correlates with changing values in a given metric.

Figure 8.5 offers the following observations. It does not really matter whether 3G or LTE connection is used in parallel with WiFi Direct. There is some negative correlation with increasing remote throughput, which repeats the earlier finding where it was found

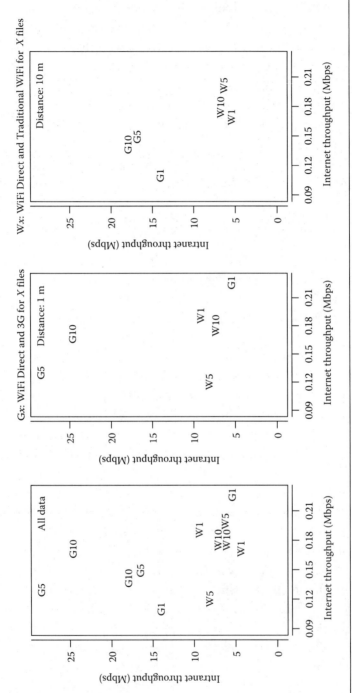

Figure 8.4   Tests of parallel connectivity in LTE/3G- and WiFi-pairs with WiFi Direct. Note that vertical scale is intrAnet (not intErnet) throughput.

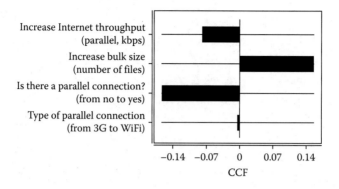

Figure 8.5 CCF analysis of various factors that can affect WiFi Direct performance.

that WiFi Direct throughput decreases when 3G or WiFi through-put increases. However, even a bigger effect is from having a parallel connection in the first place, meaning that WiFi Direct would work much better if it were used alone.

The take-home from the aforementioned results is as follows. Studies in [11,12] offered a large number of real-life measurements. However, only this and the earlier study in [1] offer analysis of parallel connectivity. This section also established an effective range of opera-tion for MultiConnect/GroupConnect applications. These practical findings are used as basis for build-up further in the chapter.

8.2 MultiConnect as Network Virtualization

Having established MultiConnect and GroupConnect concepts, as well as the baseline understanding of the local versus remote argu-ment, it is now possible to focus on how to use MultiConnect in the-ory and practice. This section focuses largely on theory, while the next section discusses a valid practical usecase.

8.2.1 Local Wireless Connectivity

Research literature on local connectivity can be separated into several distinct areas. This section discusses the most prominent of them.

Coordination in Space is all about detecting and maintaining the topology of a group of mobile nodes that share the same physical loca-tion. User profiling and identification are among the central topics in such research. In a perfect world, 3D coordination would be most welcome. However, practice shows that 2D or even 1D maps (distance only) are more realistic. Many tools are available to end users. VirtualCompass [52] creates 2D maps using WiFi and Bluetooth. NearMe [53] uses a different method but roughly the same set of techniques. BeepBeep [54] stands out in that it creates 2D maps by exchanging sounds. Note that while the tools are available to end users, their use in practice as part of a GroupConnect technology does not yet seem feasible. This is because GroupConnect adds an additional layer of complexity on top of the 2D maps, thus rendering the entire package extremely complicated. So, GroupConnect applications would rather build communication groups by randomly selecting members.

Spectral (energy) efficiency is also found here in tools like Spectra [35], but the viewpoint is different because the main question here is not about energy efficiency but feasibility of running a code of a remote node—the so-called *offload* aspect of local wireless communication.

Computation offload itself is its own aspect and is a big part of the local versus remote argument. There are many frameworks and tools for offloading code to smartphones over wireless. Offload normally goes through several stages, namely, sending the code to a remote node, waiting for the code to complete remotely while performing control or maintenance function of the code, and finally collecting the results when the code completes remotely. Tools like Chroma [36] and Cuckoo [38] do just that. There are also wireless versions of traditional parallel processing where tools like Maui [29] perform serialization of parallel processing jobs. Another traditional topic found here in its wireless versions is message passing where MMPI [42] is one of the tools.

Although it may come as a surprise, the concept of *mobile cloud* falls inside the domain of location communication [11]. In fact, mobile cloud is a kind of computation offload but with a more complex and continuous structure. The survey in [12] covers all the possible local configurations that can be used for mobile clouds. There are many platforms for mobile clouds [37,39]. There are, in fact, wireless versions of famous distributed cloud functions like MapReduce where Hyrax [50] is its wireless version. Tools like Scavenger [44] and Satin [51] do not call themselves clouds but they work by running code in a wirelessly distributed manner that makes it similar. Security aspects of running the code in local wireless groups is covered in [43].

Note that this chapter is not on mobile clouds as such. However, from the way the local wireless groups function, GroupConnect is very similar to mobile clouds. Mobile cloud research today even discusses the same notion of remote versus local that was earlier introduced in this chapter. However, there are several major differences.

First and foremost, mobile clouds do not formulate local groups as a *virtualization of networking resource*. Instead, mobile clouds go for a simple solution to the offload problem by creating a cache of remote content locally. Such research often starts from a fully synchronized local cache and only discusses the benefits of local connectivity over direct communication with the master copy of the content in (remote) clouds. The *second* major topic in mobile clouds is *grouping methods* that is achieved by relative positioning or relative coordinate systems that can help create efficient local communication groups. The key problems tackled by mobile clouds therefore are in the area of *group forming algorithms, battery efficiency*, and *real-time communications*.

The biggest difference with this chapter is that GroupConnect is not viewed as an isolation between remote and local connectivity—where the border between the two is almost always the traffic bottleneck. In fact, the entire point of GroupConnect is to boost throughput of individual members at the expense of the entire pool of networking resources available within the group. As is shown in the next section, GroupConnect offers a much more flexible formulation in the form of *virtualization* of both remote and local networking resources of the entire group. The formulation is applicable to mobile clouds but only as one practical usecase while this chapter discusses and provider deeper analysis for more interesting practical applications such as capacity-pooling.

8.2.2 Network Virtualization Basics

Network virtualization like that found under OpenvSwitch [48] or ClickRouter [49] might be another *unexpected* topic in relation to GroupConnect. However, there are already cases when network virtualization is implemented with wireless networks in mind. For example, OpenvSwitch is already successfully implemented in [47] as an altered version of the Android operating system. However, such examples lack the main component—the MultiConnect. Traditional understanding of network virtualization is that it is a function that allows routing decisions at the level of individual packets.

Virtualization—in the same meaning as it is used further in this chapter—is important because it allows to create and maintain overlay topologies on top of a given physical network, with full applicability to wireless networks as well. The overlays are abstractions that allow for capacity sharing among paths, or capacity sharing among flows, running on top of the same physical connection. Given the abstraction, all the unnecessary technical details, like the need to maintain isolation of virtual resources from each other, are hidden from the applications running in virtual environments.

WiFi Direct can be viewed as a virtualization technology. Given the now well-established local and remote wireless settings, the abstraction is simply the merger of the two, with inability on the part of applications to tell the two apart. This is, in a manner of speaking, the basic technical description of the GroupConnect technology. More details are provided further on.

8.2.3 Resource Pooling via MultiConnect

The new role is a virtual wireless user (VWU). As mentioned in the previous section, VWU is an abstraction that virtualizes local and remote connectivities within a group, that is, an implementation of GroupConnect. Another new role is SP, where services are more commonly cloud-based but can be anything as long as services are practical [2]. SP requires a remote connection and, therefore, a 3G/LTE or WiFi connectivity. In mobile cloud or MANET settings, remote might not necessarily mean Internet but can be a smaller realm of service provisioning. Mobile clouds is a good example of such a small realm [11]. The term *remote* here means *not local*, where local connections are all one-hop paths within a local group using WiFi Direct for connectivity.

With these two new roles, Figure 8.6 is a visualization of how VWU works. VWU is an abstraction and is actually based on individual wireless users (WUs). Each WU enjoys a MultiConnect environment. It is possible that some WUs may not have remote connectivity but must have WiFi Direct to participate in GroupConnect.

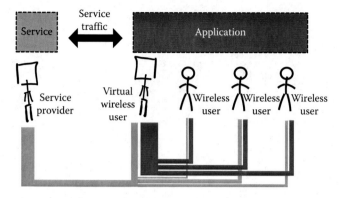

Figure 8.6 How MultiConnect can help create a group of resources that are pooled into one virtual user.

VWU, as an abstraction, aggregates remote connections supported by WUs, where WiFi Direct facilitates and maintains the abstraction locally.

How does this abstraction work based on Figure 8.6? Simply picture VWU as an application that uses the service at the SP's side. There is no need to split this *service use* into individual WUs. This is why the abstraction (virtualization) is so important in the first place. One obvious advantage in having VWU is that the role can enjoy the sum of individual remote throughputs, thus boosting the aggregate throughput.

8.2.4 *Virtual Wireless Resource*

The VWU abstraction is a good visualization of the structure of GroupConnect, but may not be immediately obvious when one needs to discuss what is happening with the traffic inside and outside of GroupConnect (local versus remote).

Figure 8.7 is another visualization, this time focusing on capacity of the network resources available to VWU.

There are three separate capacities, walking Figure 8.7 left to right.

SP Capacity, denoted as C_{SP} (leftmost), represents how much of peak throughput can be supported by SP. Note that the value is not that for individual users, but for a group of users. Many services today, especially in clouds, enforce quotas on how

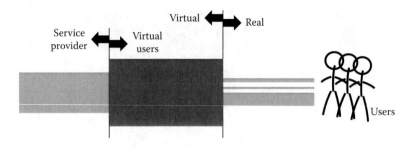

Figure 8.7 The same diagram as before but this time with virtualization component made more obvious.

much service can be used per unit of time (seconds, hours, days, months, etc.).

VWU Capacity, denoted as C_{VWU} (in the center), is simply a measure of how much throughput WiFi Direct can support within a group. Contrary to SP Capacity, this value was measured and presented earlier in this chapter. However, it still depends on the structure of data exchange within the group. Some more details on this follow shortly.

Finally, *WUs Capacities* (plural intended), denoted as C_{WUs} (rightmost), is the combination of the sum and structure of individual remote connections of WUs.

Having these three variables, we can now conduct primitive analysis based on inequalities among the values.

$C_{VWU} > C_{SP} > C_{WUs}$ is arguably the most common case. It is based on the fact that there is limited throughput on 3G/LTE Internet, much below what can be supported by SP. WiFi Direct has the biggest capacity among the three and can probably support much higher capacities both in SP and WUs.

$C_{VWU} > C_{WUs} > C_{SP}$ is the case of a poor quality or congested SP, which is not unheard of nowadays. This means that there is excess remote capacity. It is good news for WUs because each individual WU is now free to use some of the spare Internet capacity for its own activities. WiFi Direct still beats the other two capabilities.

$C_{WUs} > C_{VWU} > C_{SP}$ means that your group is either too big or WUs in the group are too far from each other, leaning toward the first case as more probable. WiFi Direct cannot fully support the VWU abstraction and starts lagging. VWU can probably compensate for some of the lag by using smart grouping—adding several hubs, hierarchy, subgroups, and so on. Yet, it is very unlikely that C_{SP} can exceed the local throughput supported by WiFi Direct, meaning that smart grouping would completely resolve the bottleneck. In this case, the new bottleneck would probably be at SP because large groups would have much more remote connectivity.

These are only several practical situations one might encounter in GroupConnect settings. While all the basic configurations are covered, practice would probably be more fuzzy, introducing complex dynamics to all variables.

8.2.5 A Practical Optimization Problem

Returning back to the local versus remote argument, we already know that WiFi Direct (local) offers much better throughput as well as overall performance compared to virtually any remote technology. It was also stated that WiFi Direct can support multiple parallel connections, thus making GroupConnect a valid reality.

The rest of this section presents two optimization formulations that put GroupConnect into a mathematical form. This section presents a specific practical usecase, while the next section offers a more generic formulation.

Let us establish all the roles first. There is *Main Client*—the main node in a wireless group. Although all other nodes also implement the same exact software, this node is in charge of delegation of tasks and collection of results, thus functioning as a representative of the entire group. Note that the *Main Client* is not a fixed role and can migrate to other nodes when the latter are in need of putting GroupConnect to practical use. Apart from the role name, nodes—users of a GroupConnect—are also numbered as A, B, and so on. Each node has access to remote (Internet) connectivity but its use is restricted by quotas, namely, packets per month, bytes per day, and so on. Quotas cannot be ignored when planning the resources of a given GroupConnect situation.

Let us denote T as an arbitrary time period. In view of modern contracts, it is convenient to use the unit of *1 month*. Given that user A is the Main Client, its networking resource (remote) can be put as N_A, or more generically for use u as N_u. Note that there are multiple users in the list $\{A, B, C, ...\}$ for each A. Also note that U is both a measure of current throughput (bps) and a monthly quota when aggregated over the entire month. N_{quota} here presents the quota, where most providers today start pacing one's traffic past 1–3 GB. Finally, N_{target} is the desired throughput of the entire GroupConnect effort, written as $N_{A,target}$ specifically for user A as the Main Client of this GroupConnect.

DTN enters the picture through a rating solution where each user u of A is assigned a rating R_u. This might be a complex variable that can take into consideration relative speeds and direction of both users, among many other metrics. The DTN part is not in focus, so a surface mention is sufficient.

The *optimization objective* is therefore to minimize the difference between desired and attainable throughput; that is,

$$\text{minimize} \quad N_{diff} = N_{A,target} - \sum_{u \in \{A,B,C,\dots\}} R_u N_u, \qquad (8.1)$$

$$\text{subject to } N_{diff} \geq 0, \qquad (8.2)$$

$$\sum_{t=1,\dots,T} N_{u,t} < N_{u,quota} \quad \forall u \in \{A,B,C,\dots\}, \qquad (8.3)$$

$$\text{count}(\{A,B,C,\dots\}) < k, \qquad (8.4)$$

where k is a reasonable limit on a number of parallel WiFi Direct connections.

Note that this formulation does not take into consideration energy efficiency and allows for any level of intensity in WiFi Direct (local) traffic. This may be a problem because people would like to conserve their batteries and, therefore, would like to participate in a process that includes energy efficiency—via establishing reasonable limits on the intensity of WiFi Direct exchange.

A more generic formulation in the next section takes this very point into consideration.

8.2.6 Toward a Generic Formulation

In order to include energy efficiency into the optimization problem, we need to include both remote and local traffic. Let us assign slightly different terms to them. Let us put remote communications as *the sky* and local (WiFi Direct) communications as *the floor*. This notation is not only useful, but is also descriptive because the terms convey the physical meanings of traffic in each region.

Note that this formulation is also directly applicable to 5G design components where energy efficiency is also viewed in the context of remote (to BS) versus local (to a local microcell or d2d) communication [9]. In fact, as mentioned earlier, the *remove versus local* approach to the energy efficiency is a general notion that exists across many wireless topics, including the mobile clouds discussed earlier [11]. Although current research and standards on 5G discuss several optimization situations [9], the formulation shown in this section is not found in current literature, while relatively simple *load balancing* techniques are common.

Further enhancing the problem, let us divide the traffic into *in* and *out*, assigning direction the packets fly on connections. Also, resources are more generic and are divided into *network* (traffic itself) and *CPU* (computation offload). Both make practical sense in GroupConnect, mobile clouds, and other settings.

Let us use U and C to denote *utilization* and *capacity*, respectively. $C_{network}$ would then denote *network capacity* and C_{cpu} CPU capacity (multicore, etc.). $C_{i,sky}$ notation can be used to denote Internet capacity of user i. L denotes *load* of a node—a generic metric describing processing overhead. The metric *diff* can be used to denote application traffic between the floor and the sky (see VWU model given earlier), measured in bytes or chunks of data to be exchanged between Main Client and the remote service. Finally, T denotes the time interval for one optimization cycle.

The optimization problem for a *GroupConnect* can then be written as

$$\text{minimize} \quad w_1 \sum_{i \in group} diff_i + w_2 \sum_{i \in group} L_i + w_3 \sum_{i \in group} \frac{U_{i,network,sky}}{U_{i,network,floor}}, \quad (8.5)$$

where

$$L_i = U_{i,cpu} + U_{i,network} \quad \forall\, i \in group, \quad (8.6)$$

$$\text{subject to } L_i \leq 2 \quad \forall \, i \in \textit{group}, \tag{8.7}$$

$$\text{size of } (\textit{merge}(\{\textit{diff}_i\})) \leq T \sum_{i \in \textit{group}} C_{i,out,network}. \tag{8.8}$$

Clearly, it is a linear problem. However, weight setting (w_1, w_2, and w_3) is a side problem that needs to be considered when looking for a solution. Note that the merged size of *diffs* is limited by the sum of capacities on Internet connections to ensure that continuous operation can be sustained. If the constraint in Equation 8.8 is not satisfied in several successive optimization cycles, backlog of unsynced changes may start growing exponentially and uncontrollably.

Note that the third term in Equation 8.5 may work to minimize remote (sky) traffic by giving higher weight to intranet communications, as long as *diffs* are property synced between local and remote parts of the service. Practical validation for the aforementioned problem is out of scope of this chapter but will be presented in future publications. However, it is important to note that the optimization problem is directly applicable to many practical problems, especially the distributed sync (group sync in this case) between a group of wireless nodes that share the same content and the remote storage (cloud drive, etc.) that provides the backend storage proper.

8.2.7 MultiConnect over MultiHop: MANETs and DTN Methods

This section repeats some of the earlier statements but now having properly established the context, they may contain more useful information.

MANETs over multihop connections are notoriously difficult [14,16], where the current trend is to introduce social aspects by grouping, flocking [22,23], and so on WUs. Basically, the trend can be split into two main directions. First, there is a clean separation of local versus remote networking. Second, grouping of WUs goes hand-in-hand with the GroupConnect spirit in this chapter.

So, how are the aforementioned statements and mathematical formulations related to MANETs and DTNs?

First, GroupConnect requires an abstraction via virtualization of network and, as shown in the previous sections, other resources of all nodes in a group. The relation to MANETs is in the virtualization parts where all the e2e QoS of multihop paths, path redundancy, and other methods central to MANET research are abstracted and hidden from the toolkit of GroupConnect functionality [16.e2e] [15]. MANET methods can help by establishing an e2e path from a user to a remote service—the remote service can in fact be within a mobile cloud several hops away—and export it upward to the GroupConnect level as a valid path. Even more important is the real-time management of e2e paths, the reliability of which in MANETs is a known fact.

The DTN approach is an alternative to MANETs in exactly the same part of functionality—remote connectivity. For example, disaster relief methods [13,27] can help by creating a route between a wireless node and a remote service. The path itself does not exist, but it is possible to use DTN methods to exchange data over multihop routes in both directions. In fact, recent research on DTN came up with several real-time methods [40,41]. Although this sounds strange, delay-intolerant DTNs are already a reality.

Having just stated that DTN is an alternative to MANETs, it should be noted that recent MANET research has branched into the so-called *opportunistic routing* that shares many features with DTN [24] but has a specific set of features that has to do with sharing the wireless medium among multiple users or multiple concurrently transmissions—an approach similar to MultiConnect but is a kind of emulation of multiconnect over one kind of connectivity. Once MAC contention and scheduling problems are resolved, a group of opportunistic users can be redefined as a GroupConnect technology because each user theoretically can communicate to multiple users in parallel. Reference [10] is an excellent and deep analysis into all the major problems and solutions in opportunistic networking.

It should be stressed again that GroupConnect is completely agnostic to remote access methods. As long as remote connectivity is established, GroupConnect can merge it into its current pool of local/remote resources for later use. Note that such an abstraction is rare in research literature, probably because e2e multihop wireless connectivity is difficult to accomplish. However, given the research on grouping/flocking and the GroupConnect in this chapter, perhaps it is just about time for such research.

8.3 GroupConnect in Practice: A University Campus

Having established

- Fundamentals and theory related to MultiConnect and GroupConnect
- Optimization targets both specific and generic

it is now time to apply these concepts to a practical application of GroupConnect. This entire section discusses one specific application— GroupConnect—in a new kind of Wireless University Campus. All stages of implementation are presented and reviewed in detail.

8.3.1 Objective: GroupConnect

Let us establish the story of this particular GroupConnect application.

The story starts with the congestion problem discussed before. The University (any university, but a role in this story) realizes that a congestion problem exists not only on 3G/LTE connections but also at the WiFi APs set up across the campus. The problem is so acute that all the plans to have wireless classes have failed after the realization that large-volume interactive content cannot be effectively served in 25+ student classes (study in [4]) where each student accesses the content asynchronously, interactively, and with little regard to resource efficiency.

So, the university decides to tackle the problem using GroupConnect. Students are allowed to bring their own remote connectivity—3G, LTE, and others—to the campus. GroupConnect is then implemented by an application distributed by the university, hopefully automatically and seamlessly as far as students are concerned. The main practical objective of GroupConnect is *throughput boosting*, as was discussed earlier.

The university conducts a preliminary feasibility study into this plan and finds that smartphone penetration across students is around 90%. The university decides that the other 10% should be covered by purchasing and handing devices to students for rent. Another study revealed that the presence of remote connection (3G/LTE) is at almost 100% penetration level, thus guaranteeing that all wireless nodes have a remote connectivity.

This is the overall story. Implementation details are presented further in this section.

8.3.2 Generic Case of Cloud APIs

Although clouds comprise many distinct and not necessarily standard technologies, access to cloud services is roughly uniform. This means that APIs are mostly standard. For example, three-party contracts are common, where OAuth is the de facto default protocol used to generate generic access tokens [34]. From the practical viewpoint, it is also common to build client-side applications as a combination of cloud services. For example, one might use storage together with map or document APIs. The GroupConnect in this section is perfectly suited to running such applications.

Yet another important feature of cloud services is the relatively small grain of content. Large content is normally split into chunks where each chunk is accessible separately. In fact, access to chunks is normally isolated, which means that they can be handled out of order. For example, one can upload or download a chunk from a larger file by specifying read/write positions as parameters in HTTP requests. As long as access tokens are correct, such HTTP requests can be sent from different devices. This, in fact, is how tasks can be delegated within a wireless group using GroupConnect.

Having established the feasibility of GroupConnect both in local wireless environments and now in cloud services, the rest of this section discusses in detail how to implement such a technology in reality.

8.3.3 Cloud API Delegation over MultiConnect

We have the same roles as before—the *Main* and *Delegated* clients. Each client, regardless of the role, has both a remote and local connectivities. Remote is 3G or LTE connections; local is always WiFi Direct.

Figure 8.8 shows the basic concept of delegation in a graphical form. As mentioned before, this is a simple form of GroupConnect where there is only one Main Client. Various other forms are possible depending on a given practical usecase. Line thickness visualizes the capacity of each kind of connection. It is natural that local connection handled by WiFi Direct is much thicker than a remote 3G/LTE connection.

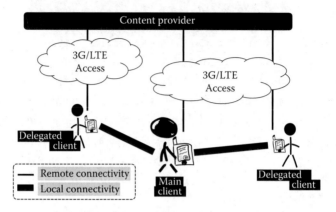

Figure 8.8 Concept of task delegation.

Figure 8.8 plays out as follows. The Main Client is aware of the granular nature of cloud services and is in a position to decide on the size of chunks, delegation configurations, and so on. The Main Client generates a *task* for each chunk of a remote service and delegates the task to one of the Delegated Clients. Note that WiFi Direct can support multiple P2P connections at the same time, which means that multiple delegations can be executed in parallel. Also note that, as mentioned once before, *parallel* here does not imply *continuous*. In reality, delegations happen as two short communication sessions: one to assign (delegate) a task and the other in the opposite direction to receive the result of a delegation. Since 3G/LTE connections are much slower, time between these two sessions can be several seconds, depending on the chunk size.

There are several practical scenarios that can be applied to this basic design. For example, this GroupConnect can be used to aggregate content via multiple parallel connections, that is, functioning a throughput boosting technology. It can also be used to distribute content among group members, where delegation would be used to ask other nodes to pass chunks along yet to other nodes.

Figure 8.9 is a generic representation of a smartphone application that implements the above GroupConnect. There is a *Virtual Client* whose entire purpose is to virtualize resources of the whole wireless group. The device's own remote connection—the Internet—is part of the pool, and remote connections of other GroupConnect members are accessed via Intranet connections that are implemented as delegations, which are in turn carried over WiFi Direct connections. This model

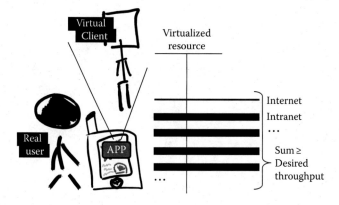

Figure 8.9   Smartphone app that implements a virtual client based on the MultiConnect.

also refers back to the optimization problems presented earlier that are now put into the context of a simple practical implementation.

Note that all members of a GroupConnect must implement the same exact Virtual Client, but the roles played by individual devices depend on a specific GroupConnect scenario and may not be the same for all devices.

8.3.4 University Coops and the App

The aforementioned implementation details lack the final component—a process that makes GroupConnect possible in university settings. After all, this section is all about building a new kind of wireless university campus based on GroupConnect. The rest of this section attempts to show that such an implementation is not only realistic but also feasible.

Figure 8.10 presents the project called "Building a Wireless University Campus based on GroupConnect" in four simple steps.

Step 1: Smartphone Application. It is possible that WiFi Direct or a similar technology is available in notebooks and desktop devices, but so far the reach seems to be limited to handheld devices, specifically smartphones and tablet devices. From the various OSes, Android appears to be the best on the market, with WiFi Direct support being a native part of this OS since version 4.1. At this step, the university develops the application plus all the group management protocols. The latter is

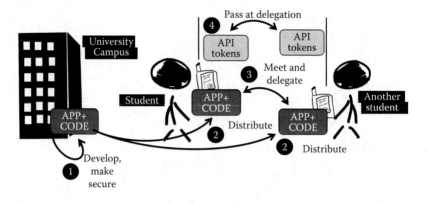

Figure 8.10 The full path one needs to walk to make GroupConnect work in practice.

obviously more important than the former because the main benefit of GroupConnect lies in the ability to put together a group ad hoc. *CODE* in the figure is a security precaution that makes sure that the application can be used only by students of a given university or only within a given university campus.

Step 2: Distribution. The best way to distribute the application is at enrollment. This can be one of many procedures a student has to undergo at enrollment. In fact, registration for Coop services is common in many universities, with good penetration levels—meaning that a very high percentage of students signs up for Coop services. The GroupConnect application is handed over to students' own devices via an installation URL or a memory card. As mentioned earlier, if a student does not have a suitable device, one can be rented for the duration of enrollment.

Step 3: GroupConnect Session. This is GroupConnect in use. When a student needs to use some remote service with a boosted throughput, he/she goes through the GroupConnect application. The application organizes a wireless group and uses its resources via delegation. However, at this step, delegation itself does not start happening yet. At this step, only the group structure is established and the Main Client gets the full knowledge of the size of the group that has become available to it.

Step 4: Services and Delegation. Having established a wireless group (or on demand, just before use), it is now possible to parallelize a cloud service by delegating portions of its own work to other members, collecting and aggregating results, and so on. As explained earlier, the delegation itself is made possible by sharing access tokens to the same service to other members.

8.4 GroupConnect Methods and Trace-Based Analysis

The material up to this point has been mostly theoretical or, when practical, descriptive in nature. This section covers the remaining area—numeric analysis of performance in realistic conditions. For simplicity, the analysis is performed within a single wireless group but two distinct group management models are analyzed. In order to make the analysis as realistic as possible, simulations are based on real mobility traces, thus avoiding the need to generate synthetic—and therefore unrealistic—mobility models. In fact, the realistic traces used in the following analysis come from a real university campus, which means that the analysis is directly applicable to the university campus example discussed in the previous section.

8.4.1 Mobility Traces

There are many synthetic mobility models in research literature [14,25]. But in several research years, there is a growing trend of trace-based analysis where the parts of simulation that need to be realistic are actually made realistic via applying real-life data. CRAWDAD [55] is one of many repositories offering mobility traces for public use. The specific trace used in this analysis comes from the KAIST campus. The trace is generated by taking GPS readings at 30 s intervals for 100 students.

There is only one minor shortage in the trace that comes from the nature of GPS itself. GPS does not work inside buildings—or at least not reliably—which means that the mobility trace cannot trace people inside buildings. If one replays the trace, then individual traces are interrupted as people go into a building and resumed when/if they come out later. This shortage does not affect the analysis too much because it can be assumed that grouping can be accomplished much easier inside buildings than outside of them, simply because people

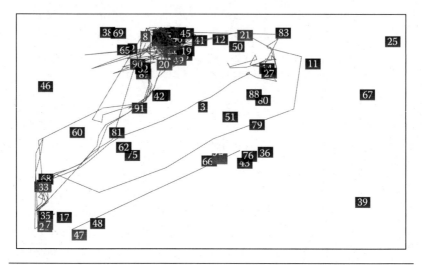

Figure 8.11 Snapshot from a CRAWDAD trace.

tend to gather inside buildings. On the other hand, people outside walk on their individual routes, which are fully captured by the trace.

Figure 8.11 shows a still snapshot of the trace, visualizing positions, and some of the routes. To avoid crowding the plot, only 10 random routes are plotted in their entirety. The rest are shown as the final positions at the end of the trace.

What can be read from this trace? First, there is obviously one large hub where most of the people crowd at and routes travel over. This is to be expected in an average university. There are two or three other locations where there is limited crowding of people and routes. It is also noticeable that many routes have shared segments—this is because students walk on the same roads between buildings.

It should again be noted that it would be hard to generate such a trace synthetically. Especially given that not only locations and routes are important but also the time when students travel between buildings. All the three components are pivotal to effective analysis of GroupConnect in realistic conditions.

8.4.2 Performance Metrics

Before the analysis itself, let us first establish some important performance metrics.

Lag is a slightly indirect metric, but it has a very practical value. Let us say that we need to download a 100 MB file in 20 s. This is a target or a deadline. The lag is then the time *past the deadline* that the user has to wait for the download to complete.

Throughput is self-explanatory. In GroupConnect settings it stands for the aggregate throughput that one can obtain via delegation within a wireless group available at the time.

Redundancy is simply the number of identical requests sent to different group members in parallel. Redundancy is necessary to overcome the problem of low reliability of P2P connections. Without redundancy, when a member you have delegated a task to goes out of range, you may lose an important chunk of your data that would invalidate the entire bulk. Redundancy is also a *safety margin* that would make achieving the target throughput a more realistic possibility, again, in view of low reliability of connections in wireless groups.

Failure rate is another way to look at redundancy. If every 10th delegation request fails for one of many possible reasons, then about 10% redundancy would be procured. In reality, 100% redundancy, that is, two requests for every chunk, is a more likely choice.

There are several other metrics but let us stick to a relatively short list to avoid confusion.

8.4.3 Practical Opportunistic Algorithms

Let us remember that GroupConnect is an opportunistic endeavor, where the various forms of opportunism are covered in depth in [24]. The basic idea is to catch whatever the opportunity one gets in regard to being able to improve one's own throughput. The bottom line is one's own remote connectivity—which, as we well know, is very slow—but one can improve this drastically by delegating tasks to other members of a GroupConnect. In GroupConnect context, opportunism directly translates into methods related to group management.

The following are the two practical models that can easily be implemented in software (the Virtual Client).

1. *Paced Model*: We have a target throughput, say, 1, 10 Mbps, and so on, which can also be expressed as a deadline (same as given earlier) based on size and preferred download time.

Then, having set the redundancy level (2×, 3×, etc.), it is now possible to manage one's group size. Group management can be dynamic as well. For example, one can keep growing the group until the aggregate throughput exceeds the targeted value. The process can be repeated every time an active member falls out of wireless range and a new member has to be added.

2. *Greedy Model*: As is already in the name, this group management creates as big a group as it can. This means that all wireless nodes that are found within the communication range are contacted and added as group members. Redundancy setting is necessary in this model as well but its operation is different. Redundancy still defines the number of identical requests issued in parallel. However, now that we can potentially have many more group members than necessary, requests can be distributed smoothly across the group.

Alternatively, in bigger groups the redundancy level can be increased as well. However, you should remember that redundancy only improves resilience of your throughput but not the throughput itself. For simplicity and in order to be able to compare results between the two models, the 2× redundancy level is applied to both models.

8.4.4 Performance Analysis

The following setting was used for the analysis. The real 3G/LTE measurements were used, distributed randomly across the population of students in the mobility trace. The simulation exists in time, which means that not only mobility is changing, 3G/LTE throughput is also changing in time. WiFi Direct (local connectivity) throughput is set to the lower performance margin at 10 Mbps, fixed and static throughout the entire simulation. Since it is much faster than any 3G/LTE connection, it has little effect on the outcome.

Groups are created using the two above models in the following way. A random time in mobility trace is selected. A random person is selected as the Main Client. All nodes within a 25 m radius are added as group members. 3G/LTE speeds have already been assigned at this point, so each group member has its point on the remote throughput timeline. It is called a timeline because remote throughput also changes in time based on throughput traces presented at the beginning of this chapter. Regardless of the size of the group, 2× redundancy is used in both Paced and Greedy models. The models themselves function in accordance to the description provided earlier.

Simulation starts and continues for 10 s. The three possible cases are SingleConnect and Paced and Greedy models of GroupConnect (the technology name is MultiConnect but the algorithm is GroupConnect). In SingleConnect, the Main Client has no group and relies only on own remote connection to download the content.

The lag needs an anchor value. In each simulation, the best remote throughput in the group is selected as the target throughput. This also applies to SingleConnect, where only one node is simulated but the same group as in multicase is defined and is available for reading metrics and parameters. The lag is then calculated relative to this target value.

Figure 8.12 shows *lag* performance. Both Paced and Greedy models are shown but their relative differences are very small compared to the case of SingleConnect, which results in both curves almost merging together. The plot is a discrete distribution of results from 1000 simulation runs.

First, we can see negative lag in Figure 8.12. This is because GroupConnect/MultiConnect was able to pool resources within the

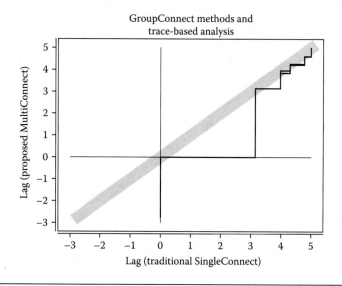

Figure 8.12   Lag performance. Both scales are seconds.

group and drastically increase throughput. The best performance here is about −3 s lag, meaning that the download took 7 s instead of the expected 10 s. There are also cases when the lag is positive, meaning that it takes the download several more seconds past the deadline to complete. However, even here the multimodels show the presence of advantage on a limited scale. Visually this translates into the curves being consistently below the diagonal line.

Figure 8.13 focuses specifically on the effect of mobility from the viewpoint of the upper limit it puts on aggregate throughput. The rules for this analysis are slightly different. Each GroupConnect session is still 10 s, but the targeted throughput is gradually increased, up to 100 Mbps. This is only a target value while the plot shows that this value was not achieved in simulation. This is in fact the limitation of mobility. Note that same 2× redundancy is retained in this case as well.

Let us read Figure 8.13 along the diagonal line. As long as targeted throughput is relatively low, both Paced and Greedy models perform nearly identically. This trend continues up to 4.05 (about 10 Mbps). Past that we see that 2× redundancy does not provide sufficient resilience to the Paced model and people going out of range cause major performance impairment. These delegations have to be re-requested to someone else, delaying the download in the process.

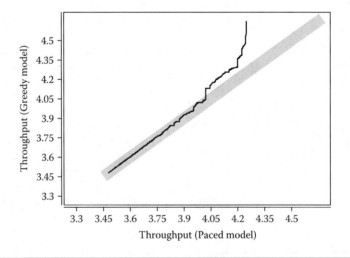

Figure 8.13   Throughput performance. Both scales are logs of throughput values in kbps.

Mobility affects the Greedy model as well, but there are many more people to whom the Main Client can send duplicate delegation requests later on, which means that recovery is fast. In fact, we can see that close to 40–50 Mbps throughput could be achieved in some cases. These are the scenarios that were played near the main building of the university where there are more people on average.

Note that changing redundancy to 3×, 5×, and so on potentially could also improve throughput. This is because having more duplicate parallel requests would increase the chance of getting at least one reply. However, following this line of inquiry complicates the analysis and is therefore avoided. The take-home lesson here is that both group size and redundancy level play a role in increasing aggregate throughput.

8.5 Conclusions

This chapter focused on two technologies: MultiConnect as a practical solution for parallel wireless connectivity and GroupConnect as a logical social usecase for MultiConnect. The two technologies share the technical core, but GroupConnect is bigger because it incorporates group management techniques and application logic. These concepts are not abstractions. Among several others, WiFi Direct is an existing technology that can be used on smartphones and tablet devices to implement MultiConnect as a connectivity feature and ultimately GroupConnect at application layer. This chapter discussed a practical usecase where GroupConnect was used by students at a university campus as a tool for boosting download/upload throughput.

This chapter presented material at several distinct layers. Theoretically, MultiConnect is a network virtualization technology where network resources of a group of wireless nodes can be merged and used collectively via an abstraction—MultiConnect is a good name for this abstraction. This formulation opens doors into a wide variety of theoretical viewpoints where this chapter presents one practical and one generic optimization problems. Potentially, there can be a separate optimization problem per each individual usecase that puts GroupConnect to practical use.

Note that the virtualization formulation is compatible with many technologies currently discussed as part of the 5G standards. SONs are a form of grouping. MIMO is a type of MultiConnect that, although commonly used in the single-connectivity multiple-paths mode similar to the MPTCP wired networks, has the potential for the true MultiConnect. One related practical scenario discussed in current research on 5G is when devices can balance or even use a combination of remote connections to BSs and those to local microcells [9]. Another compatible scenario involves d2d functionality in 5G, where, for example, device A can pool connections to base- or microcells together with those to other devices that in turn can receive content from other base- or microcells. Such a usecase can facilitate a major throughput enhancement for individual users. Note that the same goal is achieved in this chapter via GroupConnect, where, in a similar fashion, devices boost throughput by pooling cellular connections to 3G/4G BSs with local wireless connections to other devices. Also note that the two 5G scenarios can be fully described by the generic virtualization concept of GroupConnect as well as the optimization formulations presented in this chapter.

The particular usecase on which this chapter spent considerable portion of material is a new kind of Wireless University Campus. The usecase goes beyond just the utility of boosting one's throughput. The main point is that traditional infrastructure, both in 3G/LTE and local WiFi AP forms, is simply not capable of supporting classes where most traffic is exchanged wirelessly. This chapter presents results from real measurements in both wireless domains that support this harsh statement. GroupConnect here comes as a replacement technology. Students literally bring their own network resources to the campus, and GroupConnect applications running on individual students' smartphones and tablet devices work hard to put people in groups automatically and on demand. Instead of individual 3G/LTE connections or crowded WiFi APs, we get partially overlapping wireless groups where centers are students who initiate GroupConnect applications.

The University Campus example in this chapter was also presented as a manual for universities on how to implement such a large-scale project. Software that implements MultiConnect is not difficult to create. In fact, Android Development Tools has WiFi Direct example [56] already in it. In fact, this author has built a working prototype for a large-bulk download application based on WiFi Direct version of GroupConnect. The overall theme here is that students can use GroupConnect to boost throughput in cloud services, which is a generic enough definition to allow for many specific applications, including those well beyond educational use.

8.6 Future Research Directions

There are many aspects of MultiConnect and GroupConnect that were not covered in this chapter. For example, it is not discussed how to increase the scale of GroupConnect. The example in the chapter discusses a wireless university campus that has the scale limited by the size of the campus. Increasing the scale will raise several problems, among which the strongest one is arguably security and its socially acceptable solutions. The application in the chapter assumes that all smartphones in the campus have the same application as well as a security key that lets them tell on-campus devices from others. This is a simple solution to the problem that is otherwise recognized in literature as the *resource discovery* problem. Resource discovery at a large scale will require some level of manual input from smartphone users that will limit the range of automatic operation.

Issues related to the opportunistic nature of the technology and social incentives that apply are not considered in depth in this chapter because, again, they are not very important in university settings. The social aspects of group wireless connectivity is a hot topic in current literature [24] and is likely to remain one in the near future. It is likely that some kind of incentive technology will be applied, where the closest existing example is that of peer-to-peer file sharing. In case of GroupConnect, users share network, CPU, and storage resources of their own devices.

In the grand scheme of things, GroupConnect can be considered as a solution for e2e QoS problems in MANETs and DTNs. This is somewhat related to the social aspects [24] because it was shown that grouping wireless nodes together and using groups as point of routing can greatly improve resilience of e2e connectivity. However, the roles in this case are inverted—while this chapter focuses on groups and can use improved e2e connectivity in MANETs or DTNs for better remote connectivity, the *social routing* research uses groups to improve e2e connectivity. Ignoring the inverse, both research directions have similar components and can benefit from each other.

Finally, GroupConnect implements the offload function that is actively discussed as part of 5G cellular technology. In 5G, d2d, and m2m connectivity modes, femtocells and the like are all solutions to the problem of congestion in cellular network core. Congestion is partially resolved by offloading as much capacity as possible to small-scale wireless networks that, under 5G, can be a mixture of cellular and broadly defined (WiFi, etc.) wireless technologies used concurrently. It can be expected that more research will focus on such mixtures in the near future.

References

1. M. Zhanikeev, Virtual wireless user: A practical design for parallel multiconnect using WiFi Direct in group communication, in: *Tenth International Conference on Mobile and Ubiquitous Systems: Computing, Networking and Services (MobiQuitous)*, Tokyo, Japan, December 2013.
2. M. Zhanikeev, Multi-source stream aggregation in the cloud, Chapter 10, in: M. Pathan, R. Sitaraman, and D. Robinson (eds.), *Advanced Content Delivery and Streaming in the Cloud*, New York: Wiley, 2013.
3. M. Zhanikeev, Experiments with application throughput in a browser with full HTML5 support, *IEICE Communications Express*, 2(5), 167–172, May 2013.
4. M. Zhanikeev, Experiments on practical WLAN designs for digital classrooms, *IEICE Communications Express*, 2(8), 352–358, August 2013.
5. M. Rumney, *LTE and the Evolution to 4G Wireless, Design and Measurement Challenges*, New York: Wiley, 2013.
6. X. Zhang, X. Zhou, *LTE-Advanced Air Interface Technology*, Boca Raton, FL: CRC Press, 2013.
7. G. Roche, A. Valcarce, D. Lopez-Perez, J. Zhang, Access control mechanisms for femtocells, *IEEE Communications Magazine*, 48(1), 33–39, January 2010.
8. Transition to 4G: 3GPP broadband evolution to IMT-advanced, Rysavy Research Report, 3G Americas, www.3gamericas.org, 2010.
9. E. Hossain, M. Rasti, H. Tabassum, A. Abdelnasser, Evolution toward 5G multi-tier cellular wireless networks: An interference management perspective, *IEEE Wireless Communications*, 21(3), 118–127, June 2014.
10. S. Glisic, *Advanced Wireless Networks: Cognitive, Cooperative and Opportunistic 4G Technology*, New York: Wiley, 2009.
11. N. Fernando, S. Loke, W. Rahayu, Mobile cloud computing: A survey, *Elsevier Journal on Future Generation Computer Systems*, 29, 84–106, January 2013.
12. M. Satyanarayanan, Fundamental challenges in mobile computing, in: *Fifteenth Annual ACM Symposium on Principles of Distributed Computing (PODC)*, Vol. 5, Philadelphia, PA, May 1996, pp. 1–6.

13. M. Satyanarayanan, Mobile computing: The next decade, *ACM SIGMOBILE Mobile Computing and Communications Review*, 15(2), 2–10, April 2011.

14. M. Ilyas, R. Dorf, *The Handbook of AdHoc Wireless Networks*, Boca Raton, FL: CRC Press, January 2003.

15. M. Ananstapalli, W. Li, Multipath multihop routing analysis in mobile ad hoc networks, *Journal of Wireless Networks*, 16(1), 79–94, January 2010.

16. V. Borrel, M. Ammar, E. Zegura, Understanding the wireless and mobile network space: A routing-centered classification, in: *Second ACM Workshop on Challenged Networks (CHANTS)*, Montréal, Québec, Canada, September 2007, pp. 11–18.

17. E. Osipov, C. Tschudin, Evaluating the effect of ad hoc routing on TCP performance in IEEE 802.11 based MANETs, in: *Sixth International Conference on Next Generation Teletraffic and Wired/Wireless Advanced Networking (NEW2AN)*, St. Petersburg, Russia, May 2006, pp. 298–312.

18. R. Al-Qassas, M. Ould-Khaoua, Performance comparison of end-to-end and on-the-spot traffic-aware techniques, *International Journal of Communication Systems*, 26(1), 13–33, January 2013.

19. C. Chen, C. Weng, Bandwidth-based routing protocols in mobile ad hoc networks, *Journal of Supercomputing*, 50(3), 240–268, December 2009.

20. C. Chao, J. Sheu, C. Hu, Energy-conserving grid routing protocol in mobile ad hoc networks, in: M. Ilyas (ed.), *The Handbook of Ad Hoc Wireless Networks*, Boca Raton, FL, CRC Press, January 2003, pp. 399–413.

21. S. Doshi, S. Bhandare, T. Brown, An on-demand minimum energy routing protocol for a wireless ad hoc network, *ACM SIGMOBILE Mobile Computing and Communications Review*, 6(3), 50–66, July 2002.

22. A. Konak, G. Buchert, J. Juro, A flocking-based approach to maintain connectivity in mobile wireless ad hoc networks, *Journal of Applied Soft Computing*, 13(2), 1284–1291, February 2013.

23. W. Ting, Y. Chang, Improved group-based cooperative caching scheme for mobile ad hoc networks, *Journal of Parallel and Distributed Computing Archive*, 73(5), 595–607, May 2013.

24. R. Ciobanu, C. Dobre, Social-awareness in opportunistic networking, *International Journal of Intelligent Systems, Technologies and Applications*, 12(1), 39–62, July 2013.

25. A. Saha, D. Johnson, Modeling mobility for vehicular ad-hoc networks, in: *First ACM International Workshop on Vehicular Ad Hoc Networks (VANET)*, Philadelphia, PA, September 2004, pp. 91–92.

26. O. Amft, P. Lukowicz, From backpacks to smartphones: Past, present, and future of wearable computers, *IEEE Pervasive Computing*, 8, 8–13, July 2009.

27. Y. Sasaki, Y. Shibata, A disaster information sharing method by the mobile servers in challenged networks, in: *Twenty-Sixth International Conference on Advanced Information Networking and Applications (WAINA)*, Fukuoka, Japan, March 2012, pp. 1048–1053.

28. M. Satyanarayanan, P. Bahl, R. Caceres, N. Davies, The case for VM-based cloudlets in mobile computing, *IEEE Pervasive Computing*, 8, 14–23, October 2009.

29. E. Cuervo, A. Balasubramanian, D. Cho, A. Wolman, S. Saroiu, R. Chandra, P. Bahl, MAUI: Making smartphones last longer with code offload, in: *Eighth ACM International Conference on Mobile Systems, Applications, and Services (MobiSys)*, San Francisco, CA, June 2010, pp. 49–62.

30. P. Schmidt, R. Merz, A. Feldmann, A first look at multi-access connectivity for mobile networking, in: *ACM Workshop on Capacity Sharing (CSWS)*, Nice, France, December 2012, pp. 9–14.

31. Y. Chen, Y. Lim, R. Gobbens, E. Nahum, R. Khalili, D. Towsley, A measurement-based study of multipath TCP performance over wireless networks, in: *ACM SIGCOMM Internet Measurement Conference (IMC)*, Barcelona, Spain, August 2013.

32. A. Makela, S. Siikavirta, J. Manner, Comparison of load-balancing approaches for multipath connectivity, *Elsevier Journal of Computer Networks*, 56, 2179–2195, May 2012.

33. H. Pucha, M. Kaminsky, D. Andersen, M. Kozuch, Adaptive file transfers for diverse environments, in: *USENIX Annual Technical Conference*, San Diego, CA, August 2008, pp. 157–171.

34. K. He, A. Fisher, L. Wang, A. Gember, A. Akella, T. Ristenpart, Next stop, the cloud: Understanding modern web service deployment in EC2 and azure, in: *Internet Measurement Conference (IMC)*, Barcelona, Spain, August 2013, pp. 177–190.

35. J. Flinn, S. Park, M. Satyanarayanan, Balancing performance, energy, and quality in pervasive computing, in: *Twenty-Second IEEE International Conference on Distributed Computing Systems*, Vienna, Austria, July 2002, pp. 217–226.

36. R. Balan, M. Satyanarayanan, S. Park, T. Okoshi, Tactics-based remote execution for mobile computing, in: *First ACM International Conference on Mobile Systems, Applications and Services*, San Francisco, CA, May 2003, pp. 273–286.

37. G. Huerta-Canepa, D. Lee, A virtual cloud computing provider for mobile devices, in: *First ACM Workshop on Mobile Cloud Computing and Services: Social Networks and Beyond (MCS)*, New York, NY, Vol. 6, June 2010, pp. 1–5.

38. R. Kemp, N. Palmer, T. Kielmann, H. Bal, Cuckoo: A computation offloading framework for smartphones, in: *Second International Conference on Mobile Computing, Applications, and Services (MobiCASE)*, Santa Clara, CA, October 2010.

39. L. Deboosere, P. Simoens, J. Wachter, B. Vankeirsbilck, F. Turck, B. Dhoedt, P. Demeester, Grid design for mobile thin client computing, *Future Generation Computer Systems*, 27, 681–693, June 2011.

40. A. Vasilakos, Y. Zhang, T. Spyropoulos, *Delay Tolerant Networks: Protocols and Applications*, Boca Raton, FL: CRC Press, 2011.

41. A. Balasubramanian, B. Levine, A. Venkataramani, DTN routing as a resource allocation problem, in: *SIGCOMM*, San Diego, CA, October 2007, pp. 373–384.

42. D. Doolan, S. Tabirca, L. Yang, MMPI: A message passing interface for the mobile environment, in: *Sixth ACM International Conference on Advances in Mobile Computing and Multimedia (MoMM)*, Bali, Indonesia, November 2008, pp. 317–321.

43. D. Huang, X. Zhang, M. Kang, J. Luo, MobiCloud: Building secure cloud framework for mobile computing and communication, in: *Fifth IEEE International Symposium on Service Oriented System Engineering (SOSE)*, Nanjing, China, June 2010, pp. 27–34.

44. M. Kristensen, Scavenger: Transparent development of efficient cyber foraging applications, in: *IEEE International Conference on Pervasive Computing and Communications (PerCom)*, Fort Worth, TX, March 2010.

45. Wi-Fi, Peer-to-peer: Best practical guide, Wi-Fi Alliance, www.wi-fi. org, December 2010.

46. MultiPath TCP: Linux Kernel implementation [Online]. Available at: http://multipath-tcp.org (retrieved July 2014).

47. K. Yap, T. Huang, M. Kobayashi, Y. Yiakoumis, N. McKeown, S. Katti, G. Parulkar, Making use of all the networks around us: A case study in android, in: *ACM SIGCOMM Workshop on Cellular Networks: Operations, Challenges, and Future Designs (CellNet)*, Helsinki, Finland, August 2012, pp. 19–24.

48. OpenVSwitch Project [Online]. Available at: http://openvswitch.org/ (retrieved July 2014).

49. E. Kohler, R. Morris, B. Chen, J. Jannotti, M. Kaashoek, The click modular router, *ACM Transactions on Computer Systems (TOCS)*, 18(3), 263–297, August 2000.

50. E. Marinelli, Hyrax: Cloud computing on mobile devices using MapReduce, Master thesis, Carnegie Mellon University, Pittsburgh, PA, 2009.

51. S. Zachariadis, C. Mascolo, W. Emmerich, SATIN: A component model for mobile self organisation, in: R. Meersman and Z. Tari (eds.), *On the Move to Meaningful Internet Systems (CoopIS/ODBASE)*, LNCS, Vol. 3291, Berlin, Germany: Springer, October 2004, pp. 1303–1321.

52. N. Banerjee, S. Agarwal, P. Bahl, R. Chandra, A. Wolman, M. Corner, Virtual compass: Relative positioning to sense mobile social interactions, in: *Eighth International Conference on Pervasive Computing*, Helsinki, Finland, May 2010, pp. 1–21.

53. J. Krumm, K. Hinckley, The NearMe wireless proximity server, in: *Ubiquitous Computing (UbiComp)*, Nottingham, England, September 2004, pp. 283–300.

54. C. Peng, G. Shen, Y. Zhang, Y. Li, K. Tan, Beepbeep: A high accuracy acoustic ranging system using cots mobile devices, in: *Fifth ACM International Conference on Embedded Networked Sensor Systems (SenSys)*, New York, November 2007, pp. 1–14.

55. CRAWDAD Repository of Mobility Traces [Online]. Available at: http://crawdad.cs.dartmouth.edu (retrieved July 2014).

56. Android Development Tools [Online]. Available at: http://developer. android.com/tools/index.html (retrieved July 2014).

9

VIDEO STREAMING OVER WIRELESS CHANNELS

4G and Beyond

MARAT ZHANIKEEV

Contents

Video streaming is part of a larger topic of *content delivery*. The term *content delivery network* (*CDN*) will be used throughout this chapter to refer to a provider of the delivery service. Although this chapter is on wireless delivery, the network is considered a nontrivial mix of wired and wireless parts.

Content delivery went through roughly the following three stages of evolution in recent years. The *first generation* was all about *replication* and *distribution*, where *caches* of the same content were created in several locations in the global network and the load would be balanced by distributing users across the multiple locations [29–31]. The *second generation* evolved either into a peer-to-peer (P2P) delivery [32] or *adaptive streaming* [33], which solves the same problem of server-side congestion in spite of being two completely different technologies [2]. The third generation is being currently formed by research on even smarter and more dynamic streaming technologies where *multisource aggregation* [2,6,7], *scalable/layered video* (SVC part of H.264 [56] format) [19,57], and cloud-backed providers with dynamically managed populations of streaming servers [2] are the main topics.

Current research on video streaming in wireless channels is extremely complex. While basically being the combination of the aforementioned basic technologies, the complexity is increased because of the nature of the wireless delivery. Traditional CDN research does not view the network as a limited resource, instead considering the network from the viewpoint of the dynamics in its *end-to-end* (e2e) performance. For example, in P2P streaming, one can have as many as a hundred simultaneous connections to other peers. In wireless networks (WNs), such scenarios would brutally fail simply because one wireless terminal can support only a relatively small number (often only one) of parallel connections to other terminals or provider base stations (BSs). Yet the overviews on cloud-based streaming [2] and P2P streaming with SVC [5] might help establish the necessary background prior to making the adaptations for wireless delivery.

It is important to note that video streaming used to be about *source and channel coding* [16]—and still is in the wireless domain, where topics like *concealment, error resilience,* and *compression* rooted in the general area of information theory (IT) are still considered central. Today, however, one can clearly see separation of the field into purely IT-related problems (work on scalable and layered video) and the problems related to delivery. This is recently facilitated by the newly developed video formats where, for example, scalable video format (H.264 [56]) is released as a black box with a long list of configuration parameters that can be adjusted by the delivery method to suit a given e2e network environment [27]. Having been relieved of the need to worry about the video coding itself, the delivery method can now pay more attention to more abstract performance metrics like robustness to variation in e2e throughput or behaviors like *scaling content in response to decreasing throughput* [26,28]. This *separation of coding from delivery* is evident in the majority of current research on wired and wireless delivery. For example, an advanced P2P delivery with focus on multistream aggregation [5] depends and takes advantage of scalable video [5,23].

All of this applies to wireless video delivery in both 4G and 5G, but in the soon-to-start fifth generation, there are many of the same topics as are discussed in wired delivery today [15]. There is a clear shift from Mobile Ad hoc NETworks (MANETs) [11] toward small-scale local group collaboration [12], which together with *femtocells* can support P2P streaming. Multiple-input multiple-output (MIMO) is a kind of *multiconnect* technology that together with Collaborative MultiPoint (CoMP) and self-organized networks (SONs) can also support P2P streaming or, alternatively, the multisource aggregation of content [1]. There is also wireless-specific functionality that is not found in the wired domain. GroupConnect is a more general case of multipoint but with the added feature of being able to use multiple modes of connectivity [1], where WiFi Direct is the specific technology that allows for that function (also referred to as P2P WiFi) [55]. Chapter 8 focuses on GroupConnect, which should be a better source of references on the topic. Mobile cloud is a more general case of GroupConnect [35] and is also wireless specific.

Wireless video streaming is mostly standardized under the IMT-A umbrella of standards developed by 3GPP [52]. Specifically, 3G to 4G to 5G evolutions using LTE and LTE-Advanced (LTE-A) technologies are released as separate normative documents by 3GPP [53]. There are also conceptual documents that, for example, state the necessity of MultiConnect and GroupConnect functionality [54]. If limited only by mobile web, then the dynamic adaptive streaming over HTTP (DASH) [59] standard on *adaptive web streaming* covers that area as part of the HTML5 set of standards [58].

This chapter focuses on advanced video streaming over wireless channels. The focus is mostly on the delivery aspects and specifically on *e2e network performance*. The chapter will talk about quality metrics (quality of service [QoS]) as well as human perception of quality (quality of experience [QoE]). Scalable video defined by H.264 [56] is selected as the representative of the best video coding standard today and is the most common form of video content throughout the chapter. The chapter discusses recent advances under the umbrella of LTE-A such as MIMO, CoMP, and femtocell. New paradigms such as MultiConnect and GroupConnect are shown in the context of how they can be implemented using existing wireless functionality. The chapter pays special attention to *massive crowds* of wireless users and how wireless delivery can deal with such environments. The specific example of a *massive crowd* comes in the form of an Olympics stadium with the target crowd size of 100k wireless terminals. Performance analysis is done on several models, including WiFi hotspots, femtocell networks, and P2P delivery using GroupConnect. The P4P (local P2P caches) approach is applied to some of the models. The overall approach toward performance analysis is to find and visualize bottlenecks in the wireless delivery of video streams.

9.1 Advanced Video Streaming

Only a small portion of this section talks about wireless video streaming while the majority focuses on the video streaming technology proper. Specifically, this section discusses the concept of *layered video* and the coding techniques involved. Having understood the structure of the video stream, the reader is introduced to the *substream method* actively used in P2P streaming and multisource aggregation, where the total video stream is split into separate parallel substreams to be delivered over multiple e2e network paths. At the end of this section, the narrative returns back to the wireless domain and discusses the specific challenges that exist in wireless video streaming.

9.1.1 Introduction to Video over Wireless Channels

References [13–15] provide excellent overviews of recent advances in WNs where the core technologies are the current LTE and the future LTE-A. Reference [15] focuses specifically on video delivery and has some overlap in content with this chapter, mostly in the area of low-level video coding.

The history of wireless video streaming starts with MANETs [11] where some of the methods focus on multimedia delivery. The biggest problem in video delivery over MANETs is the reliability of e2e paths [46]. Current research on MANETs often focuses on e2e QoS [48] and is therefore closer in spirit to the general (wired) multimedia delivery, which is traditionally extremely concerned with e2e network performance. Among recent works, e2e TCP performance in MANETs [39,47] and bandwidth-aware routing [49] have the closest relation to this chapter. The topic of *social collaboration* is on the path toward the concept of GroupConnect and is represented in recent MANET topics of *node grouping* [51], also sometimes referred to as *flocking* [50]. *GroupConnect is defined as a technology in which wireless terminals are joined in a logical group with a clean separation between local* (*within the group*) *and remote* (*group members with the outside*) *communications*. This chapter will return to this concept on several occasions, plus one of the analytical models at the end of the chapter is specifically based on GroupConnect.

Video coding technology today has advanced to the current de facto default of the MPEG-4 format as defined by the H.264 standard [56]. The format is well studied with many real [60] and synthetic [61] traces. The QoS and QoE are also well defined and understood [60]. QoS in this context normally is an indicator of the reduced video quality due to a given level of data loss from the delivery over the network. QoE is the subjective judgment an average human makes under a given level of QoS impairment. This chapter returns to these metrics later in the analysis part. H.264 is on the way to become standard in web browsing with the DASH protocol [59], where it can be used for web-based streaming, for example, in a web-based TV [34].

In view of the aforementioned text on MANETs, it should be noted that MANETs by themselves are not directly related to content delivery. This is due to the fact that MANETs are normally assumed to operate in relative isolation from the network backbone. By contrast, the mainstream discussion on content delivery (and video streaming as a practical CDN) normally assumes that the content is delivered on long paths that start at a major content provider, continue with multihop network delivery, and end in a local (possibly wireless) reception. The real-time decision-making that is involved in such services makes it hard to apply MANETs directly. However, video streaming remains to be part of the traditional ad hoc/MANET research [11] with the QoS-aware advanced described earlier.

The default context presumed by this chapter is as follows. Content is hosted and served by a large content provider, which takes care of cache distributions, replication, and even possibly e2e request routing [29]. The first part of the network delivery happens over wired networks, but the final leg is wireless that normally involves cellular technologies like LTE and LTE-A [15], but can also accommodate local delivery over WiFi networks as part of the overall LTE-A approach toward the offload technology. In the WiFi part, specifically, modern wireless multimedia have some elements of both ad hoc and wired delivery tricks like node grouping and e2e topology control and network optimization [15].

The overall CDN technology is the same for wired and wireless domains. In the wired domain, the technology evolved from *cache management* [29] to *route optimization* [30,31] to *P2P delivery* [32] to *cloud-based multisource* [2], and so on. In the wireless domain, you do not see such evolution as clearly—some wireless-specific research still mostly talks about video coding [16]—but one can find roughly the same elements as in the wired domain. For example, P2P streaming, cloud elements, multisource aggregation, and several others all have wireless analogs. This chapter will stop at each of them along the narrative. Note that the evolution of the video coding technology itself resulting in the H.264 format is agnostic of the delivery method and can result in effective applications in both wired and wireless domains.

9.1.2 Advanced Video Streaming Concepts

Figure 9.1 showcases all the major video streaming technologies from the viewpoint of the features that are important for efficient delivery. This section first defines all the features and then discusses each technology separately.

Parallel transfer is a feature that tells you whether your content arrives over multiple smaller parallel streams or in one large stream. The obvious options are *single path* versus *multipath*. Multisource is obviously a kind of multipath [2].

Parallel connectivity only sounds the same as parallel transfer. In reality, most technologies today, including wired delivery, are *single connects*. MultiConnect is a relatively recent, natively wireless concept that describes the ability of a wireless terminal to use several distinct wireless technologies in parallel. In modern smartphones, 3G/4G/LTE, WiFi, and Bluetooth all can be used in parallel, with a relatively low level of interference [1]. Although technologically possible, there are no known implementations of MultiConnect between wired and wireless technologies.

Finally, *data rate* can be constant versus variable. Note that this feature refers to the data rate of the delivered content rather than the performance of e2e paths. However, it should be clear that variable rate content might be better suited to e2e paths that experience fluctuations in throughput. This is, in fact, the core notion of *adaptive streaming* [26].

	Parallel Transfer		Parallel Connectivity		Data Rate	
	Single Path	Multipath	Single Connect	MultiConnect	Constant	Variable
Traditional	■		■		■	
Adaptive HTTP streaming	■		■			■
P2P streaming		■	■			■
Cloud streaming			■		■	■
Traditional wireless	■				■	■
Advanced home wireless		■	■	■ (MIMO)	■	■
Femtocell				■ (MIMO)	■	■
Federated WiFi	■		■		■	■
Wireless P2P streaming		■	■		■	
5G streaming		■		■ (GroupConnect)	■	■

Wireless-specific

■ + ■ mean content-agnostic

Figure 9.1 Various video streaming technologies and their features.

Now, let us consider each potential video streaming technology.

Traditional streaming is when your terminal uses a *single connection* to receive and view constant bitrate (CBR) video. Methods that apply to this method are discussed in literature [29–31], but given the advantages of all the following technologies, this is the most unpopular option today.

Adaptive HTTP streaming still uses a *single path/connection/connectivity* but also applies *variable bitrate (VBR)*. In browsers, the DASH [61] protocol defines the active part of the adaptive streaming, in case the VBRs from H.264 itself are not sufficient.

P2P streaming is very active today. It exists in at least two separate versions: the BitTorrent-type stream aggregation [32] or the video-specific *substream method* [5,6]. Cloud streaming, which is where clouds provide the multisource environment for parallel content aggregation, is potentially the new kind of such streaming [2] and is already shown to work in practice in browsers [7]. The overlaying feature of all these methods is the *multipath* delivery. Connectivity is still singular in most cases. The methods are agnostic to the kind of bitrate, but most recent methods prefer VBR [5].

Cloud streaming is the exact copy of P2P streaming with the exception that peers are replaced with cloud-backed sources [2].

Traditional wireless streaming is mostly the same as the *traditional wired* except that wireless environments are too volatile for CBR video, which is why even relatively old technologies and methods relied on some form of VBR [11].

Advanced home wireless streaming is mostly about MIMO but recently also involves femtocell technology [15]. While MIMO poses technical difficulties when implemented between terminals and BSs—mostly because they are energy-hungry and can cause radio congestion over large areas—it already exists in wireless routers, wireless storage devices, and wireless multimedia systems at home. The main point of MIMO is to use multiple antennas in parallel. This makes the technology both MultiConnect and MultiPath, regardless of the fact that all the connections and paths are the same. Since it is used locally, the bitrate of video content does not matter. Additionally, femtocells have recently become part of the home setup with personal mini-BSs installed at people's homes.

Femtocell streaming [12]—including the one that is part of the advanced home wireless earlier—is a way for BSs to reduce congestion by offloading traffic to more numerous but smaller BSs [15,37]. Transfer is mostly single path, but femtocells can employ MIMO parallelization for traffic between itself and mobile terminals. In terms of bitrate, both are found in practice today, but large-scale video streaming services should prefer the VBR.

Federated WiFi is a close relative of the SON [15], except that here terminals are not organized ad hoc but are grouped locally by their distance to a given WiFi access point (AP) [15]. The relation to SON is mostly in the part where multiple WiFi APs (hence the term *federation*) are employed by the same overlaying service to stream the same video content to a large number of end users. User connections are single path and single connectivity, with the choice between constant and VBR. The topic of federated WiFi networks is very active in research today.

Wireless P2P streaming is listed here mostly for the sake of the argument because there are no examples of such a technology in practice today. Moreover, it is likely that this mode of streaming will become part of the 5G streaming. In wireless P2P, one would normally use MANETs (therefore single connectivity) to establish e2e paths to multiple peers. Beyond that, the standard P2P streaming method applies.

5G streaming is the umbrella technology for a number of distinct methods that can be created as the combinations of the features in the table. MultiConnect [38] (multiple connectivities) may be exploited to support GroupConnect (multiple paths) [1]. The implementation can also vary from existing WiFi Direct [55] to the femtocell + CoMP in future 5G networks.

9.1.3 Advanced Video Coding: H.264 Standards

The core terminology related to *Advanced Video Coding* (AVC) is CBR versus VBR. The latter is defined by the recent (and currently ongoing) H.264 format [56]. CBR speaks for itself—the rate of data is constant. Now, technically speaking, the size of frames may vary depending on the content but the range of variation is negligible (several bytes against frame size of several kB).

Several acronyms exist in relation to VBR and under the umbrella term of AVC. *Layered video coding* (LVC) and *scalable video coding* (SVC) are both kinds of AVC and both refer to the same basic technology that is when VBR is implemented as *layers of detail* and can be used to *scale* the video by actively controlling layers. AVC is the umbrella term, while SVC is most common in streaming research.

There is a practical reason why VBR can replace CBR. Recent research discusses the concept of *SVC smoothing* [24,28], which is when the various modes of coding defined in AVC can be used to provide various patterns of rate variability across frames in groups of pictures (GOPs). Although such a mode does not exist today, there is a practical niche for a coding mode that would emulate CBR by providing a near-constant frame size across the entire GOP. This, in fact, can already be accomplished in practice but is not yet standardized [5]. For all the common coding modes and resulting frame size distribution, refer to the study in [27].

AVC is closely related not only to QoS but also the QoE, where some coding modes may be preferred to others under certain e2e network conditions [18]. Adapting coding modes to changes in e2e network performance is a kind of *adaptive streaming* discussed in current literature [5].

Figure 9.2 shows data from one of the traces in [60] in form of the ratio between throughputs. The ratio is plotted in two pairs: SVC/CBR and VBR/CBR. There is not much difference between the two except at the left side of the plot where VBR is revealed to have higher average throughput. Vertical scale is $\log(1+y)$ for positive (CBR smaller than VBR, SVC) and $-\log(y^{-1})$ for less than 1 (CBR bigger than VBR, SVC) values, which helps put values of both signs and extremes into the same plot. Maximum I-Frame size

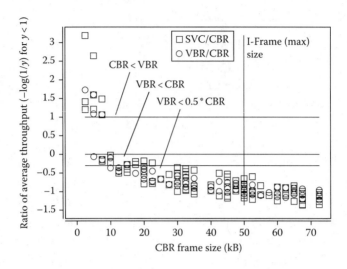

Figure 9.2   Throughput (ratio) for CBR, VBR, and baseline SVC coding formats.

(VBR, SVC) is around 50 kB. However, since bitrate is variable in SVC and VBR, the average throughput is much smaller than multiples of the I-Frame size. On the other hand, CBR is tested with various frame sizes between very small (5 kB) and much larger (70 kB). Frame sizes of SVC and VBR are not altered but are replayed directly from a trace in [60].

Let us read the plot left to right. When CBR frame size is very small, CBR throughput can be smaller than that of VBR and SVC. Here, it is immediately evident that VBR throughput is higher than SVC, on average. However, the inequality reverts to the opposite past about 10 kB, which is 20% of the I-Frame in SVC/VBR formats. Past that point, CBR always generates much higher throughput than that found in SVC or VBR formats.

The −0.5 line is a known trivia in SVC community [60] where it is assumed that SVC can achieve the same level of quality as CBR at about half the average throughput. The 0.5 line shows that that line for CBR is between 15 and 20 kB. In practical terms, this means that the alternative to a constant stream of 20 kB frames is to use SVC, which would send one large 50 kB frame per GOP but the remaining frames would be much smaller. The average SVC throughput would still be much lower than that of a CBR for the same target quality of the video stream.

9.1.4 Optimized Resource Allocation in Video Streaming

Resource allocation is about the server side of video streaming. The earliest technology in this area is the *cache distribution,* where the same content would be replicated multiple times and replicas would be stored in separate locations on the global network. There are many known *cache strategies* [29], most of which are used in CDNs today. Multiple locations help split and (preferably smoothly) distribute content viewers across multiple serves, thus not only reducing the peak load but also improving QoS and QoE on the user side.

Even if the problem of server-side congestion is resolved, the problem of delivery still remains. In CDNs, even the modern ones, servers, and network parts are often handled by different providers. What we refer to as CDN often manages the network among its multiple locations but does not have access to the network between the content and end users/viewers. In this respect, an average CDN as more a *network of servers* rather than *network and servers.* This is a known problem, which has a large body of literature on various methods for QoS-aware e2e routing [30]. The routing can be handed by CDNs by selecting the cache replica, which has the best e2e performance to a given user, or, in roughly the same way, by end users.

There are very few examples, even fewer real services, in which CDN covers both cache management and e2e network optimizations [31]. However, the concept is well understood and is achievable in terms of technology.

While there are a large number of specific resource allocation optimization methods in literature, they can be classified by implementing one of the following elements:

Cache replication and distribution is whether or not content is replicated and distributed across the delivery network. Distribution is normally done taking into consideration e2e network distance (delay) between streaming servers and viewers.

Request forwarding is used by CDNs in addition to cache replication and distribution in order to provide a seamless service. The common practice is for CDN to change global DNS dynamically to have some level of control over which requests go to which servers. The feature is used both as an offloading technology and as an optimization of average e2e distance between a community of viewers and servers for each physical location.

Delay sensitive allocation is a common element in interactive streaming but is not a requirement element of video streaming. For example, P2P streaming discussed in depth in this chapter is not a delay-sensitive technology. In fact, P2P delivery over multiple relay hops may be delayed by several seconds. However, it is not a problem for video streaming as long as the delivery is sufficiently steady—meaning that the optimization here targets *minimization of delay variation* rather than the delay itself. By contract, most interactive multimedia applications (IP telephony, gaming, etc.) have to solve a more complex problem where both delay and its variation have to be optimized within a given range of practical usability.

Source population optimization is a P2P-specific element where content is delivered over multiple substreams, in which case each peer prefers a source population that would support a steady streaming session. The simplest practical optimization target here is *minimization of variance* across sources. But there are more complex cases like those that use variable-size substreams (see the VBR methods further on).

Note that this classification lacks the wireless-specific elements, where there are additional performance goals like energy efficiency and radio interference minimization. These will be unfolded further on in this chapter.

There is also a brand new concept that can overcome the aforementioned difficulty by exploiting the natural flexibility of resource allocation in clouds. Clouds have recently been proposed as hosts for CDNs [2]. In clouds, one can have a population of streaming servers where each server can migrate across physical locations on the network. In this part, the technology is similar to that of the cache replication and management. The difference is in the scale. In clouds, one can potentially have very large populations of streaming servers with relatively smaller capacity allocated to each server. This creates an environment suitable to a P2P-like streaming method, where instead of peers, each end user would select a number of servers (sources) in the cloud [5,6]. This is a much more flexible concept that can cover QoS management in both server and network parts. At the server side, the CDN is responsible for providing a sufficient (in both cloud and geography) population of sources while users have the freedom of selecting the sources that provide the best e2e network performance at the time. This technology is already shown to work in browsers [3,7].

9.1.5 Layered Video Coding and Its Transport

Figure 9.3 shows the three distinct technologies in terms of their internal structure [26,28]. There are three data units—*frame, GOP*, and *block*—the last one is not a physical unit but it is often used by streaming methods to aggregate content.

As mentioned earlier, VBR content is, as expected, variable in size. The frame size of the I-Frame (first in GOP) is normally the biggest, while intermediate P- and B-Frames are smaller in size. VBR uses a fixed coding format that provides a relatively stable distribution across various kinds of video content.

SVC (right rightmost in the figure) has changed the overall approach to the distribution of information inside the GOP. Details in image (motion, background, etc.) are presented in layers, with each higher layer contributing more detail to the image on the frame. The size is also a decreasing distribution since smaller detail takes proportionally smaller volume of binary data to describe. The baseline of SVC is the same as in VBR, meaning that the new design can be applied to provide roughly the same frame size distribution as used to exist under VBR. This is also good for compatibility.

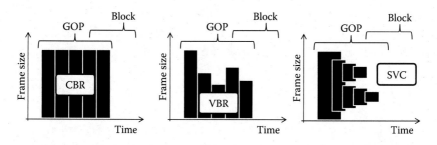

Figure 9.3 Evolution of video coding standards from CBR, via VBR, to the current scalable video.

However, the new design can accommodate alternative distributions of frame size. Some of the SVC sub-blocks can be shifted around as long as they stay within the same GOP. Some of the recent studies exploit this feature to implement a smoothing effect on bitrate [28] in order to adapt video to network environments that cannot handle high traffic peaks.

The transport of layered video exploits all the three data units. BitTorrent-like P2P delivery cuts the video into equal-size blocks and is therefore agnostic to VBR/SVC difference as well as GOP structure [32]. However, more advanced P2P streaming methods would often focus on SVC and create blocks as multiples of GOP. Bigger blocks (more GOPs) would be requested from peers with a history of better response (throughput, e2e delay), while smaller blocks would be requested from relatively slower peers [25]. Multiples of GOPs seem to be a popular method found in many recent streaming methods in literature [19,20]. It has also recently been applied to BitTorrent [22].

In [5,6], the data unit is a frame. Substreams are therefore created by frame positions in GOPs (thus pinning the number of peers to the number of frames in GOP). In this case, the difference in e2e network performance across peers can also be managed by assigning bigger frames (I- or P-Frames) to peers with better e2e network performance.

Browsers have also recently started taking advantage of layered video. Part of this development is the HTML5 in which WebSockets allow for concurrent aggregation of content over multiple parallel paths (MultiPath but not MultiConnect) [3]. All the problems and solutions found in P2P streaming given earlier apply to such methods [7]. However, a separate development is along the line of single-path delivery with a higher level of adaptation [33]. In this case, one expects that e2e network performance can experience temporary fluctuations and react to them by *scaling the video*. In the worst case, some of the upper layers in GOPs can be discarded (failed to receive), which has an effect on QoE proportional to the volume of discarded data. The proportion is nonlinear, meaning that higher levels have a proportionally smaller effect on QoE. Such methods are currently discussed as good candidates for web-based TV [34].

9.1.6 Error Resilience and Concealment

Figure 9.4 shows the standard coding–decoding process, with the network in the middle for physical delivery [16]. This is a standard way to represent and analyze the subject in IT. *Source encoding* is content-specific and can be video, audio, and so on. The main part is played by *channel encoding*, where network performance can be taken into consideration. This is where the stream can be made extremely resilient by inserting redundancy or making it less likely to be rendered undecodable via loss of its portion. If a portion of a frame is lost completely, there are methods that can help conceal the loss by inferring the content from neighboring information. The efficiency of resilience and concealment methods is normally judged using subjective scores as part of QoE evaluation.

As far as research literature is concerned, wireless video delivery still actively focuses on these subjects [16].

Note that as far as modern research on video streaming is concerned, the rest of the aforementioned methods have found their implementation in one of the coding modes of scalable video under the H.264 standard. There are also studies that verify the resilience and concealment abilities of SVC coding modes in terms of QoE [18].

9.1.7 Substream Method in P2P Networks and Clouds

Strictly speaking, P2P streaming is not done over a *distribution tree*. Figure 9.5 shows an example topology that shows why the correct term to refer to such a topology is *multiple complementary trees* [5], which is a slightly confusing term because it is not clear which trees complement which other trees. It is probably better to simply state that P2P delivery topologies are very complex.

It is interesting that complexity occurs as a result of a very simple operation. Commonly, each peer requests a given repository server for a list of

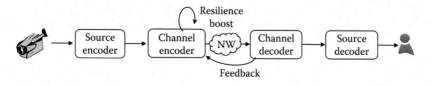

Figure 9.4 Standard e2e process experienced by video content as it travels from source to destination.

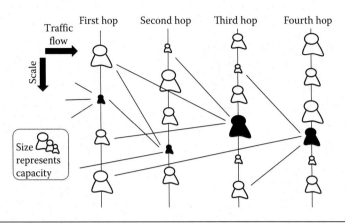

Figure 9.5 *Multiple complementary trees* topology that exists in P2P streaming.

the peers it can use to request parts of the stream. The method used by the server to generate such a list is the subject of active current research. BitTorrent traditionally (to the day) selects a random set of peers [32]. Alternatively, one can use some metrics that capture network distance (delay, throughput, etc.) among peers in which case the selection is still random but the raw list contains a subset of the entire P2P network—normally the subset of the peers that are considered to be closer to you.

In both cases, the selection is either partially or completely random. This adds the mesh-like features to the resulting connections. It is likely that a very late-arriving peer will try to connect to a peer that has been on the network for several days. If the age is represented as depths of the initial delivery tree, it would mean that peers from the bottom can connect to those at the very top.

Now, given that a peer can rotate peers (or change them based on performance history), the graph keeps growing more and more complex with time.

The *substream method* proper, in the aforementioned environment, is simply the method that aggregates the content over multiple smaller streams. For example, in a setting with 10 source peers, each peer would only send the receiving peer the 1/10th of the entire stream. The receiving peer would then aggregate and play the stream locally. More details on the substream method—specifically the reselection of peers, buffer management, and others—will be discussed further in this chapter.

Attempting to classify all substream methods, they can be roughly divided into a 2 × 2 grid of *static versus dynamic* and *synchronous versus asynchronous.* Traditional P2P methods are *dynamic and synchronous* where rapidly changing environments are counteracted by re-election of parent and/or child peers and new upstream and downstream connections have to be acknowledged by other peers or even scheduled (more common in traditional BitTorrent methods).

However, *synchronization* can also be interpreted in the global sense where CDN is the chief synchronization engine that optimizes the entire distribution network. SPPM [21] is an existing platform of this exact kind. Note that such platforms are limited in scale by definition because larger distribution networks require more management overhead.

Clouds offer an alternative example of a globally synchronized delivery without excessive management overhead [2]. While the number of cloud-based sources should be roughly the same as in traditional P2P delivery, higher management efficiency is achieved by the fact that the distribution network is flat and is ideally single-hop where each viewer is directly connected to several sources is the cloud.

9.1.8 Unique Challenges in Wireless Networks

The wireless domain poses several unique challenges but also at least one unique capability.

In terms of the challenges, the capacity limitation of APs (WiFi, femtocell, BSs, etc.) is the biggest problem. One simply cannot connect more than a relatively small number of users to a given AP. In WiFi, the recommended number is under 10 concurrent users per a commodity AP, where some studies show that after 20–25 users the network suffers major performance impairment [4]. This problem is part of the *offload* feature of the LTE-A in particular and IMT-A in general [52]. The offload here is simply a method or technology that helps avoid crowding of users at a given AP by distributing them to other APs. 3GPP discusses the offload technologies in [53] and more generally in [54].

There are technologies outside of the LTE-A or IMT-A realms. For example, WiFi Direct (also referred to as WiFi P2P) [58] is a stand-alone technology that can support MultiConnect and, by extension, GroupConnect [1,8]. Although it is only one practical use of GroupConnect, it can function as the offload technology if only one group member communicates to the BS while the rest get the content from that one member. More on other uses of GroupConnect will be considered further in this chapter.

Various grouping and flocking techniques in MANETs play roughly the same role in terms of offload [50,51]. While each mobile node used to support its own e2e path, via grouping it is now possible to reduce (focus) long-distance communication at the expense of increasing local data exchange. This approach has very high utility in video streaming simply because video content is the same for all members, making it extremely effective to appoint a single representative group member to receive the content and share it locally within the group.

Note that local grouping as the overall approach is part of the latest IMT-A [15] recommendations.

9.2 Wireless Video Streaming

This section fully focuses on the wireless aspects of video streaming. Most of the details have been briefly mentioned earlier in this chapter and are unfolded in this section.

9.2.1 Architecture

Figure 9.6 shows the detailed design of a standard technology that provides e2e video delivery over wireless. Note that only the WN is

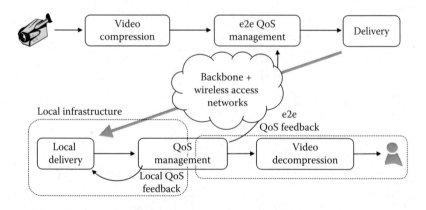

Figure 9.6   A more detailed design of e2e video streaming technology where the final legs of delivery are done over wireless.

shown because in the wired + wireless setting, performance of the wired part can normally be considered *near perfect* when compared with the wireless part.

There are several technological advances under the LTE-A umbrella. First and foremost, there is the concept of *wireless backbone + wireless access network*. Current LTE-A discussion plans to serve video streams using broadcast and/or multicast in the backbone while facilitating as much offload as possible at the local wireless access via microcells (femtocells, WiFi, etc.) [13]. In these settings, e2e QoS, assuming that the wired part is near perfect, is impaired mostly at the access part; that is, we are talking about the single-hop QoS at local APs [14]. In the offload method, local networks, SONs, CoMP [15], and femtocells [12] are the components from which one can build a working large-scale technology. Note that no one default technology yet exists in the LTE-A framework, which means that several possible combinations can be (and are) considered in research at the time of this writing. Later in this chapter, several practical technologies will be formulated and analyzed for performance.

QoS of wireless delivery is firmly a part of IMT-A [52] standard process and is already considered in several pilot implementations [15]. While the detailed design of a given technology may be complex and can involve more than two protocols, the conceptual design is simple— we are always talking about the separation of the backbone network across BSs and a large access network of small APs.

9.2.2 Streaming over Single-Hop and Mesh Networks

The main players here are MIMO [41] and femtocells [42].

MIMO is the core player in next-generation home wireless systems, where it can help boost throughput up to 100 Mbps and above. The modern 802.11n wireless routers can support up to 300 Mbps, which is achieved using MIMO. The biggest technical problem in MIMO is known as *beam forming* [40]. The methods that attempt to solve this problem propose specific configurations of antenna arrays that can result in *less radio interference*. This is still a very active area of research.

Clearly, MIMO is effective only when it is supported by both ends of a wireless connection. Since it is already implemented in home-oriented wireless equipment and some recent smartphones, home communications can benefit from MIMO directly. The advantage is not as obvious when it comes to wireless networking outdoors. Even assuming that femtocells implement MIMO for throughput boosting (which is discussed in under LTE-A), the problem of interference is much more hard to overcome in multiuser environments.

So, what happens when video is streamed over a MIMO-enabled wireless connection? From the viewpoint of the streaming application—both server and client—they have only one connection between each other. However, at the firmware/hardware level, the content is split into multiple concurrent wireless connections that deliver partitioned content to the other side of a connection where it is assembled back to the original stream. This, in a nutshell, is identical to the *substream method* in P2P streaming or *multisource aggregation* in cloud streaming. The only (but still major) difference is that all the parallel streams in MIMO are roughly identical. However, this is not a problem for MIMO because the transmission happens locally and all substreams experience (roughly) the same performance.

The MIMO with substreams pointing to different end points is called MultiConnect or GroupConnect. In fact, MIMO has no immediate future plans to implement such a functionality. However, using WiFi Direct (for example) one can easily implement concurrent streams to multiple parties [1,8]. Such a technology is better than MIMO in several aspects like the ability to balance the load across multiple parties, putting more stress on parties that have shown (through past history) to be able to handle more throughput. MIMO here would not be able to tell the difference.

GroupConnect can also be used to create local mesh networks, which can be done using MIMO. A separate chapter in this book is fully dedicated to the concept of GroupConnect and can serve as a better source of insight into the capabilities of the technology.

The role of femtocells in single-hop streaming is minor and is mostly found in its offload function [42]. Users can connect to only one femtocell AP at a time and can get only one path through the AP into the backbone WN. However, provided the AP is not crowded, the fact that femtocell APs are connected to BSs over dedicated wireless channels should guarantee that your streaming quality remains unimpaired.

9.2.3 Adaptive Video Streaming

Figure 9.7 shows example distributions of frame size between a standard VBR and SVC version of the same video. Let us assume that the SVC video is being delivered over the WN and has to adapt to changes in network performance. This section discusses the adaptation methods that are feasible in practice today. The methods are roughly split into a 2 × 2 grid by single- versus multihop and single- versus multipath types.

First, let us recall the SVC design discussed earlier. Data blocks in SVC can be moved around the GOP, thus allowing to change the distribution. Using Figure 9.7, one can interpret this ability as a manually adjusted curve roughly within the area between the two curves in the figure, where VBR is the extremely jittery and SVC is the mostly smooth extremes.

If the delivery is one-hop, then we probably implement MIMO. In this case, we normally should not care about throughput because we definitely get enough throughput for an average video bitrate. However, if we take an imaginary high-definition (HD) video stream, then MIMO could help by increasing the number of parallel wireless connections when bitrate increases. High-frequency changes within each GOP (one or two times per second) would probably be too often for the control layer, so it is probably more efficient to ignore the GOP structure and simply split the content into data block (per unit of time). This gives access to the *throughput* as the runtime indicator of bitrate, which would help MIMO decide when to increase or decrease the number of concurrent connections. This covers both one-path and multipath single-hop connections.

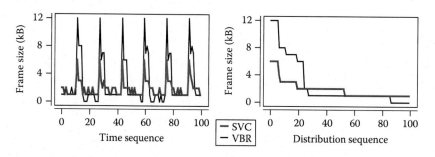

Figure 9.7 Example distributions of VBR and SVC traces for the same video at [60].

Adaptive streaming follows a relatively simple algorithm:

1. Start at a relatively (or highest) high bitrate.
2. Keep streaming until video freezes due to depleted buffer, in turn due to insufficient e2e network performance.
3. React to a freeze by switching to a lower bitrate. In practice, this can be accomplished by several methods like *switching to a lower bitrate streaming* and *scaling the video in real time* by tweaking SVC parameters (smoothing). When done, go back to (2) and resume streaming until the streaming is found too intensive that would cause another downgrade.

When you go into multihop connections, you have many more options. If you have only one path, then you almost always have to tweak the SVC coding mode when adapting to your e2e network performance. If your network can handle higher throughput peaks, the VBR-like configuration should work. However, if the network reacts poorly to rapid increases in throughput, then you should tweak SVC configuration to produce a version of the lower curve in the right plot in Figure 9.7. This method is described in the currently ongoing work on the DASH protocol for adaptive HTTP streaming [59], itself a part of the larger HTML5 [58] standardization process.

If you can support multipath streaming (P2P streaming, multi-source aggregation on the web, or GroupConnect by WiFi Direct), then you have multiple options for adaptation. First, you can move layers across frames within the GOP [5], thus, again, balancing the load across peers or sources. However, in existing research it is more conventional to use multiples of GOP [22] to create substreams of varied rates. Finally, regardless of which one of these frameworks you use, you can always adapt to changes in network conditions by changing the list of sources that provide you with substreams. This method is obviously the most compatible with a wide variety of environments and has even been implemented in browsers [3,7]. However, to provide the maximum possible resilience, most methods in P2P streaming will exploit both opportunities—they will optimize the list of sources while also adapting video content to suit the current list [5,6]. Such methods are format-agnostic but prefer H.264 [5].

9.2.4 *P2P Video Streaming*

This section provides more details on the concept of *QoS by reselection* [5,6], which is technology-agnostic and is exactly the same for both wired networks and WNs. Figure 9.8 shows the standard way to represent P2P streaming—the viewer in question in the center and parent and children peers up- and downstream. Note that *multisource aggregation in clouds* follows the same basic concept only in this case there are parents (sources in the cloud) but no children.

As mentioned earlier, the total video stream is formed from multiple substreams. Based on the various methods earlier, the substreams do not necessarily have to be the same in volume. However, even with heterogeneous substreams, it is crucial to ensure that *playback position* is the same across all substreams. The playback position is best visualized as the sequence number of a block of data in the stream. The substreams would then carry overlapping streams of blocks, that is, 1, 4, 7, …, 2, 5, 8, …, 3, 6, 9, … and so on. The playback position in all substreams has to be roughly at the same position; that is, the method would strive to minimize the variance across positions on all substreams.

This element is crucial to multipath video streaming and is the major difference between video streaming and aggregation of generic content. Video streaming is a real-time technology and crucially depends on the smooth delivery of video content to be immediately played at the user's terminal. Large difference across playback positions on substreams would eventually result in gaps in frames, in turn resulting in frozen playback.

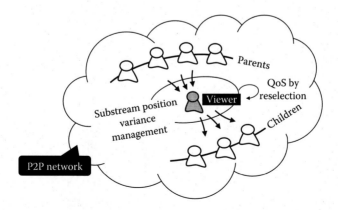

Figure 9.8 Basic concept of P2P streaming where each peer communicates with parent peers upstream and children peers downstream.

In this context, *QoS by reselection* describes a method that reselects a peer with the *bad* playback position with another peer. The *bad* here can mean both too late and too early because both extremes can eventually cause problems in playback. As will be shown further in this chapter, optimization methods normally do not care about the direction of deviation, instead optimizing for minimal overall variance across all positions.

P2P streaming follows the following general algorithm:

1. Initialize parent list randomly, commonly by sending a request to a tracker node (a centralized registry server in P2P). Also, respond to all requests from other peers, assigning as many as you can as child (downstream) nodes.
2. Monitor two *playback positions*—one as the average of positions on parent and the other on child substreams.
3. When a given playback position (either parent or child) lags too much (a reasonable value of a user parameter), that parent or child has to be reselected. In both cases, this is accomplished simply by tearing the connection with a given peer. A replacement peer is added to the respective list and the algorithm returns to (2) for continuous monitoring.

9.2.5 *Video Streaming in Wireless LANs, 4G, and Beyond Networks*

In 4G, MIMO technology is obviously one of the biggest players in wireless video streaming, as was discussed earlier [41]. The 4G (current at the time of this writing) version of LTE [52] does not offer much help in this area. There are measurement studies in literature, which show that throughputs on LTE and 3G networks are not much different [1] regardless of the fact that LTE should technologically offer a 10-fold increase in throughput. This happens mostly due to poor planning of network capacity. While the wireless part of an LTE network may actually support multi-10 Gbps throughputs, the gateway between LTE and the global network can be too narrow and experience constant congestion. This is in fact one of the findings in [1], where a new 3G entry into the market was found to be faster (in terms of e2e throughput of video streaming) than a 2-year-old LTE provider.

Streaming efficiency in the wireless domain is often considered from the viewpoint of the signal technology. The recent orthogonal frequency division multiplex (OFDM) technology—the upgrade from the time division multiplex (TDM)—has been shown to provide better efficiency in multiuser environments [9]. Recent literature on the subject proposes the coupling of MIMO and OFDM where the frequency multiplexing offers higher frequency efficiency while MIMO facilitates multistream (and therefore multiuser) video delivery. Recent studies show that such environments are compatible with SVC video where layers can be scheduled and transmitted over MIMO–OFDM separately, thus increasing the overall efficiency [10].

Wireless environments are traditionally optimized for two practical targets [9]: *maximizing quality of delivered video* and *minimizing variance across multiple users* (assuming that all users subscribe to the same level of quality). While recent studies show that MIMO and OFDM both contribute to higher efficiency in these two targets, there is an upper practical limit to this efficiency [10].

With this upper limit in mind, it should be noted that large-scale wireless services can be technically viable only if they are designed with the offload functionality in mind. Yet, even in 5G, if the LTE-A provider designs the network with too narrow backbone connections, users will experience the same problems as in today's LTE, even with the offload technologies such as femtocells and CoMP.

9.3 Mobile Multimedia Communications: 4G and Beyond

Earlier sections discussed the basics of video streaming technology and its wireless specifics up and including 4G. This section broadens the scope to general multimedia and expands into 5G WNs.

9.3.1 Various Multimedia Services

This section presents a proper classification of multimedia in which video streaming is only one of the types.

First, multimedia can be *centralized versus distributed*. CDNs are normally centralized, even though cache replication methods make it semi-distributed. However, true distribution comes with P2P network delivery or multisource aggregation of content. In these cases, you completely lose track of where the content came from originally. In fact, in P2P and cloud-based delivery, content delivery can go on indefinitely until all the copies in the network disappear completely. In the cloud environment, the copies will gradually disappear with time but in P2P networks the content may stay alive for a long time.

Multimedia can be *1-way* or *2-way*. Note that *interactive* is part of the 2-way but is not the entire group. An application can be low interactivity but still be 2-way.

Finally, there is *throughput–intensive* versus *delay-constrained* multimedia. Here, delay-constrained is always *interactive*, by definition. Video can be considered both throughput-intensive and delay-constrained because real-time playback has relatively strict deadlines on delays in arrival of fresh content.

9.3.2 5G, MIMO, and Home Networks

Figure 9.9 shows the taxonomy of all possible wireless streaming technology. Left-to-right the scenarios (top-down columns) roughly represent the evolution of wireless delivery from 3G to current. Although most technologies have already been discussed earlier in this chapter, this figure helps to review the different levels (top-down) and evolution (left–right) of wireless delivery and content provisioning. Another reference that can provide a fairly full picture of the current state of technology is [15].

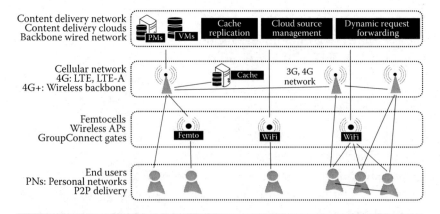

Figure 9.9 Taxonomy of all wireless delivery methods starting from 3G and up to the currently established parts of LTE-A in 5G wireless networks.

First, let us review the system top-down. At the top level, there is the CDN itself that used to be about cache management, but today it has advanced to *dynamic request forwarding* and even possibly the combination of both these technologies. Another major advance happened with the help of clouds where clouds can host much larger populations of streaming servers than are common in traditional content delivery. The major feature of the top layer is that physical locations are connected by a network that is engineered with e2e delay and bulk transfer in mind. Generally speaking, streaming servers today are well-managed and well-connected populations of either physical machines (PMs) in conventional CDNs or virtual machines (VMs) in cloud-based versions.

The second from the top layer is the network of BSs of a given cellular service provider. Topologies roughly look the same for all generations, but there can be more BS-to-BS traffic in 5G. Also, BSs in LTE-A (5G) are supposed to assume the functionally different role of the wireless backbone network. This is functionally different from current BSs that communicate with end users directly. In 5G, BSs mostly communicate with femtocells and only in the worst unavoidable cases fall back to direct communication with users. This layer can also have local caches of popular content (P4P technology). In this case, some of the traffic from the bottleneck at the wireless–wired border can be offloaded internally—users of locally cached content do not exit to the global network but get streams from local cache.

Now, let us trace all the options available to end users (bottom-up and left to right) roughly in evolutionary order. Traditionally, users connect directly to BSs and then exit to the global network. With femtocells in LTE-A (5G), users connect first to the closest femtocell, then to BSs, and then to the global network. This is how the traffic that congests BSs is distributed and offloaded to smaller and more numerous femtocells (also referred to as micro-APs). Since there is now a large network of small APs, the control that optimizes network use between BSs and APs is important and is an active area of research today [43].

In the center of Figure 9.9 is the conventional case when terminals use WiFi to connect to the network. In this case, the connection exits to the global network directly via the WiFi-wired gateway. WiFi APs are also relatively small, with the capacity roughly the same between femtocell and WiFi APs.

Finally, the rightmost scenario describes GroupConnect. This is an extremely heterogeneous environment, in which group members can communicate to an arbitrary number of distinct wireless technologies (WiFi, 3G, 4G, 5G, etc.) at the same time as communicating roughly in the mesh configuration with other members in the group.

Let us now map all the advanced parts of LTE-A to the scenarios in Figure 9.9. MIMO can help in femtocell and WiFi configurations by improving throughput between the terminal and the small AP. In both cases, throughput is much high and much more reliable between femtocell and BS (5G) or WiFi and a wired network. CoMP and SON concepts go hand in hand with the concept of GroupConnect and, more generally, P2P wireless networking. The similarity across all these technologies is that they form very small local groups of terminals network performance within which they can be easily managed.

The aforementioned femtocell scenario can also be experienced in several pilot networks like that one at [44]. Another interesting related project is a device-to-device (d2d) communication using femtocells as routers but bypassing wireless backbone [45].

9.3.3 Streaming Video over 5G Wireless Networks

As was discussed before, 4G [16] and specifically the LTE technology [14] do not offer any technical breakthroughs in the area of video streaming [13]. With the advent of 5G's LTE-A, the delivery method changes considerably [15]. First, femtocells (APs) form a new wireless hop between BSs and users [12]. Second, the WN is split into two networks—the backbone that connects BSs with APs and the access network between APs and end users (Figure 9.10).

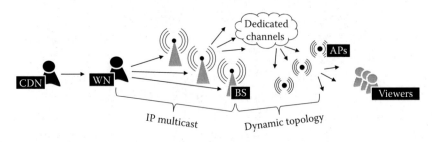

Figure 9.10 e2e process of video streaming under LTE-A in 5G.

Let us trace the entire process from left to right. We start at the CDN itself. WN gets the content from CDN (potentially having its own cache for storage) and streams it downstream into its own WN. Some of the documents on IMT-A propose that this part of the streaming be done over IP multicast [15]. In this case, performance majorly improves because the same content can be simultaneously broadcasted to all BSs. In the wireless domain, this is, in fact, easy to accomplish without increasing radio interference—broadcasts in WNs are simply the same as unicasts, provided the radio transmission is omnidirectional. BSs are not strictly omnidirectional but are configured in such a way that simultaneous broadcast is technically achievable.

Using the IP multicast, the video stream gets all the way to femtocell APs where the stream is available for single-hop access by end users. The remaining part is the offload part where users are smoothly distributed over the AP network in such a way that as few APs as possible would be congested. This part, in fact, is the most difficult from the technical point of view.

9.3.4 QoS, Energy, and Bandwidth Optimization in 5G Mobile Multimedia Applications

Figure 9.11 puts all the earlier points about 5G wireless communications in the context of the overall quality and performance guidelines. The diagram is in the form of matching performance targets to user requirements.

On the design side, the LTE-A provider has to deal with *spectrum efficiency*, *delay constraints*, and *throughput requirements*. *Spectrum efficiency* is important because you now have the backbone network in additional to the access network. Given that both the APs and

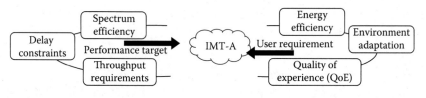

Figure 9.11 Overall guidelines of IMT-A expressed as matching performance targets to user requirements.

individual users still retain the capability of communicating with BSs directly, radio efficiency is extremely important. At the same time, one should remember that LTE-A targets 1 Gbps nominal throughputs. Finally, the added hop between backbone and femtocells cannot result in too much increased delay because this would impair e2e performance. Although delay is not linearly proportional to throughput, increase in delay as a side effect can negatively affect throughput.

Now, this all has to be matched to user requirements. Users want to be energy efficient. This means that MIMO-type multiconnect and continuous control traffic has to be minimized. This is important because handover between femtocells happens much more often for mobile users due to the small size of the cells. The same applies to environment adaptation where users want their terminals to adapt to changes in the environment (change APs, etc.) but do not want to waste too much energy on that. Finally, a due level of QoE has to be supported in seamless handover across femtocells, seamless adaptation, and so on.

9.4 Wireless Technology in Massive Crowds

This section puts all the aforementioned technology in the context of massive crowds and analyzes how they perform. The specific performance aspects considered in this section are *traffic offload*, *efficient e2e streaming*, and *ubiquitous networking*. Finally, special attention is paid to revealing and analyzing performance bottlenecks.

9.4.1 Flash Crowds and the Offload Solution

A *flash crowd* is defined as a drastic increase in the number of users accessing a given resource. In the context of this chapter, flash crowds can occur in many places. They do not necessarily coincide with performance bottlenecks but can potentially cause major performance impairment both for individual users and service providers.

Traffic offload can potentially help elevate flash crowds. The two offload scenarios widely used today are 3G-to-WiFi and P2P offloads. The former is a wireless technology, while the latter helps resolve flash crowds in CDNs.

Offload using WiFi Direct has been already mentioned and is already implemented in practice [1] as a GroupConnect technology [8]. This particular kind of GroupConnect is not part of LTE-A standards, but LTE-A discusses similar approaches via the use of CoMP and femtocells [15]. The next section considers the specific technology of femtocell-based offload.

9.4.2 Femtocell Approach

Femtocells are part of LTE-A (5G) and are actively discussed in current research literature [12]. The biggest problem femtocells have to overcome is radio interference [42]. On one hand, femtocells are expected to host under 10 concurrent terminals, which is a relatively minor load, but on the other hand, femtocells under 5G are expected to support bitrates up to 1 Gbps. Some literature proposes to use MIMO between terminals and femtocell APs, which would support the high bitrate, on one hand, but would worsen radio interference, on the other [40].

Control layer in LTE-A poses major difficulties. Some literature formulates this problem as *access control* [43], which defines which mobile users are connected to which APs. However, control technology is more complex because it has to embrace both access and backbone layers of the WN.

This is mostly because the topology among BSs and in the wireless backbone between BSs and APs in LTE-A is not expected to be fixed but should dynamically adapt in accordance to the demand. Radio efficiency is part of this requirement, but, more importantly, these adaptations can help provision multiple services—among which multiple video streaming services can exist at the same time with only partially overlaying BS–AP topologies.

While there are already pilot projects implementing LTE-A in real life [44], the complexity of such pilots is far from the practical offload requirements described in this section specifically and this chapter in general. For example, the technology that would allow terminals to exploit APs and BSs as routers on ad hoc e2e paths to other terminals (d2d [45]) so far exists only in research and is not found in any existing pilots, despite the obvious practical utility of such a function.

9.4.3 Mobile Clouds and GroupConnect

Figure 9.12 shows a realistic topology of a service that has a massive wireless user base. Note that this technology is not yet part of the official scope of LTE-A or IMT-A, but is discussed in other parts of wireless research like GroupConnect and the mobile cloud discussed further in this section.

The three groups in the figure are as follows. G1 is a WiFi AP to which some of the members of G2 are connected. However, some other members of G2 are connected to 3G/4G BSs or 5G femtocells. The G2 membership therefore is not decided based on the Internet connection but by the fact that they all are connected to each other within their local GroupConnect. The concept of GroupConnect is a fairly recent one based on the ability to support multiple simultaneous connectivities (different technologies) [38] but going beyond than with the notion of pooling and virtualizing network resources of the group [1]. Another chapter in this book is fully dedicated to GroupConnect.

G3 is even more *strange* here because none of its members have any Internet connections but have peer connections to some members of G2. The main point of a video streaming service in such an environment is to provide smooth streaming all the way to the members of G3. In fact, as performance analysis further in this chapter shows, deployment of wireless video streaming in massive crowds will fail unless a large (possibly even majority) portion of users are connected via P2P connections like the member of G3.

Note that the GroupConnect has several practical uses. It can be connection sharing where multiple users share a single Internet connection. It can also be connection pooling where multiple connections are merged (virtualized) into one in order to support larger throughput.

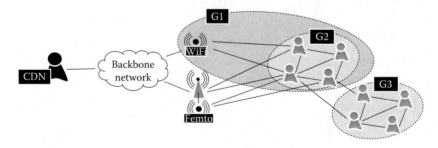

Figure 9.12 Concept of GroupConnect in the context of massive crowds.

While the *smooth delivery* of the video stream to G1 and G2 groups can be facilitated by improving WiFi and 5G technology, G3 can benefit only if its connections to other groups can support smooth streaming. This is a known problem in social networks called *community building*, which is where the larger graph is partitioned into smaller groups in such a way that all groups would retain strong connections to neighbor groups [17]. Unfortunately, space limitations cannot allow for a detailed discussion into community building, but this topic will definitely be revisited in future publications.

Although without naming the GroupConnect notion but with the offload notion in mind, there are also *mobile clouds* [35]. Mobile clouds are not created for video streaming. However, their single most important target is to offload as much traffic to local in-group communication as possible. In this respect, mobile clouds are one of the usecases of GroupConnect. The technology is extremely popular in practice with several existing implementations for a number of practical uses [36].

9.4.4 Traffic Bottlenecks

Figure 9.13 shows all the possible traffic and performance bottlenecks numbering them B1 through B5. This section explains them, while the last section in this chapter refers to them in performance analysis.

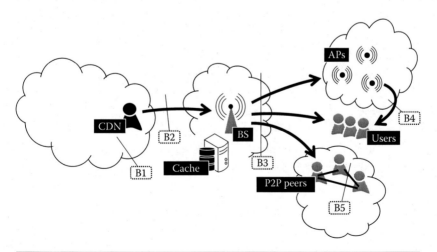

Figure 9.13 Traffic and performance (in general) bottlenecks.

B1: CDN bottleneck can happen with resource-starved or poorly managed CDNs. For example, Japan has relatively few (almost none for some providers) data centers. This means that cloud services of popular providers are delivered from data centers at very distant locations. e2e delay is high in such cases (several hundred ms), which can in turn cause major variation in e2e throughput. As a result, video delivery (even in wired environments) from such providers has limited spread in Japan, while local providers—those that have physical facilities inside Japan—enjoy very large user bases.

B2: Wired E2E path is important for wireless providers. Cellular providers are often large. With constantly increasing penetration of smartphones, the volume of traffic generated by all users cannot be neglected. B2 here is the bottleneck in terms of the maximum aggregate throughput, while it can also be a performance bottleneck if the wireless-wired gateway cannot process all the traffic in real time. Both cases exist in reality as shown by the measurements given in [1].

B3: Wireless backbone is mostly up to the implemented technology. While LTE can support up to 54 Mbps, LTE-A can potentially grow beyond 100 Mbps and even reach 1 Gbps rates [15]. The bottleneck here is the physical limit of 1 Gbps (for example), which can accommodate only about 10k users each viewing a 100 kbps video stream. Of course, this is only if we assume that each user has its own dedicated e2e connection, which is not the case in the LTE-A vision, where it is assumed that the provider will implement several offload technologies to reduce the number of e2e connections.

B4: Femtocell APs do not suffer from throughput limitation but have to resolve the social kind of congestion where the capacity of each AP is assumed to be under 10 users. In this case, the congestion is bound to occur from time to time when an extra terminal connects to an already full AP, but LTE-A framework attempts to resolve such cases by handing such terminals over to other APs or potentially connecting them directly to a nearby BS.

B5: P2P bottlenecks are the limit of P2P delivery. While it may appear that P2P has no physical limit, there are reasonable limitations that put a cap on the size of a wireless P2P crowd. First, there is the physical limitation of the number of peers per user—again, under 10 is a recommended number. Then, the number of relays—hops in e2e delivery—should not exceed a certain reasonable number. The performance study further in this chapter will try between 2 and 4 relays. A relay in P2P is when a peer gets content from another peer that itself got content from another peer and so on three or four times until you get to the origin femtocell or BS. Exceeding this number would probably result in exceedingly high volatility in the overall P2P topology that can result in disruption of service for a large number of users at once.

9.5 Performance Metrics, Analysis, and Optimization

This is the first of the two sections that analyze performance of wireless video streaming. This section establishes performance metrics, general optimization framework, and practical aspects in adaptive and P2P delivery. Optimization is first defined specifically for the P2P streaming method but then generically for any streaming technology. Note that the term *adaptive streaming* is slightly inappropriate because all existing video streaming methods implement some form of adaptation to changing network conditions. However, the term *adaptive HTTP streaming* has long been solidified as an official name of the respective method.

9.5.1 Performance Metrics

As mentioned before, performance evaluation can be split into two major groups of QoS versus QoE [18]. Any performance metric can be easily classified into one of the other group. The rule of thumb can be that if the evaluation is performed subjectively by a human then the metric belongs to the QoE groups. However, in recent years the term *objective QoE* (oQoE) has entered terminology [18]. The metric is referred to as *objective* because it is calculated from a number of underlying QoS metrics. The term *QoS–QoE mapping* specifies an equation that calculates oQoE from QoS. Needless to point that oQoE's main advantage is that it can be calculated mechanically without involving human experiments.

Peak signal to noise ratio (PSNR) is a well-known QoS metric. PSNR-QoE mapping is one of the most popular in practice today. However, the main problem is that PSNR is applicable to raw video coding [16], which is gradually disappearing from research on video streaming (see the discussion of the reasons given earlier). This is why most of the practical performance metrics today have to do with network performance.

The obvious pair of network-specific metrics is throughput and delay. However, these are only meaningful with constant bitrate because otherwise it takes some effort to separate bitrate fluctuations from those inflicted by changes in e2e network performance. For example, Skype generates a constant bitrate of very small packets and uses fluctuation in throughput as an indication of changing network conditions, to which Skype dutifully adapts.

Figure 9.14 shows several practical metrics. The best practical set of metrics appears to be the one related to software components. Let us consider the left and right sides of the figure separately.

The left side is about fluctuations in e2e throughput and how they can be resolved using a buffer. First and foremost, it is wise to implement a buffer longer than the biggest imaginary throughput drop.

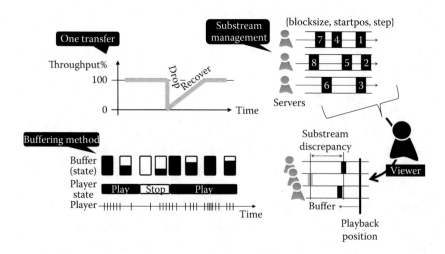

Figure 9.14   Various practical metrics that can be used to evaluate QoS of video streaming.

Naturally, there is a physical limit on the length of the buffer, part of which comes from the QoE side—humans do not want to wait for tens of seconds for their video to get buffered prior to commencing playback. The second important thing here is the buffering logic. For example, in the middle of the timeline playback stops because the buffer is empty and resumes only after the buffer is full again. It should be obvious that alternative logics are possible. For example, playback can stop at empty buffer but resume when it fills up to half its capacity (fast unfreeze). Note that the presence of a buffer makes for a better streaming service but complicates simulations and statistical analysis. This is mostly due to the fact that buffers smooth out fluctuations in throughput, letting through only the very large and/or extended drops in throughput.

The aforementioned considerations reveal the two main metrics used in practice—*playback freeze time* and *lag* [5]. Freeze time is simply the aggregate duration the playback remains frozen waiting for a slower substream to deliver its block. The lag is defined as current download position minus current playback position. Positive lag is good; negative lag means that playback is frozen waiting for the download position to catch up with playback position.

9.5.2 Lag in P2P/Cloud Substream Method

The right side of Figure 9.14 describes the substream method of P2P streaming. The same visual applies to multisource aggregation in cloud or GroupConnect environments. Since the stream is aggregated from multiple smaller substreams, playback position in all streams has to be roughly the same. When *substream discrepancy* is too high, it can result in frozen playback, which is when your player has to wait for a missing block from the slowest substream in the group. A buffer can help smooth out some of the discrepancy, but if the gap across substreams keeps increasing it will ultimately result in frozen playback.

The *lag* here is defined the same way as in the case of a single connection—the playback position of the aggregated stream is compared with the download position. Optimization by reselection of parent and child peers can improve performance, but the technique does not contribute any additional metrics.

The lag can even be calculated in BitTorrent-type streaming sessions where data blocks are not necessarily downloaded in their playing sequence. The lag and freeze time are still valid metrics because they directly represent the quality of a streaming session [22].

9.5.3 Delivery Tree Optimization in P2P Networks

Repeating an earlier statement in this chapter, P2P delivery topology is not strictly a tree but rather is a semi-mesh of multiple trees. A vast majority of P2P streaming methods do not optimize delivery topology, instead relying on runtime optimization of peer lists by reselection. The main advantage of this method is that it is completely distributed while delivery tree optimization has to be centralized.

Among the methods that do use tree optimization, the SPPM framework is arguably the most prominent [21]. Prior to its streaming session, SPPM optimizes the delivery tree and creates a schedule for intermediate nodes, which therefore serve as relays. The main disadvantage of this method is that it fails to scale. With large topologies and high-frequency arrival and departure events, it is unlikely that such a method can be feasible as a large-scale technology.

There is a recent version of SPPM that works with variable-size substreams (SVC format with multiples of GOP), but the method can work only on a small scale, such as streaming within a university campus, which, in fact, is the home environment of the platform [20].

Delivery tree optimization is part of LTE-A and will work in 5G WNs. In this case, delivery is mostly a tree-like layered topology where the upper layer is formed by BSs while the lower layer is populated with a large number of femtocell APs. The optimization problem here is to distribute active APs across BSs in such a way that the overall utility of the network is maximized.

9.5.4 P2P-Specific Optimization

In a matter of speaking, P2P streaming is a simpler technology compared to some of its rivals. Its main optimizational target is to arrive at an *optimal combination* of various observable parameters. Even more precise, parameters are divided into *observable* ones that describe network performance from a given peer to all its parent of children peers, and *known in advance* ones—these are directly read from the video stream and are mostly deterministic.

This is why, when working with P2P streaming optimizations, it is convenient to operate with *sets of numbers*. Let us assign $\{V\}_n$ to denote a set $\{v_1, v_2, ..., v_n\}$ of n values, as a generic notation for sets. The two key sets in P2P streaming are $\{S\}_m$ for distribution of frame size across the m-frame GOP and $\{R\}_m$ for distribution of end-to-end network throughput across the m available peers. Note that the count of both parameters is the same, while the optimizational objective is to find the best mapping between the two. Count m refers to the subset of *active* members of an entire available set of n. This translates well into practice where a peer can have n other peers available for contact but only needs m peers for assignment of substreams. Also remember that m is derived directly from the size of GOP, which is defined as the number of frames in a self-sufficient sequence of frames.

The following simple logic can be applied to optimize the mapping between the two sets of size m.

Step 1: Frame Ordering. In VBR (SVC is a case of VBR), frames in GOP vary in size—the first frame is normally the largest I-Frame, followed by a much smaller B-Frame, and so on. For mapping, we need a list ordered by frame size, which means that the resulting set is a mixture of frame positions, ordered by decreasing frame size. For the next step, it is sufficient to discard values for frame size and retain only frame positions.

Step 2: Peer Ordering. When implemented in practice, a peer would be continuously monitoring the network performance of all remote peers in the n set. The history of such measurements can help define the set of m peers ordered by decreasing end-to-end network performance. Note that this simple formulation does not tell us how to select the m subset from the total set of n available peers. This element is discussed further in this subchapter.

Step 3: Map frames to peers. The new *map*( ) operator is defined as:

$$map(S_i \rightarrow R_j) \quad \forall\, i, j = 1, ..., m, \tag{9.1}$$

and is used to map positions between the two m subsets. For simplicity, let us assign i to refer to positions in frame size–ordered list (i.e., $i \in \{S\}_m$) and j to refer to positions in peer-ordered list (i.e., $Zj \in \{R\}_m$). Obviously, the mapping has to be exclusive and allow that each i can be mapped to only one j, excluding it from further selection.

Note that there is a subtle relation between R and S. Denoting throughput (data rate) as $r = s/T$ where T is the unit of time, and s and r are size by bits and rate in bps, respectively, the *map()* operator can have a mathematical representation. This paper omits this discussion because of size limitations.

These steps describe the core selection and mapping process. However, they are missing the dynamics required for dealing with changes at runtime such as arrival/departure of peers (peer churn) and changes in distribution of frame size in GOP. The rest of this section first defines the simple utility function for optimization of static mapping and then reviews two example algorithms dealing with runtime dynamics.

The utility function is defined as:

$$U = eval(\{S_i \rightarrow R_j\}_k), \tag{9.2}$$

for k individual mappings $\{S_0 \rightarrow R_0, \ldots S_m \rightarrow R_m\}$, each between a frame position and a peer. The mappings do not have to follow the order of listing, that is, $i \neq j$. Moreover, the number of mappings can be lower than m, that is, $0 \leq k \leq m$.

Obviously, this utility function is static. The only parameter related to dynamics described further in this subchapter is the value k, which counts how many changes are introduced by the mapping. It is a key parameter for dynamics, as larger k values directly relate to a larger overhead required for implementing the re-mapping in practice, that is, abort connections with some existing peers and perform signaling with all the newly assigned ones.

The rest of this section describes two logics that can guide the dynamic part of the optimization.

Example 1: Re-Elect One. Each change period, select the slowest peer and replace it with a faster peer from the entire set n, if available. The notation for this operation is $R_{i,old} \rightarrow R_{i,new}$. The objective for this (dynamic) optimization is then:

$$\text{maximize} \sum_{i=0}^{k} (R_{i,new} - R_{i,old}), \tag{9.3}$$

$$\text{subject to} \quad R_{i,new} - R_{i,old} > 0, \tag{9.4}$$

$$R_{i,new} > 2S_{i,new}. \tag{9.5}$$

Note that this objective is specific to a 15-frame GOP in a 30 fps video, which is why we know that each second of video strictly has 2 GOPs—hence number 2 in (1.5).

Example 2: Change Minimization. Considering that overhead from too many changes is costly as it requires communication to old/new peers while implementing a given mapping, it makes sense to minimize the number of changes while allowing more than one (Example 1 allows for only one mapping per round). The objective function in this case is:

$$\text{minimize} \quad k, \tag{9.6}$$

$$\text{subject to} \quad R_{i,new} > 2S_{i,new} \quad \forall\, i = 1,\ldots,k, \tag{9.7}$$

$$R_i > 2S_i \quad \forall\, i = 1,\ldots,m. \tag{9.8}$$

This is a very simple form that will obviously lead to $k = 1$. In practice, one can add more constraints to allow for $k > 1$. Note that besides minimizing k the algorithm has to confirm that all substreams beyond those covered by k still possess the necessary network performance to handle their frames. Currently active peers that are found insufficient should also be made part of the k in the remapping schedule.

Note that the above formulations are generic and can be applied to both traditional CBR and VBR streams. In fact, there is virtually no difference between the two types of streams since even CBR can be rendered variable-like by cutting the stream into nonuniform pieces.

9.5.5 A Genetic Optimization Framework

Following roughly the same basic path, this section defines a generic optimization framework that can be applied to any streaming method.

Both $p\{R\}_k$ and $p\{L\}_k$ are distributions, where the only difference is that we care about the lag L only when it is positive because it would indicate that playback freezes as some point. Note that R is related to L but not obviously/proportionally because the rate of a substream S_k

can by anything from zero up to the R_k. With $F()$ meaning *function of*, the quality of your substream design can be estimated *generically* for all streaming methods as

$$Q = F\left(\frac{C}{S}, p\{R\}_k, p\{L\}_k\right),\qquad(9.9)$$

with the obvious rate ceiling of

$$\sum_{i=1k} R_i \le S \le C.\qquad(9.10)$$

Social utility in this context is the evaluation of *the response* of your design to changes in throughput. Let us define two situations (*note*: designs become situations when under a workload) A and B with respectful quality estimates Q_A and Q_B. Then, provided all the four parameters can be given scalar estimates, the utility is simply

$$U = \frac{Q_B - Q_A}{|B - A|}.\qquad(9.11)$$

Note that in this definition, the direction of the change—whether it is toward better or worse—does not matter, which is why the sign in the denominator is ignored. This is the intended behavior because the proposed method should ideally adapt to any new condition. Following this line of reasoning, utility should be zero if the method provides a perfect mapping between throughputs and frame size. Moreover, a negative change may also result in a better mapping opportunity.

The utility U is interesting from the viewpoint of its maximization in a given streaming method:

$$\text{maximize} \quad \sum_{i \in people} \sum_{j \in time} U_{i,j},\qquad(9.12)$$

$$\text{subject to} \quad \sum_{i \in people} R_i \le S \le C,\qquad(9.13)$$

$$\sum_{i \in people} \sum_{j \in time} L_{i,j} \le 0.\qquad(9.14)$$

9.5.6 Specifics of Adaptive Streaming

Adaptive streaming is confined to one e2e path that limits the choice of actions that can help the stream adapt to changing network conditions. Let us review some of the methods that are used in practice today.

Current version of HTML5 [58] defines a *VIDEO* tag with multiple sources. When the web page is generated by a web server, it would contain multiple sources that would point to several locations in CDN [3]. However, this does not mean that video is aggregated via multiple sources. The logic is simple: select one of the sources and switch to another only if streaming quality of the current source is unacceptable. *Unacceptable* normally means that the video playback freezes while minor fluctuations in throughput would not trigger the switch.

Although not yet implemented in browsers today, DASH protocol specifies a higher degree of control [59]. It is proposed that video content would be carried over WebSockets with bidirectional access to the socket from both browser and server sides. In this case, a web application in a browser can exchange control traffic with a web server, notifying on abrupt changes in throughput. The server would then react possibly by redesigning the SVC stream (video smoothing [28]).

This method reminds us of TCP connections, where control traffic can flow in the opposite direction from the main bulk effectively notifying the sending side of throughput failures. There is already research that connects HTML5 WebSockets and TCP protocol [26]. There are already prototypes of true multisource aggregation of videos streams in browsers, which also use WebSockets, but upgrade the functionality to a true multisource [3]. Unfortunately, the multisource functionality itself is not yet part of the HTML5 standard.

9.6 A Case Study. The Challenge: An Olympics Stadium

This section continues the analysis, this time from a practical point of view. While the standardization process has not yet decided on the best wireless design for massive crowds, several models presented in this section have the right to compete. This section poses a real problem in the form of the Tokyo 2020 Olympics and analyzes several wireless designs. There are three models based on 3G/4G and current vision of 5G, which are separately studied in the situation with a local cache (the P4P design). The last model is for GroupConnect, which is where majority of the user base gets the stream from other viewers.

9.6.1 A Short Background Story

The stadium for the Tokyo 2020 Olympics has not yet been built but the location is already decided in the center of Tokyo (good wired networking to the stadium). The stadium will be a huge structure for 100k+ people. It is not the only stadium that will host the Olympics, but this one is the biggest.

It is likely that by the year 2020 most smartphones will have already migrated to 5G and can support most of the technologies listed in current LTE-A documents. It is also likely that smartphones will be able to support some form of GroupConnect—both in the WiFi Direct version and in the one via CoMP and SONs in LTE-A framework. Finally, it is very likely that the ratio of people with smartphones will increase further (penetration in Japan is about 60%–70% at the time of this writing), which would mean that most people at the stadium would want an uninterrupted 5G service.

What is not clear about the stadium so far is its wireless infrastructure. There is some discussion of the stadium in media, but nothing so far about the networking. So, the main target of this section is to devise several likely wireless designs and test them in simulation.

One might ask why have a video streaming service in an Olympics stadium when you can have big screens inside the stadium itself—this is, actually, how it is done today. There are two related justifications. First, Olympics are different from conventional sports events in that multiple events can happen at different places in the stadium at the same time. Second, the *video stream* may not actually be video but come in the form of a stream of information in which video and images occupy only a relatively small fraction. To better visualize such a usage pattern, imagine a smartphone application (or glasses if you wish) that receives a continuous stream of information about the Olympics. Users can *browse* the stream by watching video snippets, images, looking up background information on players, history readings, and so on. Instead of implementing such an application as a common web application where people would browse web pages, click on links, and so on, it is more convenient for viewers to have all the information packed into a real-time stream. The best way to simplify this stream is to call it *video streaming* regardless of the contents.

From this point on, the Olympics stadium will be referred to simply as *the stadium*.

9.6.2 Overall Model

Figure 9.15 shows the overall design of the streaming service with all the components for all the models presented next. First, there is a content provider (CP), which is probably the same company that records (see small CP inside the stadium) and streams video at the stadium. In Japan, this would probably be the NHK. Although not part of the simulations given later, the CP can have viewers outside of the stadium (home or international viewers) who get the stream over CDN-to-users (CDN2U) connections. The CP at the stadium records and compiles (encodes, aggregates, etc.) content and sends it directly to the main CP or alternatively stores a copy at local cache. The parts that relate to each distinct design will be revealed at each following section where each design will stress its own weaknesses and strengths.

Figure 9.16 shows the device density inside the stadium. It is based on a likely notion that most people would be located at or near their seats at the stadium while the remaining space would have proportionally

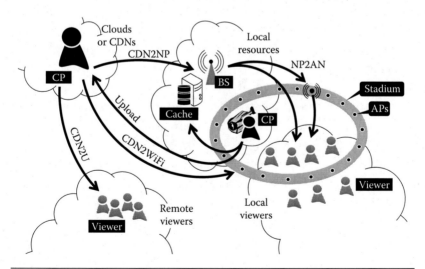

Figure 9.15   Overall design of a video streaming service at the stadium.

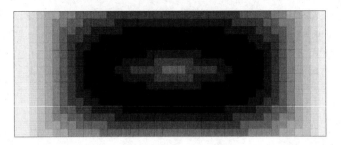

Figure 9.16   A map of device density in the stadium.

fewer devices. The space is 1000 m × 400 m in size (about the size of the spot selected for the Tokyo Olympics) with 25 m × 25 m cells (logical cells used for drawing, not the wireless cells). The total number of people is 100k, which makes it possible to allocate people into cells. This is crucial in deciding where devices and APs are located.

Note that there is no mobility in this simulation. The analysis results should therefore be interpreted as the best case, while the presence of mobility in real life would slightly decrease performance. Discussion on how much mobility one can expect from viewers inside a stadium is interesting but is too big to fit into the analysis in this chapter.

9.6.3 DIY: The Do-It-Yourself Design

DIY means that each device is on its own. It also means that no special arrangements have been made at the stadium and wireless users are expected to fare on their own. In practice, this would mean that users would come with their own 3G/4G smartphones that will connect to nearby cellular BSs for Internet. Note that even if 5G will have already been commissioned, it will not simply work out of the box without special arrangements at the stadium. As is specified in the current LTE-A standards, 5G may potentially be backward compatible with 4G (LTE) but will require a local infrastructure of femtocell APs and the proper management/control layer to get access to the full advantage of the 5G. This scenario is listed as a separate model further on.

Since we have no local infrastructure, the design can be simply analyzed in terms of the interference and its upper limit. Figure 9.17 shows the conceptual chart in which a number of people are mapped to BSs up to a certain physical limit of interference. The practical way to calculate

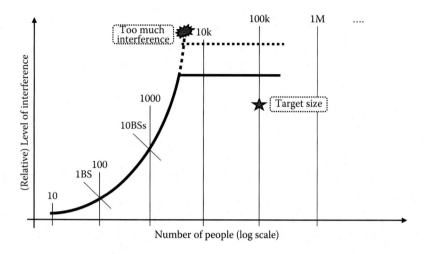

Figure 9.17 Physical limitations of a DIY design expressed in terms of interference from multiple BSs installed near the stadium.

capacity here is to remember that 1 BS can accommodate about 100 concurrent data connections [15]. The only way to increase the number of terminals is to have more BSs around the stadium, reasonably assuming that BSs will remain mostly omnidirectional as they are today.

However, there is a practical limit of interference at which increasing the number of BSs will actually decrease the total efficiency of the system. This subject is studied thoroughly on a small scale related to femtocells [42] but the results apply to large BSs as well.

Although Figure 9.17 shows that the capacity can approach 5k users (log scale), it is likely that it will not even be able to exceed 1k users. Many of the readers may have personal experience with wireless service at large gatherings like concerts today. The same basic problems will occur in the DIY design. All things considered, this is the cheapest but is also the worst possible way to handle networking at the stadium.

The provider of such a network is likely to follow a simple algorithm when managing the DIY video delivery network:

1. The provider implements its 4G or 5G network with real-time optimization in mind. In both 4G and 5G, assignment of users to BSs taking into consideration *both interference and load* is optimized. However, *interference* optimization in 4G and 5G are implemented differently—in 5G interference can be controlled by allocating frequencies dynamically.

2. At runtime (during the streaming itself), newly added users require dynamic reassignment of users to BSs based on the current situation. The algorithms in (1) are here put in action.

9.6.4 Traditional: A Network of WiFi Hotspots

The traditional way to handle wireless access at large infrastructure objects today is to install WiFi hotspot networks. In Figure 9.15, this is the case when CP headquarters get content from its local representative (CP at the stadium) and immediately send the stream to the WiFi hotspot network at the stadium. The network may not in fact be the same. CP–CP data transfer can happen over dedicated satellite links, while CDN2WiFi transfers may happen over dedicated wired connections. The wired part here is important because it is very unlikely that congestion will occur at this leg of the e2e paths. It is also very easy to plan for this capacity simply by adding enough capacity and switching equipment for 100k simultaneous connections (a 40 Gbps link or several 10 Gbps links with load balancing should do it).

At the stadium, WiFi hotspots are distributed at equal distance from each other starting from the most dense areas of the map in Figure 9.16. A visual representation is shown in Figure 9.19. As is expected, WiFi hotspots are first installed at the densest ring of the stadium, but, if more hardware and backbone capacity is available, the installation can extend into less dense rings or even to the area outside of the stadium.

Figure 9.18 shows the expected performance for the AP–AP distances of 25, 50, and 75 m. For the sake of the argument, several population sizes are tested starting from 500 people and up to 100k.

Figure 9.18a shows the overlap (vertical scale) between APs installed at various distances across each other (horizontal scale). Each curve is for a given effective range of AP. The curve for 25 m has zero overlap because of the cell structure of the map—allocation of APs is optimized for not less than 25 m, which means that the resulting distance is a higher value rounded to the nearest cell in the map. The overlap is drastic for the cases of 50 and 75 m, which is a sign to be cautious—too densely popularized WiFi networks can suffer quality degradation from excessive interference.

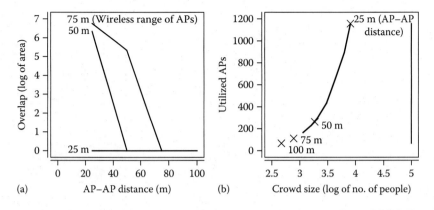

Figure 9.18 Performance of the WiFi hotspot model in terms of radio coverage overlap (a) and crowd size (b).

Figure 9.18b shows how the capacity changes with the number of deployed APs, depending on the AP–AP distance (labels on curves). Cross bullets show the maximum number of allocated people. The maximum exists first because each AP can serve at most 10 clients and secondly because certain cells may have fewer people that APs can handle (the density map). It is assumed that each user has a terminal with the effective radio radius of 50 m. People are located randomly inside the cell to avoid having a neat and predictable layout.

The results show that the network can accommodate at most about 10k users with about 1k APs installed in a 25 m grid. Note that if the range of each AP is around 50 m (same as each user terminal), then the interference in a 25 m cell grid is extremely high, requiring careful planning to avoid service degradation.

9.6.5 Smart: A Network of Femtocells

Returning to the designs in Figure 9.15, in this core 5G design the CP sends the data to the wireless network provider (NP) at the stadium via CDN2NP connections, which are likely to be dedicated optical lines with enough capacity to avoid congestion. The NP has one or more BSs at the stadium, which are used to stream—probably multicast—content to femtocell APs (NP2AN) to which users connect to get their individual streams.

The same logic as in the previous traditional model (same distribution pattern as in Figure 9.19) is applied to this design as well. In fact,

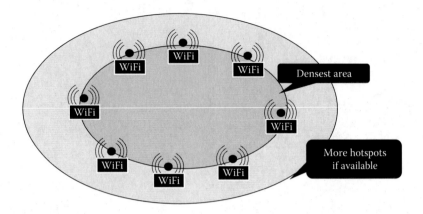

Figure 9.19 Visual representation of how WiFi hotspots may be allocated based on the density map in Figure 9.15.

the models are nearly identical and differ only in distances and ranges. Figure 9.20a shows radio overlap for 10–50 m grids and with radio coverage ranges between 15 and 50 m. These numbers come from existing pilots that show that femtocells can be considered as *slightly smaller WiFi hotspots* [44]. Overlap is still high and drops only after 25 or 30 m distances between neighbor APs.

Figure 9.20b shows that, as is expected according to the *femtocells are smaller WiFi hotspots* notion, we cannot reach the 10k mark even with 10 m grids and 1100 APs. Note that the number of APs is about the same as in the previous model, but the range of femtocells (and therefore devices) is 30 m. However, the distance between femtocells is 10 m here, which, as the plot shows, is no different from the

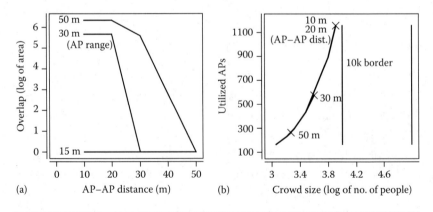

Figure 9.20 Performance of the 5G design based on femtocells in terms of radio coverage overlap (a) and crowd size (b).

case of 20 m. This is an interesting finding that speaks of the upper limit on the capacity of any large-scale 5G deployment. The fact the 10 and 20 m results are qualitatively the same is rooted in the density distribution on the map. On one hand, the most populated areas on the map do not benefit from a higher density of APs because even a twofold increase in APs cannot accommodate all the terminals that try to connect via nearby APs. At the same time, dense grids of APs in sparse areas on the map remain underutilized and fail to bring in more people as the density of the grid increases.

At the same time, radio interference does not allow for even denser AP grids. This means that one cannot simply bring all the underutilized APs and install them next to overutilized APs. Such solutions would definitely result in very high interference and would eventually lower the overall efficiency of the system.

9.6.6 P4P: All the Aforementioned with a Local P4P Cache

Figure 9.15 has the local cache installed at the stadium. This section analyzes its effect on the performance of the three aforementioned models. In the process, the discussion also recalls all the traffic and performance bottlenecks listed earlier in this chapter.

Figure 9.21 is a conceptual chart indicating how capacity is affected by various bottlenecks in which local cache plays its own role. Let us read the chart left to right gradually increasing the size of the crowd.

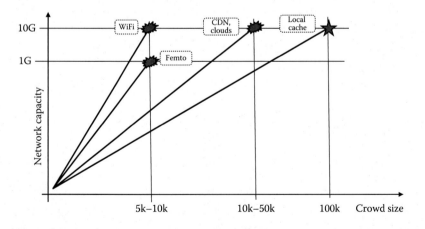

Figure 9.21 Analysis of capacity relative to technologies and specifically the effect of the presence of local cache at the stadium.

At the range of 5k–10k crowds, we saw in the aforementioned analysis that the WiFi hotspot or femtocell networks are the bottlenecks. The infrastructure literally cannot accommodate more users, which makes all the other bottlenecks irrelevant. A solution to this problem is presented in the last model further in this section. Note that there is a difference between femtocell and WiFi networks in the backbone as well. LTE-A assumes that the IP backbone across BSs will run at 1 Gbps rates. At the same time, WiFi networks can be connected directly to 10G and above networks with very little effort. This means that this particular bottleneck is twofold—while there is a cap on crowd size, there is also a separate cap on the total traffic generated by the crowd. Putting this in practical form, there is a difference between a 10k crowd of 50 versus 100 kbps video streams.

If somehow the wireless crowd is allowed to grow to 10k–50k devices, then CDNs and/or clouds can be the bottleneck. This is the classic case when CDNs get a flash crowd that can potentially disrupt the entire service. In the research community, flash crowds are often viewed as a kind of denial-of-service (DoS) attack because when the intent (malicious or benign) is set aside there is very little difference between the two. This is where the local cache can enter the picture as the offload solution to the flash crowd. The CDN in this case would send copies of its content to a local cache at the stadium from which it would be streamed locally. This small technical change can potentially increase the capacity of the system up to the target number of 100k. Note that this solution is already actively employed in P2P streaming services where Internet service providers (ISPs) install their own caches of popular content that helps drastically reduce traffic in/out of the ISP. This solution is often referred to as P4P.

9.6.7 Social: APs as Peers in a Massive Delivery Tree

This final model brings in the GroupConnect. It is called the Social Model because it is mostly about building a video streaming network from a structureless crowd of devices. The traditional and smart models show that the maximum achievable crowd size is about 10k.

GroupConnect increases the crowd by connecting the rest of the devices in P2P streaming topologies. All the P2P streaming methods described earlier are applicable to this model.

There are four combinations in which such a technology can be implemented. First, it can be added to either the WiFi hotspots or cellular networks, forming the two base cases. Additionally, the top level of the P2P distribution tree can be selected randomly or based on the crowd density in the stadium. To save space, these four cases are packed in to two plots in Figure 9.22 while the other two combinations can be deducted from the results.

The common rules in all the cases are as follows. One device can handle at most 10 parallel connections, of which one has to be used for receiving while the rest can be used for sending the video stream. The relatively simple one-path case is considered without the complexity of the substream or multisource aggregation methods. Radio range for each device is 50 m, meaning that each user can reach any other users in all neighboring cells. Downstream peers are selected randomly from a list of all reachable peers. In densely populated areas, this would mean that there can be multiple P2P connections in the same cell. As in all previous models that can potentially cause interference problems, let us set the interference component aside for the sake of modeling simplicity.

Figure 9.22a is the case when the distribution tree starts from the most densely populated areas in the map. Each WiFi hotspot or femtocell serves as a peer that gets at most nine other peers downstream, which in turn find their peers, and so on, until there are no more peers

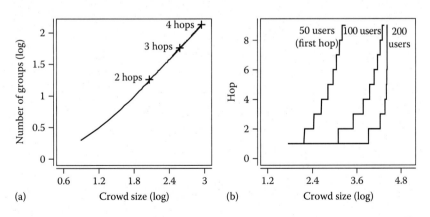

Figure 9.22 Results of the social model: (a) group count; (b) hop count.

to find. The only additional rule in this submodel is that peer search has to stop at most at the fourth hop. Obviously, this puts a major restriction on the size of the P2P network, which can increase the crowd only by 1k additional devices. With 2 hops, only 100 users on average are added.

Figure 9.22b shows an alternative approach. This time, a random number of users (50, 100, 200) are selected as delivery tree heads from the entire stadium. Obviously, this will not work with WiFi hotspots or femtocell APs simply because they have fixed locations. However, such a selection can be done using BSs that can reach any user at the stadium directly.

The head devices then start to recruit peers in the normal way, allowing at most nine downstream connections per peer. The results show that with 50 head users we can increase the crowd by 1k new devices. With 100 and 200 initial users, the increase can reach 25k new devices. In the present model, the crowd size could not grow beyond this level. Note that the maximum number of hops is 8.

It is important to see the dynamics of crowd growth relative to the number of hops. The figure shows that most of the users are brought in at the first 1–3 hops. This means that the crowd size could be increased further if the same process is repeated several times from the start—that is, selecting a new set of random users at distribution tree heads and building a brand new tree. Due to size limitations, such complex scenarios cannot be discussed in this chapter.

9.7 Conclusions

This chapter discussed the topic of video streaming over wireless channels in the fourth generation and beyond. The *beyond* part refers to the currently ongoing standardization process whose goal is to build a new level of advanced wireless technology under the name of LTE-A. In fact, current normative documents show that LTE-A will introduce several major changes in the way WNs operate.

The biggest change under LTE-A is that the cellular network will be separated into the high-speed and high-performance backbone network connecting BSs to each other, on one hand, and a large number of femtocell APs, on the other. Given that femtocells have a small range, the topology of this network will look like a meshed tree with very many leaves (BS–AP connections) at the lower level. In addition to the increased technical challenge of controlling such a network, the major shift of paradigm is in the way devices will access the network. While currently devices access their cellular provider via BSs—the capacity of which is roughly 100 concurrent users—in LTE-A the users will connect via small but more numerous femtocell APs where each AP will have a relatively small capacity of up to 10 devices, but the high density of such APs would allow for seamless handover of devices across neighboring APs.

This chapter discussed several other 5G technologies in relation to femtocells. First, MIMO is discussed in its main functionality of a throughput-boosting technology, which makes it perfect for high-capacity home multimedia streaming. Second, LTE-A contains references to several technologies like CoMP and SONs, which can all be classified as *traffic offload solutions*. This chapter discusses in detail how they can help offload traffic in practical situations.

When put into the context of video streaming, this chapter shows that current and future WNs are different from the wired ones in several ways. The low reliability of wireless connections poses unique challenges for wireless video streaming, which is traditionally counteracted by improved resilience and error concealment at the low level of video coding. However, WNs also have unique features that may be exploited to achieve much better throughputs.

The one specific such feature on which this chapter spends considerable effort is MultiConnect and GroupConnect. To the day, only wireless devices have the ability to support multiple connectivities like WiFi + Cellular. This can be exploited to implement the GroupConnect technology in which multiple devices share their Internet connections while at the same time enjoying high-capacity local connections among each other. This functionality allows for a unique kind of P2P networking in WNs and is extremely useful for video streaming.

This feature is also useful in view of one of the main problems raised by this chapter—*wireless video streaming in massive crowds*. While the offload concept is a big part of current standardization under the LTE-A framework, massive crowds pose additional challenges even to the current set of 5G technologies.

These challenges are solidified in this chapter in form of a practical usecase—the challenge of building a suitable infrastructure for the main stadium for the Tokyo 2020 Olympics. Four separate designs are formulated and analyzed for capacity and performance bottlenecks. It was found that even with the 5G installation that would involve a dense grid of femtocells, the maximum size of the crowd would not exceed 10k devices. The fourth model—the social model based on the concept of GroupConnect and the P2P video streaming implemented on top of it—shows that it is possible to increase the crowd size to 35k devices if, in addition to the 5G infrastructure, devices are encouraged to form P2P delivery trees that would be used to stream content to the portions of the crowd that cannot be covered by the main 5G infrastructure.

Due to volume limitation, the models had to be relatively small and simple. Future research on this topic will include mobility (there are many public traces like [62]), algorithms for large-scale multicasting and network optimization in 5G settings, and, finally, the distributed algorithms that can be used by devices to participate in P2P delivery of video streams. Special attention will be paid to the radio efficiency, which is where the algorithms will be evaluated in terms of how much local interference they generate.

9.8 Future Research Directions

The first obvious problem that remains unsolved in current 5G literature is capacity. While the rates are expected to increase 1k-fold or even 10k-fold, capacity in terms of the number of parallel connections is expected to experience a relatively meager 10- to 100-fold increase. This means that 5G technologies may be capable of supporting small crowds with very high bitrates but will not be able to support large crowds, regardless of the bitrate.

P2P delivery can help in two ways.

First, P2P networks offload the majority of delivery effort to the edges of the delivery network, that is, the viewers themselves. Devices can provide portions of content to other devices that can aggregate and play video from multiple sources. This solution is shown to work well in wired networks and is successful in WNs as well. However, local radio interference puts a hard physical limit on the size of such P2P networks. OFDM and MIMO–OFDM hybrids can somewhat increase density but not at the level of an order of magnitude.

Therefore, it is likely that increased attention will be paid to hybrids of frequency and time multiplexes. While MIMO and OFDM can facilitate higher efficiency radio, TDM can be emulated in software simply by avoiding continuous access to radio. Such emulation is simple—software simply has to minimize the time it spends using a given radio channel, instead of sending requests or receiving payloads from peers in short bursts on demand. With opportunistic wireless networking methods in mind, such environments can offer drastically higher capacity in terms of the number of users that can be supported concurrently.

In general, more research is expected on the topic of interference avoidance in massive wireless crowds where video delivery in general and the Olympics usecase specifically discussed in this chapter can be viewed as efficiency benchmarks.

References

1. M. Zhanikeev, Virtual wireless user: A practical design for parallel MultiConnect using WiFi direct in group communication, in: *Tenth International Conference on Mobile and Ubiquitous Systems: Computing, Networking and Services (MobiQuitous)*, Tokyo, Japan, December 2013.
2. M. Zhanikeev, Multi-source stream aggregation in the cloud, Chapter 10, in: M. Pathan, R. Sitaraman, and D. Robinson (eds.), *Advanced Content Delivery and Streaming in the Cloud*, New York, Wiley, 2013.
3. M. Zhanikeev, Experiments with application throughput in a browser with full HTML5 support, *IEICE Communications Express*, 2(5), 167–172, May 2013.
4. M. Zhanikeev, Experiments on practical WLAN designs for digital classrooms, *IEICE Communications Express*, 2(8), 352–358, August 2013.
5. M. Zhanikeev, A method for reliable P2P video streaming using variable bitrate video formats, *JSSST Journal of Computer Software*, 31, 234–245, May 2014.
6. M. Zhanikeev, A method for extremely scalable and low demand live P2P streaming based on variable bitrate, in: *First International Symposium on Computing and Networking (CANDAR)*, Matsui, Japan, December 2013, pp. 461–467.
7. M. Zhanikeev, A practical software model for content aggregation in browsers using recent advances in HTML5, in: *IEEE Conference on Computers, Software and Applications (COMPSAC)*, Kyoto, Japan, July 2013, pp. 151–156.
8. M. Zhanikeev, WiFi direct with delay-optimized DTN is the base recipe for applications in location-shared wireless networking virtualization, *IEICE Technical Report on Radio Communication Systems (RCS)*, 113(386), 25–28, January 2014.

9. S. Guan-Ming, H. Zhu, W. Min, K. Liu, A scalable multiuser framework for video over OFDM networks: Fairness and efficiency, *IEEE Transactions on Circuits and Systems*, 16(10), 1217–1231, November 2006.

10. L. Maodong, C. Zhenzhong, T. Yap-Peng, Scalable resource allocation for SVC video streaming over multiuser MIMO-OFDM networks, *IEEE Transactions on Multimedia*, 15(7), 1519–1531, October 2013.

11. M. Ilyas, R. Dorf, *The Handbook of AdHoc Wireless Networks*, Boca Raton, FL: CRC Press, January 2003.

12. J. Zhang, G. Roche, *Femtocells: Technologies and Deployment*, New York: Wiley, 2010.

13. E. Dahlman, S. Parkvall, J. Skold, *4G LTE LTE-Advanced for Mobile Broadband*, Oxford, UK: Elsevier, 2014.

14. M. Rumney, *LTE and the Evolution to 4G Wireless, Design and Measurement Challenges*, New York: Wiley, 2013.

15. X. Zhang, X. Zhou, *LTE-Advanced Air Interface Technology*, Boca Raton, FL: CRC Press, 2013.

16. H. Wang, L. Kondi, A. Luthra, S. Ci, *4G Wireless Video Communications*, New York: Wiley, 2009.

17. P. Thilagam, Applications of social network analysis, in: B. Furht (ed.), *Handbook of Social Network Technologies and Applications*, Springer, May 2010, pp. 637–649.

18. J. Ruckert, O. Abboud, T. Zinner, R. Steinmetz, D. Hausheer, Quality adaptation in P2P video streaming based on objective QoE metrics, in: *NETWORKING*, Prague, Czech Republic, 2012, pp. 1–14.

19. N. Capovilla, M. Eberhard, S. Mignanti, R. Petrocco, J. Vehkapera, An architecture for distributing scalable content over peer-to-peer networks, in: *Second International Conference on Advances in Multimedia (MMEDIA)*, Athens, Grece, June 2010, pp. 1–6.

20. P. Baccichet, T. Schierl, T. Wiegand, B. Girod, Low-delay peer-to-peer streaming using scalable video coding, in: *Packet Video*, Lausanne, Switzerland, November 2007, pp. 173–181.

21. J. Noh, P. Baccichet, E. Hartung, A. Mavlankar, B. Girod, Stanford peer-to-peer multicast (SPPM)—Overview and recent extensions, in: *Picture Coding Symposium (PCS)*, Chicago, IL, May 2009, pp. 1–4.

22. C. Gurler, S. Savas, A. Tekalp, Variable chunk size and adaptive scheduling window for P2P streaming of scalable video, in: *Nineteenth IEEE International Conference on Image Processing (ICIP)*, Orlando, FL, October 2012, pp. 2253–2256.

23. Z. Liu, Y. Shen, K. Ross, S. Panwar, Layer P2P: Using layered video chunks in P2P live streaming, *IEEE Transactions on Multimedia*, 11(7), 1340–1352, November 2009.

24. M. Mushtaq, A. Toufik, Smooth video delivery for SVC based media streaming over P2P networks, in: *Fifth IEEE Consumer Communications and Networking Conference (CCNC)*, Las Vegas, NV, January 2008, pp. 447–451.

25. C. Kong, Y. Lee, Slice-and-patch an algorithm to support VBR video streaming in a multicast-based video-on-demand system, in: *Ninth International Conference on Parallel and Distributed Systems*, San Francisco, CA, 2003, pp. 391–397.

26. R. Kusching, I. Kofler, H. Hellwagner, An evaluation of TCP-based rate-control algorithms for adaptive internet streaming of H.264/SVC, in: *First ACM SIGMM Conference on Multimedia Systems (MMSys)*, Phoenix, AZ, 2010, pp. 157–168.

27. A. Undheim, Characterization and modeling of slice-based video traffic, Doctoral dissertation, Norwegian University of Science and Technology (NTNU), Trondheim, Norway, May 2009.

28. M. Fidler, Y. Lin, P. Emstad, A. Perkins, Efficient smoothing of robust VBR video traffic by explicit slice-based mode type selection, in: *IEEE Consumer Communications and Networking Conference (CCNC)*, January Las Vegas, NV, 2007, pp. 880–884.

29. S. Sivasubramanian, G. Pierre, M. Steen, G. Alonso, Analysis of caching and replication strategies for web applications, *IEEE Internet Computing*, 11(1), 60–66, January 2007.

30. K. Walkowiak, QoS dynamic routing in content delivery networks, in: S. Papavassiliou and S. Ruehrup (eds.), *NETWORKING*, LNCS, Vol. 3462, Springer, Wroclaw, Poland, May 2005, pp. 1120–1132.

31. C. Chen, Y. Ling, M. Pang, W. Chen, S. Cai, Y. Suwa, O. Altintas, Scalable request-routing with next-neighbor load sharing in multi-server environments, in: *Nineteenth IEEE International Conference on Advanced Information Networking and Applications*, Piscataway, NJ, March 2005, pp. 441–446.

32. C. Stais, G. Xylomenos, Realistic media streaming over BitTorrent, in: *Future Network and Mobile Summit*, Berlin, Germany, July 2012, pp. 1–5.

33. S. Lederer, Ch. Muller, Ch. Timmerer, Dynamic adaptive streaming over HTTP dataset, in: *Multimedia Systems Conference (MMSys)*, Chapel Hill, NC, February 2012, pp. 89–94.

34. F. Daoist, Adopting HTML5 for television: Next steps, Report by W3C/ERCIM Working Group, 2011.

35. N. Fernando, S. Loke, W. Rahayu, Mobile cloud computing: A survey, *Elsevier Journal on Future Generation Computer Systems*, 29, 84–106, January 2013.

36. G. Huerta-Canepa, D. Lee, A virtual cloud computing provider for mobile devices, in: *First ACM Workshop on Mobile Cloud Computing and Services: Social Networks and Beyond (MCS)*, Vol. 6, San Francisco, CA, June 2010, pp. 1–5.

37. M. Satyanarayanan, Fundamental challenges in mobile computing, in: *Fifteenth Annual ACM Symposium on Principles of Distributed Computing (PODC)*, Vol. 5, New York, May 1996, pp. 1–6.

38. P. Schmidt, R. Merz, A. Feldmann, A first look at multi-access connectivity for mobile networking, in: *ACM Workshop on Capacity Sharing (CSWS)*, Nice, France, December 2012, pp. 9–14.

39. Y. Chen, Y. Lim, R. Gobbens, E. Nahum, R. Khalili, D. Towsley, A measurement-based study of multipath TCP performance over wireless networks, in: *ACM SIGCOMM Internet Measurement Conference (IMC)*, Barcelona, Spain, August 2013.

40. W. Malik, B. Allen, D. Eswards, A simple adaptive beam former for UWB wireless systems, in: *IEEE International Conference on Ultra-Wideband*, Waltham, MA, September 2006, pp. 453–457.

41. A. Martini, M. Franseschetti, A. Massa, Capacity of wideband MIMO channels via space time diversity of scattered fields, in: *Forty First Asilomar Conference on Signals, Systems and Computers (ACSSC)*, Pacific Grove, CA, November 2007, pp. 138–142.

42. D. Lopez-Perez, A. Valcarce, G. Roche, J. Zhang, OFDMA femtocells a roadmap on interference avoidance, *IEEE Communications Magazine*, 47(9), 41–48, September 2009.

43. G. Roche, A. Valcarce, D. Lopez-Perez, J. Zhang, Access control mechanisms for femtocells, *IEEE Communications Magazine*, 48(1), 33–39, January 2010.

44. J. Weitzen, L. Mingzhe, E. Anderland, V. Eyuboglu, Large-scale deployment of residential small cells, *Proceedings of the IEEE*, 101(11), 2367–2380, August 2013.

45. L. Jin, S. Ju Bin, H. Zhu, Network connectivity optimization for device-to-device wireless system with femtocells, *IEEE Transactions on Vehicular Technology*, 62(7), 3098–3109, April 2013.

46. M. Ananstapalli, W. Li, Multipath multihop routing analysis in mobile ad hoc networks, *Journal of Wireless Networks*, 16(1), 573–575, January 2010.

47. E. Osipov, C. Tschudin, Evaluating the effect of ad hoc routing on TCP performance in IEEE 802.11 based MANETs, in: *Sixth International Conference on Next Generation Teletraffic and Wired/Wireless Advanced Networking (NEW2AN)*, St. Petersburg, Russia, May 2006, pp. 298–312.

48. R. Al-Qassas, M. Ould-Khaoua, Performance comparison of end-to-end and on-the-spot traffic-aware techniques, *International Journal of Communication Systems*, 26(1), 13–33, January 2013.

49. C. Chen, C. Weng, Bandwidth-based routing protocols in mobile ad hoc networks, *Journal of Supercomputing*, 50(3), 240–268, December 2009.

50. A. Konak, G. Buchert, J. Juro, A flocking-based approach to maintain connectivity in mobile wireless ad hoc networks, *Journal of Applied Soft Computing*, 13(2), 1284–1291, February 2013.

51. R. Ciobanu, C. Dobre, Social-awareness in opportunistic networking, *International Journal of Intelligent Systems, Technologies and Applications*, 12(1), 39–62, July 2013.

52. Framework and overall objectives of the future development of IMT-2000 and systems beyond IMT-2000, ITU-R Recommendation M.1645, June 2003.

53. System architecture evolution: Report on technical options and conclusions, 3GPP TR 23.882 V8.0.0, 2008.
54. Transition to 4G: 3GPP broadband evolution to IMT-advanced, Rysavy Research Report, 2010.
55. Wi-Fi peer-to-peer: Best practical guide, Wi-Fi Alliance, December 2010.
56. Advanced video coding for generic audiovisual services, ITU-T Recommendation H.264, January 2012.
57. H. Schwarz, D. Marpe, T. Weigand, Overview of the scalable video coding extension of the H.264/AVC standard, *IEEE Transactions on Circuits and Systems for Video Technology*, 17(9), 1103–1121, September 2007.
58. HTML5, W3C working draft [Online]. Available at: http://www.w3.org/TR/html5/ (retrieved current).
59. Dynamic adaptive streaming over HTTP (DASH), Normative Document ISO/IEC 23009-1, 2012.
60. P. Seeling, M. Reisslein, B. Kulapala, Network performance evaluation with frame size and quality traces of single-layer and two-layer video, *IEEE Communications Surveys and Tutorials*, 6(3), 58–78, 2004.
61. M. Dai, Y. Zhang, D. Loguinov, A unified traffic model for MPEG-4 and H.264 video traces, *IEEE Transactions on Multimedia*, 11(5), 1010–1024, August 2009.
62. CRAWDAD Repository of Mobility Traces [Online]. Available at: http://crawdad.cs.dartmouth.edu (retrieved July 2014).

Index